NON-LINEAR VARIABILITY IN GEOPHYSICS

NON-LINEAR VARIABILITY IN GEOPHYSICS

Scaling and Fractals

edited by

D. SCHERTZER

EERM/CRMD, Météorologie Nationale, Paris, France

and

S. LOVEJOY

Physics Department, McGill University, Montreal, Canada

KLUWER ACADEMIC PUBLISHERS

DORDRECHT / BOSTON / LONDON

Library of Congress Cataloging-in-Publication Data

Non-linear variability in geophysics : scaling and fractals / edited by D. Schertzer and S. Lovejoy.
p. cm.
ISBN 0-7923-0985-5 (alk. paper)
1. Geophysics. 2. Atmospheric physics. 3. Scaling laws (Statistical physics) 4. Fractals. I. Schertzer, D. II. Lovejoy, S. (Shaun), 1956-
QC806.6.N66 1991
550--dc20 90-48007

ISBN 0-7923-0985-5

Published by Kluwer Academic Publishers,
P.O. Box 17, 3300 AA Dordrecht, The Netherlands.

Kluwer Academic Publishers incorporates
the publishing programmes of
D. Reidel, Martinus Nijhoff, Dr W. Junk and MTP Press.

Sold and distributed in the U.S.A. and Canada
by Kluwer Academic Publishers,
101 Philip Drive, Norwell, MA 02061, U.S.A.

In all other countries, sold and distributed
by Kluwer Academic Publishers Group,
P.O. Box 322, 3300 AH Dordrecht, The Netherlands.

Printed on acid-free paper

Printed in the Netherlands

TABLE OF CONTENTS

PREFACE:

This book contains a collection of papers from a wide variety of geophysical specializations. They share a preoccupation with a basic aspect of geophysical phenomena: the existence of highly variable fractal structures over wide ranges of scale. While variability has always been an important subject of geophysical research, in the last 10 years, a series of developments in chaos, non-linear dynamics, turbulence and fractals - some of which were specifically stimulated by this basic geophysical problematic - have given it a more precise focus and have lead to profound changes in our outlook. These developments were encouraged by a number of specialized workshops and symposia in particular: "Scaling, Fractals and Non-linear VAriability in Geophysics" (NVAG 1, Montreal 27 -30/8, 1986 , NVAG 2, Paris, 27/6-1/7 1988) and the Joint Session "Chaos, Turbulence and Nonlinear Variability in Geophysics" (European Geophysical Society XIV General Assembly, Barcelona, 13-17/3 1989).

For a long time, this extreme spatial (and often temporal) variability routinely encountered by geophysicists was considered at best a nuisance or at worst an insurmountable obstacle; it can now be understood as natural consequence of the multiple scaling and can be treated directly with new data analysis and modelling techniques. Furthermore, the ideas of scaling and scale invariance have now been greatly generalized well beyond the restrictive geometric frameworks involving sets of points to directly deal with fields. Furthermore, the anisotropic nature of most geophysical fields requires the use of scaling notions which are considerably more general than the now familiar: self-similar or self-affine notions. This anisotropic, generalized scale invariance coupled with the discovery of multifractal measures provides such a broad framework that it is quite possible -a priori- that virtually all geophysical phenomena (including those exhibiting differential rotation such as cyclones or spiral galaxies) respect some form of scaling symmetry over significant ranges of scale. These developments provide a broad enough framework to allow scaling to become a unifying geophysical problematic: extreme variability, structures spanning wide ranges of scale produced by non-linear interactions, the phenomena we collectively identify with the term "nonlinear variability".

Furthermore, geophysical systems have always been important in stimulating new ways of thinking in chaos, turbulence and nonlinear dynamics. It is quite obvious, that advances in understanding non-linear variability will have immediate and far reaching practical implications in numerous areas where progress has been blocked -sometimes for long periods- by lack of appropriate theoretical notions (and hence of appropriate analysis and modelling techniques). A specific example from our own experience is the area of remote sensing where in spite of colossal financial investments, there has been a paucity of quantitative applications of the data. At the same time, the quantity and quality of the data- which is now routinely available- makes it potentially a unique source for developing and experimentally testing new notions in dynamical systems. A number of the papers in the remote sensing section help to bridge this gap, since multifractals provide an obvious framework for analyzing resolution-dependent fields. Geophysical systems such as the atmosphere could provide unique testing and development grounds for new ideas in physics and non-linear systems. No other non-linear systems involve variability over ranges of scale even approaching the ten or more decades that are routinely available in geophysics. For comparison, three-dimensional numerical models are currently limited in ranges of scale to factors of about 100, while controlled laboratory experiments (such as wind tunnels) are limited to factors of 1000 to 10,000 at most.

As the table of contents indicates, the contributors came from a wide variety of geophysical backgrounds. We have provided some structure by assigning papers to the following categories: turbulence and geophysical fluid dynamics, data analysis techniques, modelling and analysis of clouds, rain and other atmospheric fields, modelling and analysis of the climate, oceans, and solid earth geophysics, and remote sensing. One of our goals was specifically to encourage research in these relatively new areas, so this book is clearly a research book, although some efforts have been made to make research papers as accessible as possible.

The section of turbulence and geophysical fluid dynamics includes two papers (by Cho and by Van Atta and Poddar) that investigate the problem of extreme fluctuations and scaling in semi-geostrophic and wind tunnel flows respectively. Levich and Shtilman examine scaling cascade processes in which the consequences of broken symmetry -here parity- is studied. In this model, turbulence is dominated by a hierarchy of helical (corkscrew) structures. The authors stress the unique features of such pseudo-scalar cascades as well as the extreme nature of the resulting (intermittent) fluctuations. Intermittent turbulent cascades was also the theme of a paper by us in which we show that universality classes exist for continuous cascades (in which an infinite number of cascade steps occur over a finite range of scales). This result is the multiplicative analogue of the familiar central limit theorem for the addition of random variables. Finally, an interesting paper by Pasmanter investigates the scaling associated with anomolous diffusion in a chaotic tidal basin model involving a small number of degrees of freedom.

Although the statistical literature is replete with techniques for dealing with those random processes characterized by both exponentially decaying (non-scaling) autocorrelations and exponentially decaying probability distributions, there is a real paucity of literature appropriate for geophysical fields exhibiting either scaling over wide ranges (e.g. algebraic autocorrelations) or extreme fluctuations (e.g. algebraic probabilities, divergence of high order statistical moments). In fact, about the only relevant technique that is regularly used -fourier analysis (energy spectra) - permits only an estimate of a single (power law) exponent. If the fields were mono-fractal (characterized by a single fractal dimension) this would be sufficient, however their generally multifractal character calls for the development of new techniques. This is the theme of the papers in the section on "data analysis techniques".

A paper by Essex addresses the non-trivial problem of accurately estimating dimensions with finite sample sizes. This work was motivated by a number of attempts to estimate dimensions of strange attractors (see e.g. Lookman et al. this volume), where empirical values as high as 10 have been reported. Essex finds that the sample size necessary for estimating dimensions are far greater than has been commonly realized, and may explain away the many (surprisingly) low values of dimension cited in the literature. Lavallée et al. discuss in detail some of the problems associated with multiplicative (multifractal) cascade processes - in particular, the problem of estimating the basic parameters (such as the co-dimension function which characterizes the sparseness of the increasingly intense regions). A paper by us re-examines the application of several primarily mono-dimensional data analysis techniques (such as area-perimeter relations) to multifractal fields. The example of the cloud and rain fields over the range 10^{-3} to 10^6m is studied in detail using blotting paper, lidar, radar and satellite data. Mahrt discusses an empirical method for analyzing complex atmospheric structures, and Papadakis, a method of piecewise linear interpolation applied to fractal signals.

The section on modelling and analysis of clouds, rain and other atmospheric fields includes a paper by Austin et al. on CO_2 fluxes measured by aircraft one of the conclusions being that the extreme variability is consistent with the hyperbolic distribution of fluctuations predicted by cascade models. Tsonis describes an interesting (Laplacian) model of dielectric breakdown for lightning. His model not only yields realistic looking lightning strokes, but also permits the author to investigate why fractal stokes are more likely than non-fractal ones. Ever since the early 1960's when Kolmogorov suggested that turbulent energy dissipation might be log-normally distributed, a series of multiplicative cascade models have been developed in order to produce scaling fields with log-normal statistics. We now know that such models yield at best approximately log-normal statistics with extremes being generally hyperbolic; Waymire and Gupta add to this picture a mathematical argument showing that log-normal statistics are incompatible with a simple kind of scaling involving only a single fractal dimension. Finally, Wilson et al. discusses continuous multiplicative cascades in which an infinite number of cascade steps occur in a finite range of scales. He relates his models to the (non-linear) dynamical equations governing passive scalar advection and shows how to implement the process numerically.

The section on the modelling and analysis of the climate, ocean and solid earth geophysics includes a valuable paper by Glazman which relates the standard oceanographic phenomenology of ocean wave spectra with a geometrical fractal description. Perrie describes a statistical closure technique for obtaining scaling spectra. The paper by Jensen et al. discusses the special role of 1/f noises in linear models of geophysical processes, and Todeuschuk et al. applies them to the problem of seismic deconvolution, supporting this approach with mine-core data of propagation speed variations with depth. Ladoy investigates a number of empirical climate series using energy spectra, and probability distributions to investigate the scaling and extreme temperature and rainfall fluctuations. Nemberg et al. investigate 500mb atmospheric pressure data for evidence of the existence of low-dimensional strange attractors, as part of the on-going debate over whether atmospheric fields can be represented by a small number of degrees of freedom rather than the nearly infinite number usually assumed in the turbulence literature. Finally a paper by Viswanathan investigates the scaling and fluctuation properties of global climate models.

The section on remote sensing features two papers by Cahalan and Yano and Takeuchi dealing with the empirical analyses of cloud fields from satellite data using primarily area-perimeter relations. Finally, a paper by Gabriel et al. examines the radiative properties of fractal clouds in which scattering occurs only in a finite number of directions. Various scaling exponents are calculated-both analytically and numerically, and the results compared to the usual (non-fractal) plane-parallel radiative transfer results which assume that the clouds are completely homogeneous in the horizontal direction.

We are greatly indebted to a number of colleagues and collaborators who gave valuable help during various stages of the editorial process, from proof-reading, typing as well as various useful comments and suggestions. In particular, we acknowledge D. Lavallée and J. Wilson. We also thank M. Edwards for typing much of the manuscript.

-The editors

1 - TURBULENCE AND GEOPHYSICAL FLUID DYNAMICS

ENERGY SPECTRUM AND INTERMITTENCY OF SEMI-GEOSTROPHIC FLOW

Han-Ru Cho
Department of Physics
University of Toronto
Toronto, Ont, Canada M5S 1A7

ABSTRACT: The semi-geostrophic theory of atmospheric flow is used to study the spectra and intermittency properties of a frontogenetic flow. Prior to the formation of the frontal discontinuity, the spectra of meteorological variables are found to decay exponentially in the short wavelength limit. At the time of front formation, the spectra obeys power laws for large wave numbers. The power laws give a good approximation only when the flow field is fairly close to the formation of discontinuity. At the time of front formation, the probability for large values of certain meteorological variables, such as vorticity, is also derived. It is found to be asymptotically hyperbolic for large values.

1. INTRODUCTION

There has been considerable interest in the statistical properties of atmospheric flow at the mesoscale, due to the emphasis on mesoscale meteorological research in recent years. Empirical studies of these properties using observational data have become possible because of the increasing availability of remotely sensed and modern in situ data with sufficient temporal and spatial resolutions. Results of these studies seem to suggest that mesoscale atmospheric flow is highly intermittent with most of the fluctuations contained in sparsely distributed active regions. Furthermore, the energy spectrum of the wind in the horizontal appears to be in the form of a power law. Valuable reviews of these studies can be found in Lilly (1983) and in Lovejoy and Schertzer (1986).

These observational studies have led to the suggestion that atmospheric processes over a wide range of spatial scales are scale-invariant (e.g., Lovejoy and Mandelbrot, 1985; Lovejoy et al. 1986), provided that the effects of vertical stratification of the atmosphere is taken into account. Schertzer and Lovejoy (1985) developed the formalism of generalized scale invariance (GSI) as a mathematical tool which may be used to describe these phenomena.

While the possibility of using the scale invariance principle to describe the statistical properties of atmospheric process is very attractive, the validity of such an approach has yet to be established rigorously. The basic question remains whether atmospheric processes are self-similar, despite many supporting observational data.

The purpose of this paper is to review and to examine the properties of air flow described by the semi-geostrophic system of equations. This system is capable of describing many mesoscale dynamic processes, including the formation of discontinuities in the atmosphere (atmospheric fronts). In particular, we shall examine the spectrum and the property of intermittency implied by the system of equations, as these properties, particularly in the short wavelength limit, depend on the character of the discontinuities which may form in the dynamic system. A brief review of the semi-geostrophic theory is first presented in the next section.

D. Schertzer and S. Lovejoy (eds.), Non-Linear Variability in Geophysics, 3–11.

2. SEMI-GEOSTROPHIC DYNAMICS

It is well known that large-scale atmospheric flows are approximately in geostrophic balance. The formation of frontal type of discontinuities in such flows is caused by the ageostrophic flow which is required to maintain geostrophic balance. When an ageostrophic flow brings into contact air parcels of different properties, discontinuities form. One of the simplest sets of dynamic equations which is capable of describing these processes is known as the semi-geostrophic system.

The dynamics of the semi-geostrophic system has been discussed by Hoskins and Bretherton (1972), and Hoskins (1975). It is described by the conservation of potential temperature θ

$$\frac{D\theta}{Dt}=0 \tag{1}$$

and the conservation of potential vorticity q

$$\frac{Dq}{Dt}=0 \tag{2}$$

The essential nonlinear dynamics in such a system can be made implicit by transformation into the so-called geostrophic coordinates. In the context of a two-dimensional problem in the (x, z) plane. The geostrophic coordinates (X, Z) are given by

$$\begin{aligned} X &= x + v_g/f \\ Z &= z \end{aligned} \tag{3}$$

where v_g is the y-component of the geostrophic wind velocity, and f the Coriolis parameter. From eq. 3 one obtains readily $DX/Dt = u_g$ where u_g is the x-component of the geostrophic wind. The geostrophic coordinate X can be interpreted therefore as the position of an air parcel if it moves with the geostrophic wind. Or, in other words, in the geostrophic coordinate, air parcels move with the geostrophic wind. Thus the displacement of air parcels by ageostrophic wind which causes the formation of discontinuities is made implicit through the coordinate transformation.

The Jacobian of the coordinate transformation is given by

$$\frac{\partial X}{\partial x}=\frac{\zeta}{f} \tag{4}$$

where ζ is the vertical component of absolute vorticity. At the time of formation of discontinuity two parcels of different origin are brought into contact in physical space, i.e., $\Delta x = 0$, while they remain separate in the geostrophic coordinate, i.e., $\Delta X \neq 0$. At this time, $\zeta \rightarrow \infty$ at the point of discontinuity.

Under the Boussinesq approximation and the assumption of constant Coriolis parameter f, the thermal wind relation in the geostrophic coordinates has the form

$$f\frac{\partial v_g}{\partial Z}=\frac{g}{\theta_0}\frac{\partial \theta}{\partial X} \tag{5}$$

where θ_0 is a constant potential temperature giving a typical value of θ at the earth's surface. The potential vorticity is given by

$$q=\zeta\frac{\partial \theta}{\partial Z} \tag{6}$$

From eq. 6 one can derive readily that

$$\frac{1}{f^2}\frac{\partial^2\theta}{\partial X^2}+\frac{\partial}{\partial Z}\left[\frac{f\theta_0}{g}\frac{1}{q}\frac{\partial\theta}{\partial Z}\right]=0 \tag{7}$$

Provided that q is positive everywhere, the distribution of θ in the interior of the atmosphere is completely determined by the elliptical eq. 7 and the values of θ on the boundaries of domain of interest.

As an example, consider atmospheric flow forced at the synoptic scale by a deformation wind field $\mathbf{V}_d$:

$$\mathbf{V}_d=\hat{x}(-\alpha x)+\hat{y}(\alpha y) \tag{8}$$

α here is the rate of deformation which is assumed to be a constant. The total geostrophic wind can be decomposed into two components:

$$\mathbf{V}_g=\mathbf{V}'+\mathbf{V}_d \tag{9}$$

where $\mathbf{V}' = (u',v')$ is the component of $\mathbf{V}_g$ other than the imposed deformation flow. In a two-dimensional problem on the (x, z) plane in which horizontal temperature gradient exists only in the x-direction, we may assume $u' = 0$. The horizontal projection of trajectories of air parcels in the (X, Z) space then becomes easy to determine. Since in the geostrophic coordinate, air parcels move with the geostrophic wind,

$$\frac{DX}{Dt}=u_g(x=X)=-\alpha X \tag{10}$$

For a parcel located at X_0 at t=0, its position at t>0 is given by

$$X=X_0e^{-\alpha t} \tag{11}$$

This contraction along the X-axis is caused entirely by the deformation flow.

For the temperature field in the (X, Z) space, we assume

$$\theta=\theta_0+\frac{\theta_0}{g}N^2Z+\theta'(X,Z) \tag{12}$$

θ' represents a baroclinic field needed to induce significant flow features when forced by the deformation flow. N is the Brunt-Väisälä frequency. If the horizontal temperature gradient is very weak at the initial time t=0, the potential vorticity at that time is approximately given by (Hoskins and Bretherton, 1972)

$$q=f\theta_0\frac{N^2}{g} \tag{13}$$

Equation (2) then implies that q will remain at this constant value everywhere.

Assuming this constant value of potential vorticity, the distribution of θ' in the interior of the atmosphere is determined completely by the boundary values of θ' through eq. 7. We will use $\theta' \rightarrow 0$ as $Z\rightarrow\infty$ as the upper boundary condition. In this case θ' is determined by the surface temperature distribution $\theta_s(X) = \theta'(X, Z=0)$. At any given time t, the interior values of θ' are most easily determined in terms of the Fourier components of θ_s:

$$\theta_s=\frac{1}{2\pi}\int_{-\infty}^{+\infty}A(k)\,e^{ikx}\,dk \tag{14}$$

For each Fourier component, we have from eq. 7

$$\theta'(k, Z) = A(k) \exp\left(\frac{-kNZ}{f}\right) \tag{15}$$

The potential temperature field is then obtained by the Fourier integral

$$\theta'(X, Z) = \frac{1}{2\pi} \int_{-\infty}^{+\infty} A(k) \exp\left(\frac{-kNZ}{F}\right) e^{ikx}\, dk \tag{16}$$

The time evolution of θ' is therefore completely determined by the time evolution of θ_s.

To determine the distribution of θ_s at $t > 0$ from its initial distribution at $t = 0$, say θ_{si}, we note that prior to the formation of the frontal discontinuity, air parcels on the ground remain on the ground. Since θ is conserved following the motion of an air parcel, and the X-coordinate of a parcel varies in time according to eq. 10, we have

$$\theta_s(X) = \theta_{si}(Xe^{\alpha t}) \tag{17}$$

If the Fourier component of θ_{si} is given by $A_i(k)$, then in terms of $A_i(k)$

$$\theta'(X, Z) = \frac{1}{2\pi} \int_{-\infty}^{+\infty} A_i(k) \exp\left(-e^{\alpha t}\frac{kNZ}{f}\right) \exp(ike^{\alpha t}X)\, dk \tag{18}$$

Once θ' is determined from the initial conditions, the geostrophic wind field can also be determined by integrating the thermal wind relation eq. 5. Specifically, if we assume that the y-component of the geostrophic wind v' vanishes as $Z \to \infty$, we can obtain from the temperature field that

$$v' = -\frac{i}{2\pi} \int_{-\infty}^{+\infty} \frac{g}{N\theta_0} A_i(k) \exp\left(-e^{\alpha t}\frac{kNZ}{f}\right) \exp(ike^{\alpha t}X)\, dk \tag{19}$$

Note that at the ground surface $Z = 0$,

$$v'_s(X) = v'(X, Z{=}0) = -\frac{i}{2\pi} \int_{-\infty}^{+\infty} \frac{g}{N\theta_0} A_i(k) \exp(ike^{\alpha t}X)\, dk$$

$$= v'_{si}(Xe^{\alpha t}) \tag{20}$$

where v'_{si} is the surface wind distribution at the initial time. This relationship is similar to that given by eq. 17 for the surface potential temperature.

3. ENERGY SPECTRUM OF FRONTOGENETIC FLOW

The energy spectrum of air flow in a two-dimensional semi-geostrophic Eady Wave was studied by Andrews and Hoskins (1978). They showed that at the time of formation of frontal discontinuity at the ground surface, the energy spectrum obeys an $k^{-8/3}$. Prior to the formation of a front, the power spectrum follows the power law approximately over a large range of k, particularly at low wave numbers. In the context of more general semi-geostrophic frontogenesis models, these authors showed that at the time of formation of frontal discontinuity, the power spectrum approaches $k^{-8/3}$ in the short wavelength limit, i.e., as the wavenumber $k \to \infty$.

The power spectrum of air properties in the short wavelength limit, both at and prior to the formation of frontal discontinuity, is determined by the nature of the discontinuity. The possibility of formation of discontinuity in the semi-geostrophic dynamic system can be seen from the coordinate transformation given in eq. 3. Consider

$$\frac{\partial x}{\partial X} = \frac{f}{\zeta} = 1 - \frac{1}{f}\frac{\partial v'}{\partial X} \tag{21}$$

If a discontinuity forms in v' (and also in θ') in the physical space at x_0 (with the corresponding geostrophic coordinate X_0), then at this point, $\zeta \to \infty$, and $\partial x/\partial X = 0$. This corresponds to the point where $\partial v'/\partial X = f$. Now, from eq. 19

$$\frac{\partial v'}{\partial X} = \frac{1}{2\pi}\frac{g}{N\theta_0}e^{\alpha t}\int_{\infty}^{+\infty} A_i(k)\exp\left(-e^{\alpha t}\frac{kNZ}{f}\right)\exp(ike^{\alpha t}X)\,dk \tag{22}$$

Since v' decreases exponentially with height, the formation of discontinuity first takes place at the ground surface $Z = 0$. Consequently, we will limit our discussions in the rest of this paper to the ground level.

Due to the exponential growth of $(\partial v'/\partial X)$ in time, the condition of discontinuity first becomes satisfied at the place where $(\partial v'/\partial X)$ is a maximum. Therefore, at the time of formation of discontinuity t_0, we have at X_0

$$\frac{\partial x}{\partial X} = 0$$

$$\frac{\partial^2 x}{\partial X^2} = 0 \tag{23}$$

Taylor series expansion then gives

$$(x - x_0) = A_3\,(X - X_0)^3 + 0(\,|\,X - X_0|^4\,) \tag{24}$$

where

$$A_3 = \frac{\partial^3 x/\partial X^3}{6}$$

Prior to the formation of discontinuity, $\partial x/\partial X \neq 0$ everywhere on the X-axis. However, if we replace X by the complex variable $C = R + i\,D$ and consider x and v' as complex functions of C, the condition $\partial x/\partial C = 1 - (\partial v'/\partial C\,)/f = 0$ may be satisfied at a point $C = C_0 = R_0 + iD_0$ on the complex plane. Clearly, as $t \to t_0$, $R_0 \to X_0$ while $D_0 \to 0$.

As to the functional form of v'(C), we note from eq. 20 that at $t<t_0$,

$$v'(X, t) = v'(Xe^{\alpha(t-t_0)}, t_0) \tag{25}$$

Except for the change in scale by a factor which varies exponentially in time, v' has the same functional form in X, and therefore in C, for all time $t<t_0$. Therefore, by analytic continuation, Taylor series expansion around C_0 should take the same form as in eq.24:

$$(x - x_0) = A_3 (X - C_0)^3 + 0(|X - C_0|^4) \tag{26}$$

where

$$x_0 = x(C_0) \text{ and } A_3 = \frac{\partial^3 x/\partial C^3}{6}.$$

As is well known in the theory of Fourier transformation, the asymptotic behaviour in the short wavelength limit of the Fourier transform of a function h(x) depends on the singularity of h or its derivatives. Consider the Fourier transform of h(x):

$$h(k)= \int_{-\infty}^{+\infty} h(x)\, e^{-ikx}\, dx \tag{27}$$

If h(x) has a singularity at a complex location $x = r_0 + i\delta_0$ in the neighbourhood of which it is characterized by

$$h(x) = b\,(x - x_0)^\rho \text{ as } x \to x_0 \tag{28}$$

then the asymptotic behaviour of h(k) as $k \to \infty$ is given by (Frisch, 1983)

$$h(k) \underset{k\to\infty}{\approx} \frac{2 \sin(\pi/\rho)\, \Gamma(\rho + 1)}{k^{\rho+1}} e^{-i\pi\rho/2}\, \varepsilon\, b\, e^{ikx_0} \tag{29}$$

where ε is a determination factor. If the singular point x_0 is on the real axis, i.e., $\delta_0 = 0$, then the spectrum obeys the power law $k^{-(\rho+1)}$, as $k\to\infty$. On the other hand, if $\delta_0 > 0$, then $h(k) \to e^{-k\delta_0} k^{-(\rho+1)}$ as $k\to\infty$. It deviates from the power law by the exponential factor $e^{-k\delta_0}$.

To study the spectrum of air properties in a semi-geostrophic flow, we follow the technique introduced by Andrews and Hoskins (1977). Consider physical variable h(x) and its representation H(X) in the geostrophic coordinate. A discontinuity forms in h(x) when there is a singularity in the coordinate transformation X = X(x), assuming that H(X) is well behaved. The Fourier transform of h(x) is given by

$$h(k)= \int_{-\infty}^{+\infty} h(x)\, e^{-ikx}\, dx = (ik)^{-1} \int_{-\infty}^{+\infty} \frac{\partial h}{\partial x} e^{-ikx}\, dx \tag{30}$$

Since

$$\frac{\partial h}{\partial x} = \frac{\partial H/\partial X}{\partial x/\partial X} \tag{31}$$

and near the singular point x_0,

$$\frac{\partial x}{\partial X} \approx 3A_3 (X - C_0)^2 \approx 3A_3^{1/3} (x - x_0)^{2/3} \tag{32}$$

we have

$$h(k) \underset{k\to\infty}{\approx} -i\frac{2}{3}\Gamma(1/3) e^{i\pi/3}\, \varepsilon\, A_3^{-1/3} \left[\frac{\partial H}{\partial X}\right]_{C_0} e^{ikr_0}\, k^{-4/3}\, e^{-k\delta_0} \tag{33}$$

provided that $\partial H/\partial X$ is well-behaved. Prior to the formation of discontinuity, the spectrum of h(k) decays exponentially for large k. At the time of formation of discontinuity $t = t_0$, the spectrum decays as $k^{-4/3}$. In particular, if we let $h(x) = v'$, the spectrum of v' decays as $k^{-4/3}$ for large k, at $t = t_0$. It follows that the spectrum of $(v')^2$ decays as $k^{-8/3}$ and the spectrum of absolute vorticity decays as $k^{-1/3}$ at the instant of front formation.

The deviation from the power law prior to the formation of the front depends on the magnitude of δ_0. To obtain an estimate for its magnitude, consider one Fourier component:

$$\theta_{si}(X) = \Delta\theta \sin(l\,x) \tag{34}$$

From (18) we obtain for the temperature field

$$\theta'(X, Z, t) = \Delta\theta \exp\left(-e^{\alpha t}\frac{l\,NZ}{f}\right) \sin(l\,e^{\alpha t}X) \tag{35}$$

The wind field is then obtained from (19)

$$v' = -\frac{g}{N\theta_0}\Delta\theta \exp\left(-e^{\alpha t}\frac{l\,NZ}{f}\right) \cos(l\,e^{\alpha t}X) \tag{36}$$

At the ground level Z = 0, we have

$$v' = -\frac{g}{N\theta_0}\Delta\theta \cos(l\,e^{\alpha t}X) \tag{37}$$

The absolute vorticity is given by

$$\frac{f}{\zeta} = \frac{\partial x}{\partial X} = 1 - \frac{g\,l\,\Delta\theta}{fN\theta_0} e^{\alpha t} \sin(l\,e^{\alpha t}X) \tag{38}$$

Discontinuities form on the ground when $\partial x/\partial X = 0$. This happens at the time t_0 when $\beta(t_0) = 1$ where

$$\beta(t) = \frac{g\,l\,\Delta\theta}{fN\theta_0} e^{\alpha t} \tag{39}$$

The locations of the discontinuities are given by

$$l\,e^{\alpha t_0} X_0 = (2n + 1/2)\pi \quad \text{for } n = 0,1,2,... \tag{40}$$

Prior to t_0, $\beta < 1$, $\partial x/\partial X = 0$ only for complex values of $X = C_0 = R_0 + iD_0$ with

$$l\,e^{\alpha t} R_0 = (2n + 1/2)\pi \quad \text{for } n = 0,1,2,...$$

$$l\,e^{\alpha t} D_0 = \cosh^{-1}\left(\frac{1}{\beta}\right) \tag{41}$$

Corresponding to these complex values of X, the value of x is given by

$$x = x_0 = r_0 + i\delta_0 = R_0 + i\left(D_0 + \frac{\sqrt{1-\beta^2}}{e^{\alpha t}}\right) \tag{42}$$

The exponential decay of the spectrum at large wavenumbers prior to the formation of discontinuity is determined by

$$\delta_0 = \frac{\left[\cosh^{-1}\left(\frac{1}{\beta}\right) + \sqrt{1-\beta^2}\right]}{l\, e^{\alpha t}} \tag{43}$$

We note that $l\, e^{\alpha t}$ is the wavenumber of the basic Fourier Component at time t in the X space. The wavenumber at which the spectrum deviates from the power law by a factor of e , in the unit of $l\, e^{\alpha t}$, is

$$\frac{k}{l\, e^{\alpha t}} = \left[\cosh^{-1}\left(\frac{1}{\beta}\right) + \sqrt{1-\beta^2}\right]^{-1} \tag{44}$$

This ratio approaches infinity as $\beta \to 1$. The values of this ratio for $\beta = 0.99$ and 0.90 are 3.57 and 1.10 respectively. Therefore the power law gives a good approximation for the spectrum only when the development of the system is fairly close to the formation of discontinuity.

4. INTERMITTENCY

At the time of formation of discontinuity, the absolute vorticity ζ becomes infinity at x_0. Large values of ζ occur in a small neighbourhood about x_0. It is of interest to estimate the size of the region in which the value of vorticity exceeds, say, a large value ζ_0. From eq. 32, the values of vorticity in a small neighborhood around x_0 is given by

$$\frac{\zeta}{f} = \frac{1}{3} A_3^{-1/3} (x - x_0)^{-2/3} \tag{45}$$

The region in which $\zeta/f > \gamma$ for large values of γ is given by

$$| x - x_0 | < (3A_3^{1/3}\, \gamma)^{-2/3} \tag{46}$$

If the sinusoidal temperature field given by eq. 34 is assumed as the initial field, then all flow variables will remain periodic in the physical coordinate x. The periodicity can be determined from eqs. 3 and 36:

$$\Delta x = \frac{2\pi}{l\, e^{\alpha t_0}} \tag{47}$$

The probability for $\zeta/f > \gamma$ may be defined in this case as

$$\Pr(\zeta/f > \gamma) = \frac{(3A_3^{1/3}\, \gamma)^{-3/2}}{\Delta x} = \frac{1}{2\pi}\, l\, e^{\alpha t_0} (3A_3^{1/3}\, \gamma)^{-3/2} \tag{48}$$

Since

$$A_3 = \frac{1}{6} \frac{g \Delta\theta}{f N \theta_0} (l\, e^{\alpha t_0})^3, \tag{49}$$

We have

$$\Pr(\zeta/f > \gamma) = \frac{1}{2\pi}\left(\frac{9}{2}\,\frac{g\Delta\theta}{fN\theta_0}\,l\,e^{\alpha t_0}\right)^{-1/2}\gamma^{-3/2} \tag{50}$$

The probability is asymptotically hyperbolic for large γ.

5. ACKNOWLEDGEMENTS

This research was supported by research grants from the Natural Sciences and Engineering Research Council of Canada and Canadian Atmospheric Environment Service. The author wishes to thank Ms. Anna Reale for typing the manuscript.

6. REFERENCES

Andrews, D. G., and B. J. Hoskins, 1978: Energy spectra predicted by semi-geostrophic theories of frontogenesis. J. Atmos. Sci., 35, 509-512.

Frisch, U., 1983: Fully developed turbulence and singularities. Chaotic Behaviour of Deterministic Systems. G. Iooss, R.H.G. Helleman, R. Stova, Eds., North-Holland, Amsterdam, 665-704.

Hoskins, B. J., and F. P. Bretherton, 1972: Atmospheric frontogenesis models: Mathematical formulation and solution. J. Atmos. Sci., 29, 11-37.

Hoskins, B. J., 1975: The geostrophic momentum approximation and thesemi-geostrophic equations. J. Atmos. Sci., 32, 233-242.

Lilly, D. K., 1983: Mesoscale variability of the atmosphere. Mesoscale Meteorology - Theories, Observations, and Models. D.K. Lilly and T. Gal-Chen, Eds., D. Reidel, New York, 13-24.

Lovejoy, S., and B. Mandelbrot, 1985: Fractal properties of rain and a fractal model. Tellus, 37A, 209-232.

Lovejoy, S., and D. Schertzer, 1986: Scale invariance, symmetries, fractals and stochastic simulations of atmospheric phenomena. Bull. Amer. Meteor. Soc., 67, 21-32.

Lovejoy, S., D. Schertzer, and P. Ladory, 1986: Fractal characterization of inhomogeneous geophysical measuring networks. Nature, 219, 43-44.

Schertzer, D., and S. Lovejoy, 1985: Generalized scale invariance in turbulent phenomena. P.C.H. Journal, 6, 623-635.

HELICITY FLUCTUATIONS AND COHERENCE IN DEVELOPED TURBULENCE

E. Levich and L. Shtilman
The Benjamin Levich Institute for
Physico-Chemical Hydrodynamics
The City College of
The City University of New York 10031

ABSTRACT. Arguments are presented indicating the existence of large fluctuations of helicity and reflectional asymmetry of turbulence. Available numerical and experimental data to support this conjecture is reviewed.

1. INTRODUCTION

The last decade has been a time of changes in generally conservative and unrewarding studies of turbulent flows. Partially this may be seen in the light of aggressive optimism emanating from the computational fluid mechanics community. Indeed, the authors have often heard the opinion that the advances in computer capabilities should be able to solve the puzzles of turbulence in a relatively near future. Although to the best of our understanding, such high expectations tend to be inflated, still direct numerical simulations present splendid possibilities for experimentation in turbulence arising in flows with simple geometries and low Reynolds numbers. This is especially true for the quantities containing velocity fluid derivatives such as the vorticity field, or the rate of energy dissipation at a point. Clearly these quantities are at the core of turbulent phenomena. This is recognized by a majority of fluid dynamicists. Yet neither the vorticity field nor the local rate of energy dissipation had been the subject of scrutiny in laboratory measurements. This is because of the significant technical difficulties associated with the measurement of all nine velocity derivatives. It is only recently that the experimental situation has changed for the better and the vorticity field and other related quantities have been measured first time in turbulent flows (Balint et al., 1987; Wallace, 1986; Kit et al., 1986). On the other hand, direct numerical simulation of turbulence is simply a solution of the discretized Navier-Stokes (NS) equation and hence allows the calculation of the velocity field derivatives with reasonable accuracy, provided that the numerical resolution is sufficient. Thus it allowed the testing of certain very novel concepts with respect to the topology related properties of the vorticity field.

The vorticity field was known long ago to possess a fascinating topology. In inviscid fluids, the lines of vorticity field are "frozen in", i.e., they are rigidly attached to fluid particles. Hence, their topology, simple or complex, is invariant under all smooth deformations caused by the flow dynamics. Some, although not all, of the topology of vorticity lines is described by the helicity invariant, a quantity well known in magnetohydrodynamics and independently recovered in fluid mechanics by Moreau (1961) and Moffatt (1969). The helicity invariant can be interpreted as a certain averaged asymptotic knottedness of the vorticity lines. In difference to the real mathematical knots, the asymptotic knots are not necessarily between the closed vector field lines. This is advantageous, since the vorticity field lines are not necessarily closed. They can be ergodic or wound about the torus, or any other similar surface. The rigorous mathematical analysis of helicity was given by Arnold (1974). Mathematically, helicity is given by the following expression

D. Schertzer and S. Lovejoy (eds.), Non-Linear Variability in Geophysics, 13–29.

$$H = \int_D \mathbf{v} \cdot \omega \, d^3 r \tag{1}$$

where the vorticity field ω = curl v and D is any domain closed by the vorticity lines surface, e.g., $\mathbf{v} \cdot \omega |_{\partial D} = 0$, in particular over infinite space. It should be noted that helicity is a specifically vorticity related quantity. Indeed, any gauge transformation to a nonphysical field $\mathbf{v}' = \mathbf{v} + \nabla\phi$ leaves the value of H invariant. A particular consequence of gauge invariance is the Galilean invariance of H.

Due to the rarity of invariant of inviscid flows, helicity attracted immediate attention, as a potentially important quantity for turbulence (André et al., 1977; Kraichnan et al., 1973). However, interest quickly dissipated with a rather unsubstantiated conclusion that helicity is irrelevant (Brissaud et al., 1973). Partially, this happened, as we view it, as the result of a wrong attitude to helicity as a quantity on a par with energy. The question asked was mainly like this. Is it possible that helicity may impose a serious constraint on the energy cascade picture, as for example, the enstrophy conservation does in the case of 2-D turbulence (Kraichnan et al., 1980). The answer was, as a rule, no (Brissaud et al., 1973), although in André et al. (1977) it was demonstrated that helicity can retard the turbulence decay quite significantly.

It should be noted that helicity is a pseudoscalar, and as such is zero for reflectionally symmetric flows. In particular, for a statistical ensemble of realizations in isotropic turbulence, the ensemble averaged helicity density is

$$\lim_{V(D)\to\infty} \frac{\langle \int_D \mathbf{v} \cdot \omega \, d^3 r \rangle}{V(D)} = \langle \mathbf{v} \cdot \omega \rangle = 0. \tag{2}$$

Similarly, in Fourier space the ensemble averaged spectral helicity density in isotropic turbulence is

$$h(k) = \langle H(\mathbf{k}) \rangle \, k^2 = 0 \tag{3}$$

where for H(k) one can derive easily (Levich et al., 1987):

$$H(\mathbf{k}) = \frac{1}{k^2} \varepsilon_{ijl} \, k_i \, \omega_j(\mathbf{k}) \, \omega_l(-\mathbf{k}) \tag{4}$$

and evidently in each realization

$$H = \int_D \mathbf{v} \cdot \omega \, d^3 r = \int_D H(\mathbf{k}) \, d^3 k \tag{5}$$

The conclusion made was that if helicity is important it is only for reflectionally asymmetric turbulence with a non-zero average helicity density. But then it was supposed to be concentrated at certain large scales, e.g., $\int H(k)\,dk$ to be determined by the small values of wave number $|\mathbf{k}|$. At small scales helicity was seen as possibly convected by the velocity field, similarly to the passive scalar additive, and not influencing the small scale dynamics. The above reasoning is, however, quite dangerous due to the same simple reason that helicity is not a scalar and takes both positive and negative values. Consequently, conclusions obtained from dimensional consideration, highly successful for the scalar quantities, such as the mean energy spectral density E(k), or the mean enstrophy spectral density $\Omega(k) = k^2 E(k)$, cannot be readily applied to the helicity spectral density H(k). Even when the mean H(k) = 0, the nonaveraged quantity H(k) can be large in various domains of k-space in typical realizations of turbulence. Moreover, it may happen that the total helicity $\int H(\mathbf{k})\,d\mathbf{k}$ in particular realizations is not zero, though $\langle H \rangle = 0$. In other words, helicity can fluctuate strongly from realization to realization, and also spatially and temporatily, within each of the realizations.

2. LARGE SCALE HELICITY FLUCTUATIONS

To illustrate precisely what is meant by the fluctuations of helicity, we start from the latter case. Let us consider the helicity density balance equation (Moffatt et al., 1978):

$$\frac{D\gamma}{Dt} = \frac{D(\mathbf{v}\cdot\boldsymbol{\omega})}{Dt} = \nabla\cdot\boldsymbol{\omega}\,(-P + v^2/2) + \text{viscous term}$$
$$= \nabla\cdot\mathbf{F} + \text{viscous terms} \tag{6}$$

where

$$\frac{D}{Dt} = \frac{\partial}{\partial t} + (\mathbf{v}\cdot\nabla)$$

Applying eq. 6 at two points **x** and **x** + **r**, averaging and making use of homogeneity, one obtains

$$\frac{D}{Dt}\langle\gamma(\mathbf{x})\gamma(\mathbf{x}+\mathbf{r})\rangle = \nabla_r\left[\langle\gamma(\mathbf{x})\mathbf{F}(\mathbf{x}+\mathbf{r}) + \gamma(\mathbf{x}+\mathbf{r})\mathbf{F}(\mathbf{x})\rangle\right] + \text{viscous terms} \tag{7}$$

and after integrating over the infinite volume we arrive at the following (Levich et al., 1983)

$$I = \int\langle\gamma(\mathbf{x})\gamma(\mathbf{x}+\mathbf{r})\rangle\, d^3r \equiv \text{invar.} \tag{8}$$

with the accuracy of viscous terms. In other words, the integrated helicity density two point correlation function is conserved by the nonlinear terms of the NS equation, similarly to the averaged energy density E.

To estimate the I-invariant, it is advantageous to consider $\langle\gamma(\mathbf{x})\gamma(\mathbf{x}+\mathbf{r})\rangle$ in Fourier space of wave numbers. We can observe readily that this is a local quantity in **k**-space. The fourth order correlation function can be vastly simplified if the quasi-normal assumption is made with respect to the statistical properties of **v**(**k**). Then it can be shown that if the angle $\psi_\mathbf{k}$ between Re **v**(**k**) and Im **v**(**k**) is a random quantity then

$$I = A\int[E(k)]^2\, dk \qquad \text{where } A = \text{const.} \tag{9}$$

What is the necessary requirement for the I-invariant not to be zero in terms of the total helicity in various realizations? It can be shown that the alternative representation of the I-invariant for homogeneous turbulence is as follows (Levich et al., 1983, 1987), with random $\psi_\mathbf{k}$:

$$I = \lim_{V\to\infty}\frac{\langle H^2\rangle}{V} \equiv \lim_{V\to\infty}\frac{\int\langle[H(k)]^2\rangle\, dk}{V^2} \tag{10}$$

Thus it is not zero if typically $\int \mathbf{v}\cdot\boldsymbol{\omega}\, d^3r \propto V^{1/2}$. This latter is not in contradiction with the condition of ensemble average reflectional symmetry.

Indeed,

$$H = \lim_{V \to \infty} \frac{\int \mathbf{v} \cdot \omega \, d^3r}{V} \propto \lim_{V \to \infty} V^{-1/2} \to 0. \tag{11}$$

where we assumed that by the order of magnitude we can estimate the ensemble average as the volume average. The conclusion is that the value of I can be large, although the ensemble averaged helicity is zero, and even helicity in typical realization divided by the volume, the volume averaged helicity, tends to zero, but relatively slowly, as $V^{-1/2}$.

We observe that in fact only certain largest scales contribute significantly to the I-invariant value. Indeed, let us filter the high wave number harmonics from the velocity field at each realization constructing the filtered fields $\mathbf{v}^F(\mathbf{k})$ and $\omega^F(\mathbf{k})$. The phenomenology of Kolmogorov theory indicates that the large scale harmonics $\mathbf{v}^F(\mathbf{k})$ are the energy containing ones. On the contrary the average enstrophy density $\langle \omega^2 \rangle = \int \Omega(k)\, dk$ is mainly determined by the small scale harmonics, the ones filtered out. At least this is true for high Reynolds number flows with an extensive inertial range of space scales. Hence we can conclude quantitatively that in a typical realization the $\omega^F(\mathbf{k})$ field amplitudes are small. Still, certain topological properties of the nonphysical $\omega^F(\mathbf{k})$ field lines are conserved on average by the nonlinear terms of the NS equation. This follows from the observation of locality of the I-invariant in **k**-space which follows from the definition (9)

$$I = \int \langle \gamma(\mathbf{x})\gamma(\mathbf{x} + \mathbf{r}) \rangle \, d^3r \;\cong\; \int [E^F(k)]^2 \, dk \tag{12}$$

where we have used the same quasi-normal assumption for the $\mathbf{v}^F(\mathbf{k})$ correlator, as we did previously for the $\mathbf{v}(\mathbf{k})$ itself. We again assumed that the angle between Re $\mathbf{v}^F(\mathbf{k})$ and Im $\mathbf{v}^F(\mathbf{k})$ is random.

We conclude from the above that the existence of the I-invariant signifies generation of large scale helicity fluctuations. In physical space such fluctuations should be characterized by a certain coherence, e.g., alignment of $\mathbf{v}^F$ and ω^F. This can be seen in terms of filtered ralative helicity

$$H^F_{rel} = \frac{\int \mathbf{v}^F \cdot \omega^F \, d^3r}{\left[\int (\mathbf{v}^F)^2 \, d^3r \int (\omega^F)^2 \, d^3r \right]^{1/2}} \tag{13}$$

provided that the fluctuations are bounded by the filtered vorticity surfaces $\omega^F \cdot \mathbf{n} = 0$. This latter is not problematic when the scale of fluctuations is of the same order as that of the flow itself. Alternatively, we can expect that the helicity spectral density H(k) should exhibit significant fluctuations at the corresponding small wave number values $|\mathbf{k}| \sim L_I^{-1}$. In fact, a much stronger statement can be made with respect to H(**k**). Indeed, by virtue of Kraichnan (1980) we observe that the maximum of the I-invariant is reached when $|H(\mathbf{k})|$ is maximal for all **k** in contributing realizations of turbulence. This is not at all in contradiction with the fact that the ensemble $\langle H(\mathbf{k}) \rangle = 0$.

The important consequence of the I-invariant is a possibility of the "inverse cascade" of helicity fluctuations (Moffatt (1978) and Levich et.al. (1985)). This follows from the fact that it is incompatible to have both the I-invariant (9) and the mean energy density $E = \int [E(k)]dk$ conserved in the Kolmogorov inertial range $E(k) \propto k^{-5/3}$. This is similar to the incompatability of having the energy and enstrophy densities to be conserved in one inertial range in 2-D turbulence and subsequent inverse energy cascade. In the case of 3-D turbulence, however, the energy cascades to small scales as it should, but the energy cascade does not allow the I-invariant to cascade to small scales. Hence the conclusion that the I-invariant cascades upscale. It should be noted that the terminology of inverse cascade is quite inappropriate for a quantity such as the I-invariant. One should rather speak in terms of large scale instability and the growth of helical correlation length. The subject of large scale instabilities is very popular these days. Some of the results seem supportive of our concept of the growth of helical correlation length and subsequent generation of large scale helical structures (Levich and Tzvetkov (1985)). Still the problem of 3-D "inverse cascade" is far from being resolved.

In particular in the papers Baily et al. (1986) and Frisch et al. (1987), the claim is made that the large-scale instability may not take place in quasi-isotropic flows even with nonzero total helicity, in particular in quasi-isotropic homogeneous turbulence with nonzero mean helicity. We remark that most probably the inverse cascade of the I-invariant is terminated by the appearance of a certain coherence, e.g. the coherence in the angle between Re **v**(**k**) and Im **v**(**k**) and subsequent invalidation of (10) (Levitch and Shtilman, 1988; Kit et al., 1988).

Summarizing the above considerations, the assertion can be made that the large scales of turbulence are likely to show a well-defined coherence, e.g., the large scale fluctuation of helicity, exhibited either in the spectral helicity density or equivalently through the alignment of filtered velocity and vorticity fields. We would like to note that the large-scale fluctuations of helicity contain a hierarchy of small scale excitations, as well. These small scale excitations bury the coherence of larger scales. Thus the problem is to extract the buried coherent features from the vast sea of chaos consiting of the small-scale excitations. This is easiest to do by the filtering out of the small scales in the Fourier space of scales.

We should like to point out a rather obvious similarity of the above ideas with the concept of coherent structures in turbulence, in particular with the definition perception and eduction scheme of coherent structures pursued by Hussain (1986).

3. COHERENT STRUCTURES

So far no one has come up with a totally satisfactory definition of coherent structures in turbulent flows, despite the enormous experimental effort in seeking and identifying these structures. Basically it is usual to understand coherent structures as a certain large-scale underlying motion in turbulence with distinct coherent characteristics as opposed to the prevailing chaos of small scale motion in the interior and outside of the structures. The coherent structures are usually associated with the turbulent shear flows and the possibility of their existence is silently disregarded in "homogeneous isotropic" turbulence. This, in our view is mainly a prejudice leaving its origin in an imperfect definition of the meaning of coherent structures.

After works of (Hussain, 1986), it has become much clearer that coherent structures can be defined as regions of large scale coherent vorticity recurrently present in typical realizations of turbulent flows. This definition allowed a successful methodology for educing the coherent structures in certain turbulent flows. The difficulties, however, are significant. Indeed, the above definition of coherent structures presumes the full knowledge of the 3-D large-scale vorticity field. This should be provided by experimental data on the vorticity field $\omega(\mathbf{k})$ and subsequent filtering out of the small-scale harmonics in Fourier space in each realization of turbulence. The next step is to seek a coherent pattern of the filtered large-scale vorticity field $\omega^F(\mathbf{k})$ similar in typical realizations. If such a pattern is recognized it is argued to represent a coherent structure (Hussain, 1986). In practice the situation is complicated. Notwithstanding the difficulties of measuring the vorticity field, we shall discuss the recent successes in this regard shortly, the comparison of the 3-D $\omega^F(\mathbf{k})$ pseudo-vorticity field in various realizations may present serious difficulties.

To illustrate these, consider the vorticity field in the regions of a turbulent shear flow where the spanwise vorticity component is dominant. There, even though the genuine vorticity field is 3-D, the spanwise vorticity component may be sufficient as a first approximation to identify the domains of large-scale correlated vorticity in various realizations simply as regions of, say, clockwise and anticlockwise rotation vorticity contours in the plane. Such identification is the first step in an elaborate procedure of coherent structures education. It provides data on the 2-D projection of the structure footprints. If structures are 3-D, and the evidence is that they almost always are, then 2-D vorticity contours are not sufficient. The 3-D structures may have a very complicated topology of the ω^F-field and this would not be revealed in 2-D projections. It is likely that the ω^F-field should have what is intuitively perceived as a very knotted structure of tangled and twisted vorticity lines. In two realizations this structure can be quite different, but at the same time the 2-D contours could appear similar. Such domains clearly cannot be considered as footprints of the same structure. The above simple example illustrates the point that investigation of the ω^F-field should start from its topology. The question is: given two different realizations in a certain bounded flow domain what is the set of rules for determining whether ω^F-field lines in these two realizations are similar, i.e., reflect the presence of footprints of the same structure(s). The conjecture made in Tsinober and Levich (1983) was that coherent structures should possess a significant coherent helicity. It was pointed out in Hussain (1986) that in fact the relevant quantity is the coherent helicity corresponding to the filtered $\omega^F(r)$ pseudo-vorticity field. It is quite clear that one is concerned with the coherence of the large scale vorticity harmonics buried in the sea of small scale chaotic motions. Summarizing, we should like to stress our belief that turbulent coherent structures are likely to be associated with the large scale fluctuations of helicity, in the sense defined in Section 1.2. Although the generation of structures may be amplified in turbulent shear flows, still in our view, the coherence defined above should be an intrinsic feature of all turbulent flows.

4. HIERARCHY OF HELICAL FLUCTUATIONS

At this point it is reasonable to critically ask ourselves whether the large amplitude fluctuations of helicity and associated coherence are the property of large scales only, or may these be present at all scales of turbulence. The importance and intrinsic nature of large amplitude helicity fluctuations in turbulent flows was conjectured in a number of papers (Balint et al., 1987; Wallace, 1986; Kit et al., 1986; Moreau, 1961; Moffatt, 1969; Arnold, 1974). Moreover, it was conjectured that these form a self-similar hierarchy at the scales in the inertial range. In its initial form, the conjecture about self-similar helicity fluctuations at various scales was presented as follows (Levich et al., 1984; Levich, 1984; Moffatt, 1984, 1985; Berger and Field, 1984): any volume $V(L_0)$ of turbulent flow such that L_0 is a scale from the inertial range may contain helical fluctuations of all scales $L_i < L_0$ belonging to the inertial range, like the Russian "matrejca" and different signs of helicity, so that the average helicity is zero, or small. Evidently, the helical fluctuations of all scales with the volumes $V(L_{i+1})<V(L_i)...<V(L_0)$, and corresponding lifetimes $\tau(L_{i+1})<\tau(L_i)$,etc. form a self-similar hierarchy. It was also conjectured that they are separated from each other by convoluted boundaries, the seats of intensive dissipation and the energy cascade is retarded in the hierarchy of fluctuations. The presence of boundaries is important, since if we want to assign helicity to a fluctuation during its life-time, this should be bounded by a pseudo-vorticity surface. Without such a boundary, the helicity integral and the mean helicity density are not invariant under the gauge transformation $v \rightarrow v + \nabla\phi$, and hence contains spurious ϕ dependent surface terms. This kind of difficulty is well known to exist for the helicity related problems in magneto-hydrodynams (Berger and Field, 1984).

The conjecture of a hierarchy of helical fluctuations in turbulence was inspired and influenced by the experimental observations in the solar wind of magnetohydrodynamical helicity fluctuations (Matthaeus et al., 1982). It was found that although the main contribution to the mean magnetic helicity comes from the large scale Fourier harmonics, the magnetic helicity spectral density fluctuates strongly at all small scales in the inertial range. The conclusion made in Matthaeus et al. (1982) was that the solar wind contains "clumps" of helicity of various scales and opposite signs.

It is quite a subtle matter to try to educt the small scale helicity fluctuations should they exist at all. The difficulty is quite similar to that in identification of coherent structures. Indeed, each of the helical fluctuations should contain the hierarchy of fluctuations of smaller scales. Hence the underlying coherence may be completely buried. The education scheme should closely follow the same pattern as for the large scale helicity fluctuations. Let us consider a domain characterized, for simplicity by one scale L_0, as before from the inertial range. We introduce the auxiliary field $\tilde{\omega}$ such that $\tilde{\omega} = \omega(r)$ inside D, but $\tilde{\omega} \rightarrow 0$ exponentially outside D, on the scale much smaller than L_0. The helicity intergral corresponding to the $\tilde{\omega}$-field is

$$\tilde{H} = \int \tilde{v} \cdot \tilde{\omega}\, d^3r \approx \int_D \tilde{v} \cdot \tilde{\omega}\, d^3r \qquad (14)$$

The corresponding helicity spectral density $\tilde{H}(|k|)$s given by the substitution of $\tilde{\omega}$ instead of $\omega(\mathbf{k})$ in eq. (5). If the domain D contains the large amplitude helical fluctuation of the scale L_0, then by analogy with the above reasoning the $\tilde{H}(|\mathbf{k}|)$spectral function should exhibit a large amplitude value for $|\mathbf{k}| \sim 1/L_0$. Alternatively, one can expect that the corresponding $v^F(r)$ and $\tilde{\omega}^F$ should be strongly aligned in physical space. The same procedure can be iterated for the subdomains of D. It should be pointed out that contrary to the large scale fluctuations where it was not difficult to identify the boundaries made by the filtered vorticity field, $\omega^F \cdot \mathbf{n} = 0$, say at infinity, for the small scale fluctuations in a given domain, a priori, we do not know whether they exist or not. Thus the necessity of introducing the pseudo-vorticity field $\tilde{\omega}$. For a more detailed discussion of this difficulty and various ways of its resolution, we refer to Levich (1987). Another even more essential difference is that the large scale fluctuations of helicity can be associated with the I-invariant and a certain scale singled out by this quantity. On the contrary, the small scale fluctuations are likely to be self-similar for all scales in the inertial range. Consequently, the large scale helicity fluctuations, or structures may be expected to have an enhanced stability, whereas the small scale helicity fluctuations may not. We note also, that for any flow domain D one can introduce a quantity similar to the I-invariant but build on the basis of the $\tilde{\omega}$-field. Then a relation between the spatial fluctuations of helicity and the fluctuations of $\tilde{H}$, similar to (10) can be derived.

5. NUMERICAL EXPERIMENTS

It is timely to discuss the degree to which the above conjectures can be supported by experiment. The attention will be given to numerical experiments.

The first relevant observations (Shtilman et al., 1985; Brachet et al., 1983) were made in numerical simulations of the decaying Taylor-Green vortex (Code originator, M. Brachet). The first quantity computed in the experiment was the probability distribution function (Pdf) of the relative helicity density, i.e., Pdf of the cosine of the angle between the v and ω fields, P(cos θ) where

$$\cos\theta = \frac{\mathbf{v} \cdot \omega}{|\mathbf{v}|\,|\omega|} \tag{15}$$

We note that the helicity density $\mathbf{v} \cdot \omega$, and cos θ are obviously not gauge invariant and generally not Galilean invariant. This is in difference to the spectral helicity density H(k) which is invariant under these transformations. The Galilean invariance is restored, however, for the product $(\mathbf{v} - \langle \mathbf{v} \rangle) \cdot \omega$ and the corresponding cos θ. In the case of isotropic (not necessarily reflectionally symmetric) flow $\langle \mathbf{v} \rangle = 0$.

It turned out that P(cos θ) shows two distinct peaks for the values cos $\theta = \pm 1$, instead of being flat as the simple minded intuition would tend to expect. This latter would undoubtedly be the most probable case should we consider a well constructed random velocity field, not the solution of the NS equation. Intuitively we felt that the peaks signify the presence of helicity fluctuations in the flow, though we did not have exact understanding as to how to relate the two convincingly. The calculation of E(|k|) showed that it is almost exactly zero, as it should be for the T-G vortex due to the imposed symmetry conditions. The detailed analysis of nonaveraged H(k) and H(|k|) in various domains of the flow was not done, since we were under the pressure of proving the generality of the peaks in P(cos θ). This was done for a variety of flows in particular for the homogeneous decaying turbulence in (Pelz et al., 1986). The results were quite similar in all the above experiments. Briefly they can be formulated as follows

1) In all the cases the final shape of P(cos θ) corresponding to the developed stage of numerical turbulence showed a good deal of similarity in having very similar peaks at cos $\theta = \pm 1$. However, in some of the cases like in Kerr and Gibson (1985) only one peak at cos $\theta = +1$ showed up. The P(cos θ)'s for different runs of quasi-isotropic homogeneous turbulence also showed a noticeable but different extent of asymmetry for different runs at cos $\theta = 1$ and cos $\theta = -1$ values.

2) In the case of quasi-isotropic homogeneous turbulence all runs exhibited a noticeable global reflectional asymmetry, e.g. the mean helicity density $\langle \mathbf{v} \cdot \omega \rangle_V \neq 0$. The $\langle\ \rangle_V$ means the averaging over the mesh points, as in usual numerical simulations. We are reminded that numerical simulations produce

one or at best few realizations of turbulent flows. Therefore no genuine ensemble averaging can be done. The dynamical generation of reflectional asymmetry was obvious even in a specially devised run corresponding to an almost complete absence of the mean helicity density for the initial flow, at t = 0 (Pelz et al., 1986).

3) The shape of P(cos θ) appeared to be remarkably similar when considered in the partial domain of the flow volume. In the case of homogeneous turbulence no conditional sampling with regard to the rate of energy dissipation and vorticity squared showed a noticeable difference, even when the corresponding domains were only fractions of the total flow volume (Pelz et al., 1986; Kerr and Gibson, 1985).

Lately, arguments have been heard above the results with respect to P(cos θ) and the role of helicity in general. The arguments partially stem from a confused understanding of various quantities such as helicity, helicity density, etc. An example of a sometimes confused understanding of the ideas expressed in this paper results in the following point of view. Since the peaks of P(cos θ) are small and do not depend on the value of energy dissipation the effect is not important for turbulence. Such a point of view is not constructive and in our view is not worthy of a prolonged discussion. The fact is that the effect is persistent and seemingly manifests some intrinsic physics of turbulence. Our task is to try to understand whether it reflects the physics compatible with our conjectures, such as the self-similar helicity fluctuations, or not. If not then we have to understand the underlying physics anyway. A relation with other dynamical processes should be sought in what follows.

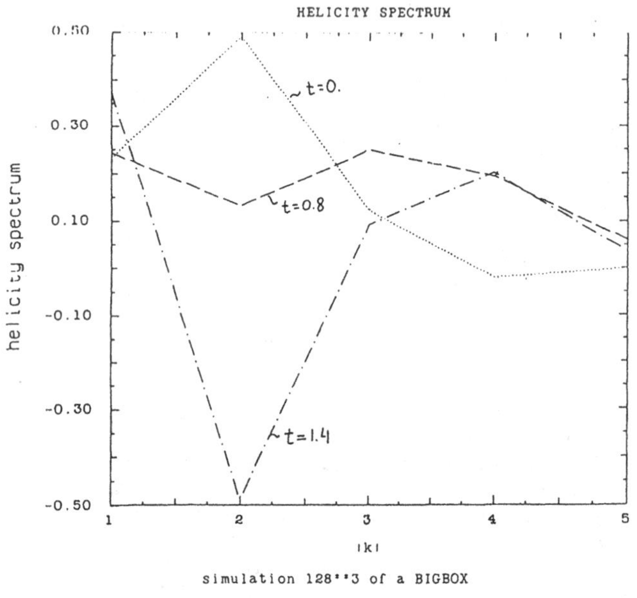

Fig. 1: Helicity spectrum (only modes with |k| ≤ 5 are shown).

To clarify some of the questions posed above we considered yet another run of quasi-isotropic homogeneous turbulence. We used BIGBOX code modified to enable 128x128x128 resolution, and the data base described in Orszag and Patterson (1972) and Shtilman (1987). The Taylor microscale Reynolds number corresponding to the time $t_F = 1.4$, at which time the average rate of dissipation $\langle\varepsilon\rangle$ and the enstropy $\langle\omega^2\rangle$ reach their maximal values, is about Re = 65. All scales in this simulation were resolved. The initial state at t = 0 is characterized by a reflectional asymmetry resulting in a nonzero positive value of the mean helicity density $\langle \mathbf{v} \cdot \boldsymbol{\omega} \rangle_V$. In fig. 1 we show the plots of averaged helicity spectral density

$\langle H(\mathbf{k})\rangle = H(k)$ for the times $t = 0$, intermediate time $t = 0.8$ and $t = t_F = 1.4$ in the interval $|\mathbf{k}| \leq 5$. The average, as is usual in numerical simulations, is obtained by the shell integration in **k**-space. Considering fig.1, we observe that although at $t = 0$, the maximum of H(k) is located at $k = 2$ and positive, at $t = t_F$ this becomes as large and negative. In the same time a positive maximum of H(k) has grown at the lowest value of $k = 1$. We can see the monotonous growth in time of this maximum. We believe that the incidence of helical inverse cascade has been detected. We should like to point out that the energy spectrum E(k), at the same time has suffered a decay at $k = 1$ and $k = 2$ in comparison with its values at $t = 0$, but the maximum of E(k) shifted from $k = 2$ at $t = 0$ to $k = 1$ at $t = 1.4$ (Levich and Shtilman, 1987). This shift is natural to associate with the well documented experimental phenomenon of integral scale growth (Monin and Yaglom, 1975). The relation between the latter and the helical inverse cascade seems to be a distinct possibility.

The next observation is the persistence of asymmetry of H(k) at all $k > 2$, which at $t = 1.4$ corresponds to the positive values of H(k) even for the smallest resolved values of k. This observed reflectional asymmetry of small scales of turbulence is compatible with the results of numerical simulations of forced turbulence (Kerr and Gibson, 1985), decaying turbulence (Pelz et al., 1986) and laboratory experiments of turbulence in the flow past the grid (Kit et al., 1986). Finally, we note that the decrease of H(k) for even large values of k is not monotonous and is very slow.

We deal now with a phenomenon interesting in its own right. Indeed, it seems to indicate the dependence of small scales of turbulence in terms of reflectional asymmetry on the large scale reflectional asymmetry. This would be a very nontrivial effect. Indeed, we all are used to the idea of complete statistical independence of large and small scales of turbulence. We provide here a simple explanation of the effect. The balance equation for H(k) in the absence of sources can be derived as follows:

$$\frac{\partial}{\partial t}\int H(\mathbf{k})\, d\mathbf{k} = -2\nu \int k^2 H(\mathbf{k})\, d\mathbf{k} + \text{surface terms} \tag{16}$$

where the nonlinear coupling except from the surface terms has vanished owing to the conservation of total helicity. For $t \to \infty$, the left hand side is likely to tend to zero, asymptotically, not slower than the energy. This means that in the limit $\nu \to 0$ (or $Re \to \infty$) the integral $\int k^2 H(\mathbf{k})\, d\mathbf{k}$ should diverge at $|\mathbf{k}| \to \infty$ and be of the same sign as $\int H(\mathbf{k})\, d\mathbf{k}$ itself. It can, however, oscillate on the time scale much faster than the decay time. Hence,

$$\lim_{\substack{t\to\infty \\ k\to\infty \\ \nu\to 0}} \left\{\left[\int k^2 H(\mathbf{k})\, d\mathbf{k}\right] \cdot \left[\int H(\mathbf{k})\, d\mathbf{k}\right]\right\} > 0 \tag{17}$$

the asymmetry of $\int k^2 H(\mathbf{k})\, d\mathbf{k}$ in the above limit explains the reflectional asymmetry of H(k) and subsequently of H(k) for large values of $|\mathbf{k}|$. The sign of this asymmetry is determined by the sign of $\int H(\mathbf{k})\, d\mathbf{k}$ at some instant of time, in particular in the case of our numerical experiments at $t = 0$. However, generally for a finite time helicity can grow or oscillate, depending on the value of $\int k^2 H(\mathbf{k})\, d\mathbf{k}$. For instance, we can start from $\int (\mathbf{v} \cdot \omega)\, dr^3 = \int H(\mathbf{k})\, dr = 0$. Then generally the absolute value of helicity will start to grow. The growth, in principle may continue unless it is arrested by the general decay of turbulence. This explains the growth of helicity pointed out in Pelz et al. (1986). It is interesting that if one starts from the initial condition $H(\mathbf{k}) = 0$, this condition is absolutely unstable. The nonlinear coupling term will start to generate nonzero H(**k**) generally for all **k**. Then the integral $\int k^2 H(\mathbf{k})\, d\mathbf{k}$ will become nonzero and will trigger the growth of integral helicity. The conclusion we would like to make is that nonhelical flows are likely to be generally unstable with respect to the growth of helical fluctuations in **k**-space and in physical space (Levich and Shtilman , 1988; Kit et al., 1988).

Why is it that H(k) in our numerical experiment although positive for all $k > 2$ is not monotonously decreasing? This is easy to understand if we take a position that the nonaveraged pseudo-scalar spectral density H(**k**) fluctuates much more strongly than the scalar nonaveraged quantities do. To see how the fluctuations of H(**k**) are hidden in the averaged density of H(k) we consider an idealized example of

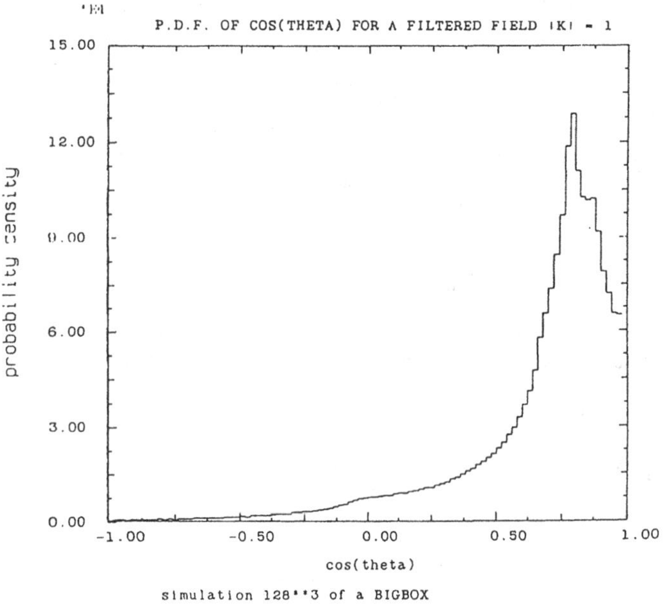

Fig. 2: P. D. F. of cos(θ) for a filtered field, |k| = 1.

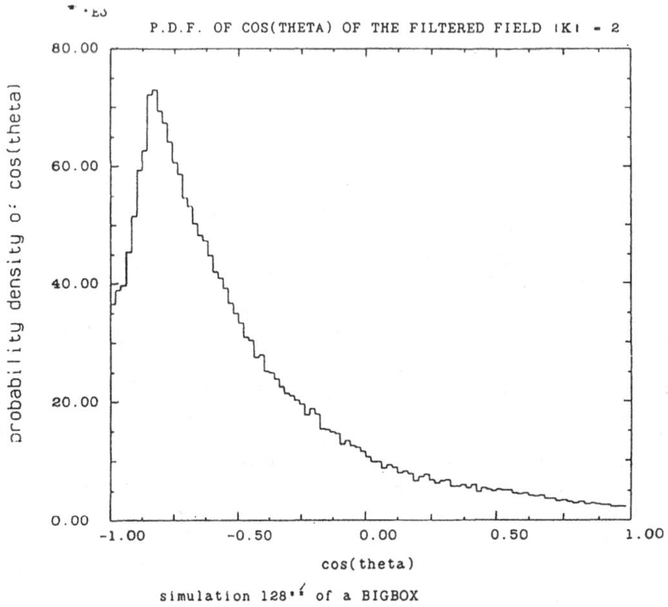

Fig. 3: P. D. F. of cos(θ) for a filtered field, |k| = 2.

reflectionally symmetric flow. Then $\langle H \rangle_V = 0$, and hence $\langle H(k) \rangle = 0$. By continuity it is clear that even if $\langle H \rangle \neq 0$, as is the case in all numerical investigations and laboratory experiments we are familiar with, still the fluctuations of H(k) are reduced. This is easy to understand, since the helicity fluctuations of opposite signs and comparable scales from the different domains in k-space cancel each other. To estimate the amplitudes of helicity fluctuations we start from the lowest values of k-numbers, k = 1 and k = 2. At these scales the fluctuations of H(k) are large even after the averaging, as is clear from the consideration of the spectrum of H(k) in fig. 1. Moreover, at these scales we observe indications of the large scale instability or helical inverse cascade. It is instructive to plot the probability distribution function for the cosine of the angle between the pseudo-velocity $v^F(r)$ and pseudo=vorticity $\omega^F(r)$ in physical space obtained by the inverse Fourier transformation of filtered $\omega^F(r)$, separately for modes with k = 1 and k = 2. These are shown respectively in fig. 2 and fig. 3. Comparing fig. 1 with figs. 2 and 3, we observe an obvious correlation between the peak in the p.d.f. of cos θ for k = 1 (fig. 2) and a positive maximum of H(k = 1) and respectively between the peak in the p.d.f of cos θ for k = 2 (fig. 3) and negative maximum of H(k) = 2). The conclusion we draw is that the large peaks in the p.d.f. of cos θ between the pseudofields $v^F(r)$ and $\omega^F(r)$ is simply a manifestation of large helicity fluctuations exhibited by the spectrum H(k). A total alignment between $v^F(r)$ and $\omega^F(r)$ would have meant that the corresponding H(k) has its maximal possible value 2kE(k) (Moffatt, 1978).

We note that the problem of exact relation between the fluctuations of $\tilde{H}(|k|)$ and the topological fluctuations at various scales in physical space is a very complicated one, and should be the subject of a study interesting in its own right. Still certain deductions can be made from the experimental material we have presently at our disposal. Let us consider the figs 4 and 5. They represent P(cos θ) for the real unfiltered $v^F(r)$ and $\omega^F(r)$ in the whole flow volume and P(cos θ) for fields $v^F_{k\leq3}$ and $\omega^F_{k\leq3}$ filtered in such a way that all Fourier harmonics with k > 3 were taken out. The origin of the peaks at cos θ = ±1 in both these figures is now apparent. Indeed, the peak at cos θ = -1 is due primarily to the large negative value of H(k) at k = 2. The peak at cos θ = 1 is due to all other harmonics which as we have seen as a rule have H(k) > 0. In fig. 5 the left hand side peak (l.h.s.) is more prominent than in fig. 4. This is because the contribution of harmonics with positive H(k) is relatively less prominent for $v^F_{k\leq3}$ and $\omega^F_{k\leq3}$ than for the total v and ω. Still the figs. 4 and 5 were remarkably similar. The reason for this is that the main contribution to the v and ω fields come from relatively small wave number Fourier harmonics. For example, the maximum of E(k) is at k = 1, and the maximum of Ω(k) is located at k = 4. For the high Reynolds number flows the situation could have been different in a sense that the maxima of E(k) and Ω(k) would have been separated much more significantly. In any case the conclusion we draw is that the peaks in P(cos θ) between the total v and ω fields observed in our numerical experiments unambiguously can be put in correspondence with the large helicity fluctuation at large scales. To illustrate further this argument we show P(cos θ) for $v^F_{k>4}$ and $\omega^F_{k>4}$ in fig. 6. We still can see a small elevation at the r.h.s. reflecting the fact we have mentioned above that H(k) is positive at all high wave vector numbers. Still it is quite different from the p.d.f. in figs. 4 and 5. On the other hand, it was noted above that the shape of P(cos θ) stays surprisingly the same, or sometimes the peaks amplify, even if only a fraction of the total volume of the flow is considered. We conclude that the subdomains of the flow contain "large scale" helical fluctuations, as well. However, the "large scale" is relatively diminished, since if a subdomain D_i has a typical scale L_i , the harmonics $k < 1/L_i$, do not contribute significantly to H(k) the quantity defined in section 4. Decreasing L_i, we decrease the scale of educted helicity fluctuations. Thus we demonstrated that the total flow volume is characterized by the presence of large amplitude fluctuations of H(k) at k = 1 and k = 2. For larger k the fluctuations of opposite signs cancel and we have to consider nonaveraged H(k). For smaller flow subdomains we shall discover that the maxima of $\tilde{H}(k)$ is located at some larger values of k, etc. Therefore the explanation of a lack of sensitivity of P(cos θ) to conditional sampling made on the basis of ω^2 and the dissipation rate ε is a very important indication in favour of the concept of turbulence as a hierarchy of helicity fluctuation at all scales in the inertial range.

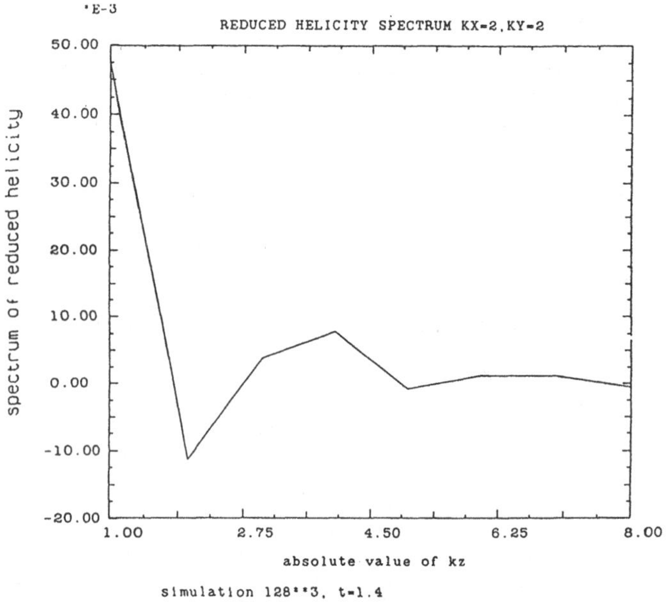

Fig. 4: P. D. F. of cos(θ) in whole space.

Let us consider now in more detail how the coherence exhibited by the nonzero value of H(k) can naturally evolve during the decay of turbulence. If the phase ψ_k are not random and only scaling is assumed with respect to I(k), then its most general form compatible with scaling assumption is

$$I(\mathbf{k}) = \lim_{\substack{V\to\infty \\ \eta\to 0}} \frac{[H(\mathbf{k})]^2}{V\delta(\eta)} (kL)^{\mu_4}, \tag{18}$$

where the parameter μ_4 determine the extent of deviation from the random phase approximation, and L is a certain integral scale kL » 1. As long as $\mu_4 < 5/3$ the conservation law (12) hold (Levich and Shtilman, 1988). It is convenient to interpret the meaning of $\sup\{[H(\mathbf{k})]^2\}$ in physical space.

We construct pseudo-velocity and pseudo-vorticity fields as follows. We assume that only two modes $\pm k_0$ are present. Then

$$\mathbf{v}^F(\mathbf{r}) = \mathbf{v}(\mathbf{k}_0)\exp(i\mathbf{k}_0\cdot\mathbf{r}) + \mathbf{v}(-\mathbf{k}_0)\exp(-i\mathbf{k}_0\cdot\mathbf{r}),$$
$$\omega^F(\mathbf{r}) = \mathrm{curl}.\mathbf{v}^F(\mathbf{r}).$$

Evidently,

$$\int \mathbf{v}^F\cdot\omega^F\, d^3r = H(\mathbf{k}_0) = H^F.$$

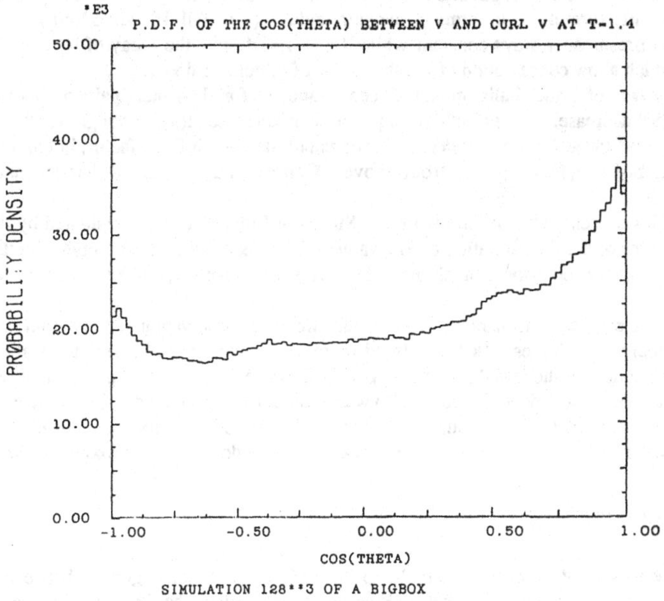

Fig. 5: P. D. F. of cos(θ) for a filtered field, |k| ≤ 3.

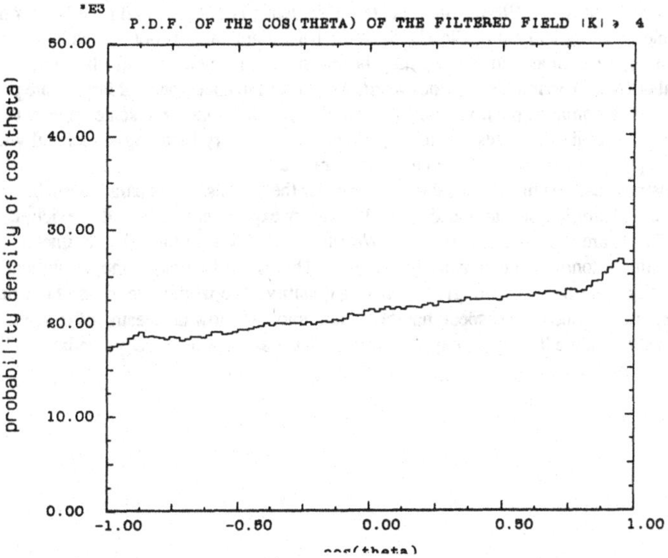

Fig. 6: P. D. F. of cos(θ) for a filtered field, |k| ≥ 4.

The maximum of $|H^F|$ is reached among all fields with a given $|v(k_0)|^2$ for a field $v^F(r)$ such that $v^F = \lambda(r)\omega^F$. Consider some implication using the example of decaying turbulence. The decay process can be conveniently formulated in terms of the amplitudes $|v(k,t\to\infty)|\to 0$, while $Re(k) = |v(k)|k^{-1}/\nu\to\infty$. The I-invariant presumably stays constant while $Re(k)\to\infty$, despite the decay of $|v(k)|$. There are three mechanisms which allow conservation of I in the course of turbulence decay:

(1) The growth of induced alignment between v^F and ω^F, for elementary pairs of modes $\pm k$ while the amplitudes $|v(k)|$ decrease. The growth of alignment may increase H(k), although $|v(k)|$ suffers a decay. This, however, cannot be for long. Indeed, the alignment may win a finite factor, but at a certain instant $t = t_{crit}$, inevitably $\sup\{I(t_{crit})\}^R \to I$ from above. Consequently exists another mechanism keeping I=invar. Namely:

(2) The "inverse cascade" of the invariant. Since the I-invariant is determinated by the amplitudes $|v(k)|$ at low values of $|k|$, the excitation of low values of $|k|$ modes while contributing a certain amount to the energy $E = \int E(k)dk$, contribute much more decisively to the I-invar, and keeps it constant despite the energy decay.

(3) At last an attractive mechanism which would allow the conservation of I-invariant notwithstanding decreasing velocity amplitudes $|v(k)|$, is a build up of probably universal correlations among the phase $\psi_k(t\to\infty)$, independent of the initial statistics $\psi_k(t=0)$. In particular if $\psi_k(t=0)$ are random, the build up of coherence would terminate independence of H(k) within certain groups of modes, i.e. create wawe packets of coherent helicity in turbulent realization. The H(k) would cease to be a δ-correlated field and subsequently the correlator $\langle H(k)H(k')\rangle$ should be able to grow beyond its random phase approximation value

$$\langle [H(k)]^2\rangle \frac{\delta(k+k')}{\delta(0)} .$$

In physical space a sign of the coherence build up should be an enhanced alignment between pseudo-fields v^F and ω^F constructed for the wave packets in **k**-space, the same way it has been done above for elementary pairs of modes $\pm k$. If scaling is assumed then the extent of correlations or coherence build up among the phase ψ_k, determines a value of the parameter of intermitence μ_4. The extreme coherence would mean a large value of μ_4. In particular if $\mu_4 \geq 5/3$, the I-invariant can be dissipated at $|k| \sim k_d$, as it occurs wiyh energy (Levich and Shtilman, 1988). Such acute intermitence is unlikely and may be inconsistent with a limited available experimental data. Clearly since the I-invariant value is dominated by the low $|k|$ modes, among these modes the phase coherence may be the most conspicuous and obseved, even in the low Reynolds number (Re) experiments. In other words we expect the appearance of large scale phase correlated wave packets. It is a common phenomenon of statistical physics, that large scale coherent fluctuations of physical fields are unstable with respect to a spontaneous symmetry breaking associated with these fields and subsequent appearance of mean of caertain order parameters.

To demonstrate that this may be the case we consider the dimensionless parameter $\alpha(k)=H(k)/2kE(k)$ in the above numerical simulations, and also in the laboratory experiment (E. Kit, E. Levich, L. Shtilman and A. Tsinober). These are depicted in figs. 7, 8. We can see that indeed the $\alpha(k)$ parameter, is nit zero and only slowly nonmonotonously decreasing for large k. This is the first experimental indication for us that indeed the helicity spectrum is a strongly fluctuating quantity. We would like to point out that although in a numerical experiment one may suspect that the nonmonotonous slow decreasing of $\alpha(k)$ may be the effect of a freak realization, in the laboratory experiment $\alpha(k)$ is the average over many hundred realizations.

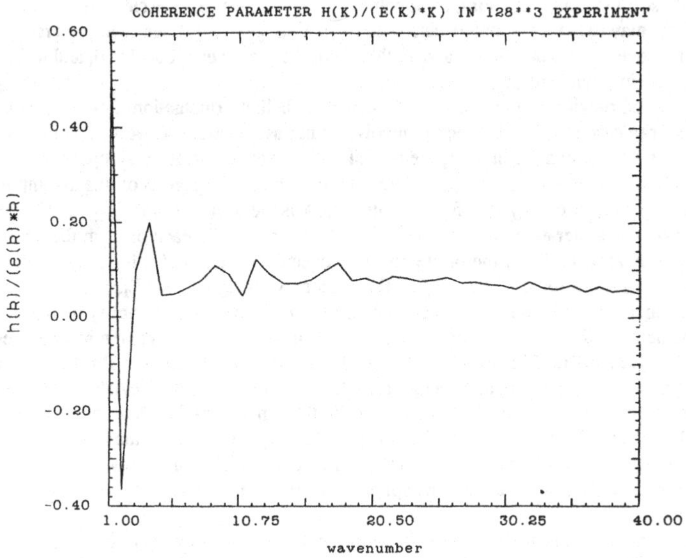

Fig. 7: Coherence parameter - numerical experiment.

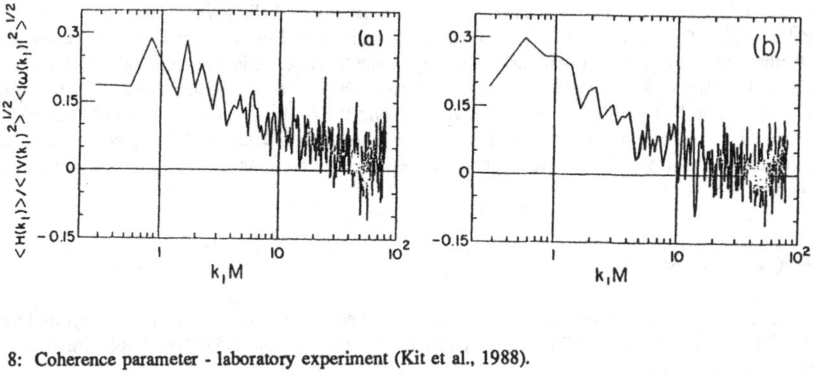

Fig. 8: Coherence parameter - laboratory experiment (Kit et al., 1988).

6. DISCUSSION

What may be the significance of various aspects of helicity and their relation to other dynamical processes in turbulence? The answer to this cannot be given with certainty at this stage. With regard to the large scales of turbulence, we believe that coherent structures and large scale instabilities are intrinsically related to the fluctuations of helicity. The generation and the swing of helicity fluctuations at small scales we believe, is a manifestation of the competition between the two processes, the stretching of vorticity lines at all scales and their breaks and reconnections. A subtle and difficult point is that in fact we have to deal not with the genuine vorticity lines but rather with some mythical lines of pseudovorticity field obtained by filtering out unneeded harmonics in the space of scales. Again at the largest scales the lines of pseudovorticity seem to conserve their quasitopology in a certain sense, as is clear from the existence of the I-invariant. This may explain the maintenance and relative longevity of large scale coherent structures. At small scales, however, as we mentioned above, the fluctuations are very short-lived, and in fact "virtual" as explained in detail in Levich (1987).

We conjectured previously that the role of small scale helicity fluctuations is in that they may be able to retard the energy cascade. The conjecture involves other assumptions as well. The main assumption is the locality of the energy transfer in the space of scales. In other words, the assumption that energy cannot significantly transfer between two disparate scales. One of the consequences of this assumption is that the characteristic time scale of energy transfer from the scale k is the turnover time τ_{turn} $\bar{\varepsilon}^{-2/3}k^{-2/3}$. If, indeed turbulence consists of a hierarchy of virtual coherent structures then in each of them the nonlinear transfer would be relatively reduced by virtue of alignment between $v(L_i)$ and $\omega(L_i)$ for any i^{th} fluctuation of the scale L_i. This is repeated for the hierarchy of fluctuations with $L_i.>L_{i+1}.> L_{i+2}$ ... Thus the effect may significantly accumulate. Should this be true, the nonlinear transfer should anplify owing to the coupling of helicity fluctuations of comparable scale, but not inside each of them. A great many degrees of freedom thereby may be eliminated. The picture breaks at the boundaries of coherent fluctuations, which is the natural seat of intensive interaction and energy transfer (Levich et al., 1984; Levich, 1987). We emphasize that as the structures themselves are virtual, with a life time much smaller than the turnover time, so the boundaries should also be virtual. The fast time of helicity fluctuations may be associated with the sweeping propagation of these surfaces destroying and creating helical structures along their path (Levich, 1987). The same time the energy transfer transpires on a larger time scale, of the order of the eddy turnover time.

Some of the above ideas are, of course, speculative, in that they have not been proved directly in the experiments described in this paper. This applies especially to the identification of short-lived fluctuations of helicity and the short living boundaries between them. These latter can naturally enough be associated with the fractal set(s), or intermittency but the concept is likely to remain theoretical for some while, without decisive proof by experiment (Levich, 1987). It can be shown that the growth and decay of helical fluctuations, that is the growth and decay of $\int \langle \tilde{H}(\mathbf{k})\, \tilde{H}(\mathbf{k}_1) \rangle d\mathbf{k} d\mathbf{k}_1$ in any flow domain D takes place only through the nonlinear coupling at with the boundary D. In particular, for the total flow domain the generation and dissipation of the I-invariant and thereby of large scale helicity fluctuations can proceed only owing to the wall effects (Levich, 1987). Indirectly, the existence of virtual fluctuations at various scales can be further tested in numerical and laboratory experiments by conditional sampling in physical space followed by filtering in Fourier space, as was described above. Also, the large scale fluctuations of helicity, clearly, can be investigated with relative ease, and their relation to what is usually perceived as coherent structures can be established. This and the subsequent phenomenon of 3-D "inverse cascade" may be the most important aspects of the concept of helicity fluctuations in application to geophysical phenomena.

7. ACKNOWLEDGEMENTS

We are indebted to A. Tsinober for numerous valuable comments. The work was partially supported by the US DOE Grant DE-FG0288ER 13837 and by Grant 85-00347/1 from USA-Israel Binational Science Foundation, Jerusalem, Israel.

8. REFERENCES.

André, G. D., and M. Lesieur, 1977: Fluid Mech., 81, 187.
Arnold, V., 1974: The Assymptotic Hopf Invariant and Its Application, (in Russian) Proc. Summer Scool in Differntial Equations, Erevan, Armenian S. S. R. Aca. Sci.
Baily, B., and V. Yakhot, 1986: Phys. Rev., A34, 381.

Balint, G.-L., P. Vukoslavcevic, and G. M. Wallace. 1987: Proc. of 1st European Turbulence Conference. To be published in Lecture Notes in Engineering (Springer-Verlag).

Berger, M. A., and G. B. Field, 1984: J. Fluid Mech., 147, 1339.

Brachet, M., D. Meirou, J. Orszag, B. Nickel, R. Morf, and U. Frisch, 1983: J. Fluid Mech., 130, 411.

Brisseau, A., U. Frisch, G. Leorat, M. Lesieur, and A. Mazure, 1973: Phys. Fluids, 16, 1366.

Frisch, U., Z. S. She, and P. L. Sulem, 1987: Large Scale Flow Driven by Aminstropic Kinetic Alpha Effect, submitted to Physica D.

Hussain, A. K. M. F., 1986: Fluid Mech., 173, 303, and references therein.

Kerr, R. M., and C. H. Gibson, 1985: Bull. Am. Phys. Soc., 30.

Kit, E., E. Levich, L. Shtilman, and A. Tsinober: Coherence and Symmetry Breaking in Turbulence: Theory and Experiment, submitted to the special issue of the PCH Journal.

Kit., E., A. Tsinober, G.-L. Balint, G. M. Wallace, and E. Levich, APS meeting, Columbus, Ohio, november 1986: submitted to Phys. Fluids.

Kraichman, R. H., 1973: Fluid Mech., 59, 745.

Kraichman, R. H., and D. Montgomery, 1980: Rep. Prog. Phy., 43(5), 547.

Levich, E., and A. Tsinober, 1983: Phys. Lett., 93A, 293.

Levich, E., B. Levich, and A. Tsinober, 1984: Proc. IUTAM Symp. Kyoto Sept. 1983, North-Holland, Amsterdam.

Levich, E., and E. Tzvetkow, 1985: Phys. Rep., 120, 1.

Levich, E., and L. Shtilman, 1987: Inverse Cascade and Helicity Fluctuations in Homogeneous Turbulence, to be published.

Levich, E., 1987: Certains Problems in the theory of Developed Hydrodynamical Theory, Phys. Rep. to appear.

Levich, E., and L. Shtilman, 1988: Phys. Lett. A, 126, 243.

Lipscombe, T., 1986: Thesis, Oxford University.

Matthaeus, W. H., M. L. Goldstein, and C. Smith, 1982: Phys. Rev. Lett., 48, 1256.

Moffatt, H. K., 1969: Fluid Mech., 35, 117.

Moffatt, H. K., 1978: Magnetic Field Generation in Electrically Conducting Fluids, Cambridge Univ. Press.

Moffat, H. K. , 1984: Proc. IUTAM Symp. Kyoto Sept. 1983, North-Holland, Amsterdam.

Moffat, H. K. , 1985: J. Fluid Mech., 159, 359.

Moiseev, S. S., R. Z. Sagdeev, A. V. Tur, G. A. Khomenko, and V. V. Yanovskii, 1984: Sov. Phys. Getp., 58, 1149.

Monin, A. S., and A. M. Yaglom, 1975: Statistical Fluid Mechanics, Vol 2, MIT Press, Cambridge, MA.

Moreau, G. G., 1961: C. R. Acad. Sci. Paris, 252, 2810.

Orszag, S., and G. S. Patterson, 1972: Phys. Fluids, 14, 347.

Pelz, R., L. Shtilman, and A. Tsinober, 1986: Phys. Fluids, 29, 3506

Shtilman, L., E. Levich, A. Tsinober, R. Pelz, and S. Orszag, 1985: Phys. Lett., 113A, 32.

Shtilman, L., 1987: On One Spectral Property of the Homogeneous Turbulence, to be published.

ter Haar, D., and T. Lipscombe, 1987: Concerning the Large Scale Instabillity of Helical Flows, to be published.

Tsinober, A., and E. Levich, 1983: Phys. Lett., A99, 99.

Wallace, G. M., 1986: Experiments in fluids, 4, 61.

Yakhot, V., S. Orszag, U. Frisch, A. Yakhot, and R. Kraichnan, 1986: Report at the APS meeting, Columbus, Ohio (Nov. 1986).

DETERMINISTIC DIFFUSION, EFFECTIVE SHEAR AND PATCHINESS IN SHALLOW TIDAL FLOWS

Ruben A. Pasmanter
Tidal Waters Division, Rijkswaterstaat
P.O. Box 20904, 2500 EX The Hague
The Netherlands

ABSTRACT. In order to better understand the observed variability of dispersion processes in the sea, we study the passive advection of particles by non-random, tidal (i.e., oscillatory) flows with spatial inhomogeneities due to the bottom topography of the basin. The model velocity field we use is two-dimensional, incompressible, and periodic in space (Zimmerman, 1976, 1978). A very broad spectrum of possibilities is uncovered. Four basic mechanisms can be defined:

1) Non-dispersive and semi-dispersive patches,
2) Chaotic trajectories with properties similar to those of a random walk,
3) Position dependence of the drift velocity leading to shear stretching,
4) Trapping of chaotic trajectories by regular (≡ non-chaotic) ones leads to anomalous diffusion and anomalous stretching.

The importance of each mechanism for the dispersion process varies with the values of the parameters characterizing the velocity field. Shear stretching can be generated even by velocity fields that are homogeneous after being averaged over the tidal oscillation. When the effects of turbulent diffusion are included, all the above-mentioned phenomena become (long-lived) transients. Numerical calculations done with more realistic velocity fields in the Wadden Sea and Western Scheldt estuary,confirm the presence of chaotic trajectories; the existence of patches is a well-known experimental fact. This simple model is capable of predicting essentially all the surprising properties of dispersion in shallow seas.

1. INTRODUCTION

The spreading of substances in the sea and in coastal areas presents us with a remarkable variability of the dispersion process. One outstanding feature of this process is the presence of patches, i.e., parts of the solute cloud that disperse much slower than the rest of the cloud. Another aspect of the varibility manifests itself when one compares log-log plots of the time evolution of the average squared size of solute clouds in different areas: one observes that the measurements corresponding to similar areas (similar in geometry, bottom topography, etc.) fall approximately on straight lines, the slope of the lines being different for different areas. The slope may be as small as 1 in shallow seas and as large as 3 in in deep open seas (see fig. 1).

It does not seem to be an easy task to develop a convincing model of so many distinct dispersion regimes. In the present work we study the specific case of dispersion in shallow seas and show that an extremely simple non-random oscillatory, i.e., tidal, flow is sufficient in order to obtain most of the striking features of dispersion as observed in shallow seas.

D. Schertzer and S. Lovejoy (eds.), Non-Linear Variability in Geophysics, 31–40.

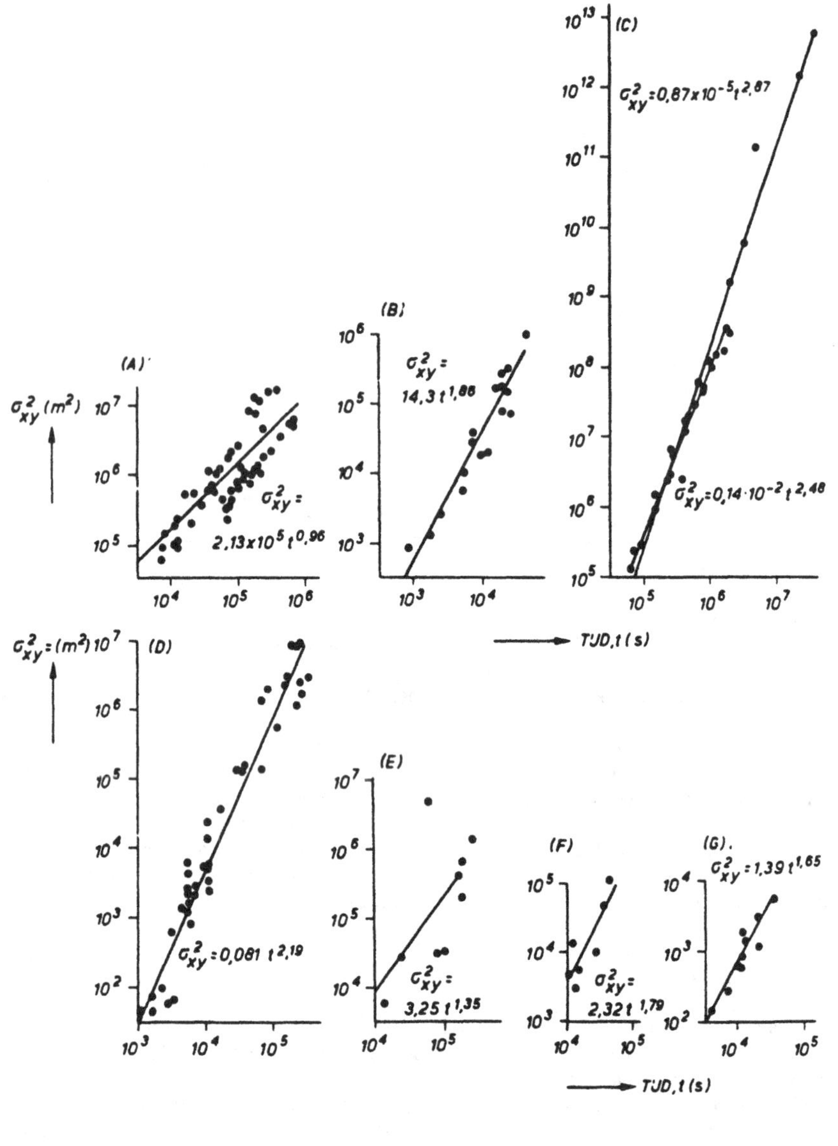

Fig. 1: Logarithm of the average squared size of pollutant clouds as function of the logarithm of time.(A) In shallow seas, (B) in English estuaries, (C) in open sea, (D) in American coastal waters, (E) in American estuaries, (F) in the Baltic coast and (F) in Fjords. Taken from Talbot 1974.

Essential for the understanding of these results is the existence of chaotic paths covered by the advected particles. This means that the separation distance between two particles that are initially (infinitely) close to each other, grows exponentially in time, (see eq. 3 for more precision). Since the initial positions are always known within a finite accuracy, the implication is that we can predict the position of a particle only for a relatively short span of time. Remember that the velocity field is a simple smooth non-random function of space and time. One of the surprising consequences of the presence of chaotic trajectories is that a cloud of particles may spread like a cloud of Brownian particles.

Another essential point is that the incompressibility of the velocity field implies that we are dealing with a conservative or Hamiltonian system (for the precise meaning of these words, see the references on dynamical-systems theory). In such a case, it is possible to observe for one and the same velocity field chaotic and regular (i.e., non-chaotic) trajectories coexisting side by side and occupying different regions of space (the so-called KAM theorem proves that such a situation is possible). An important consequence of this is that when a finite area is enclosed by regular trajectories, then all particles inside the area will remain inside for always: this part of the solute cloud is a non-dispersive patch. (Irreversible processes like turbulent diffusion will lead, of course, to leaky patches).

2. A MODEL VELOCITY-FIELD

The velocity field that we use in our calculations is essentially the one proposed by Zimmerman (1976, 1978) and other investigators for shallow tidal basins like the Wadden Sea. For our purposes, the field can be approximated by a two-dimensional incompressible flow with components given by

$$v_x = R_x\,(t) + A\,(t)\sin y$$
$$v_y = R_y\,(t) - A\,(t)\sin x \qquad (1)$$

where the time dependent amplitudes R(t) and A(t) contain not only the tidale-frequency (fundamental components) with amplitudes R_1 and A_1 but also the tidal harmonics, in particular, time-independent components with amplitudes R_o and A_o. The first terms on the r.h.s. of eq. 1 describe a spatially homogeneous tide while the second terms correspond to oscillatory eddies that are attached to the bottom topography. The presence of the harmonic components reflect the non-linear nature of a flow in a shallow basin; for more details, see the above-mentioned references. We have used the tidal period as a natural unit of time and a typical bottom-irregularity wave-length as a natural unit of length in order to write eq. 1 in dimensionless form.

In the Wadden Sea, a typical wave-length associated with the bottom irregularities along the main tidal direction is approximately equal to 5 km and 2.5 km or less along the perpendicular direction, the tidal velocity amplitude is 0.9 m/s and the amplitudes of the time-independent and fundamental components of the eddies are both approximately equal to 0.1 m/s; since the tidal period is 4.5 10^4 s one has the following typical values of the dimensionless parameters characterizing the velocity field: $R_1 \approx 8.$ and $A_o \approx A_1 \approx 0.9$.

The time-dependence of eq. 1 is essential for the presence of chaotic behaviour: if the time-dependence can be eliminated (e.g., by redefinition of the time variable), then we are left with an autonomous two-degree of freedom system; such systems have exclusively non-chaotic solutions (Poincare-Bendixon theorem).

One nice property of the field eq. 1 is that it is periodic in space; this means, in particular, that it makes sense to study the long-term and long-distance behaviour of the dispersion process.

3. THE COMPUTATION OF PARTICULAR PATHS

Given an initial position [x(0), y(0)] and taking into account the periodicity in time of the flow, we plot the successive positions of the same particle after one tidal period [x(1), y(1)], two tidal periods [x(2), y(2)] etc. This is equivalent to taking a Poincare cross-section of the trajectories.

Similarly, we exploit the space periodicity of the flow in order to plot the positions modulo a wavelength; in this way we can keep track of particles that move far away from their initial positions without having to enlarge the scale of our graphs.

The set of tidal positions of a particle [x(0), y(0)], [x(1), y(1)], etc., may form a smooth one-dimensional curve. Notice that two different curves (corresponding to two particles with different initial positions) can never intersect.

4. MAIN RESULTS

In spite of the very simple form of the field eq. 1, the paths of the advected particles show an astonishingly broad spectrum of possibilities. Figures 2a and 2b show two characteristic plots for the same velocity field, the only difference between the plots being the initial positions of the particles. In fig. 2a one sees a smooth curve formed by a number of islands: the particle "jumps" from one island to the other each tidal period. By contrast, in fig. 2b one sees a path that seems to cover a two-dimensional area.

The most important observed features can be understood in terms of four basic mechanisms.

4.1. Non-dispersive and semi-dispersive patches

As already stated, different paths can never intersect. This means that the particles inside the islands in fig. 2a can never cross the boundary of such an island; i.e., they form a permanent block or patch. It should be noted that in certain cases an area is surrounded by a smooth curve with very small openings. In such cases a particle that starts in the surrounded area needs a very long time until it hits one of the small openings and leaks out. In other words, the field can also generate leaky or semi-dispersive patches. Moreover, patches may be drifting or non-drifting, they may contain only regular (≡ non-chaotic) paths or both regular and chaotic ones.

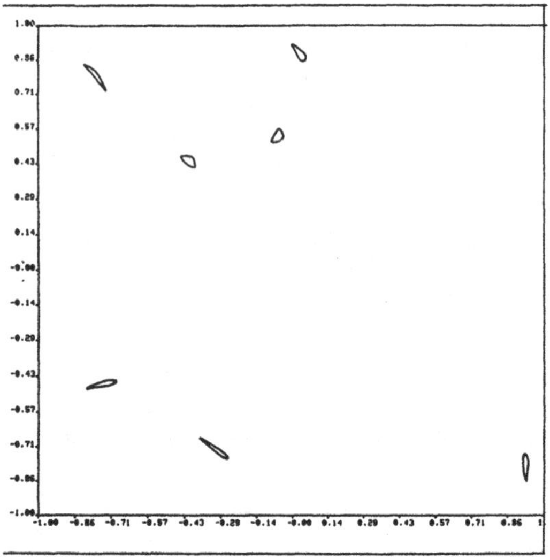

Fig. 2 (a): Eight hundred successive tidal positions of one particle form seven smooth islands. Witheach tidal period, the position jumps from one island to the next one. Velocity field characterized by the dimensionless parameters $R_1 = (8, 5)$, $R_o = (0.025, 0)$ and $A_o = A_1 = (0.9, 0.9)$.

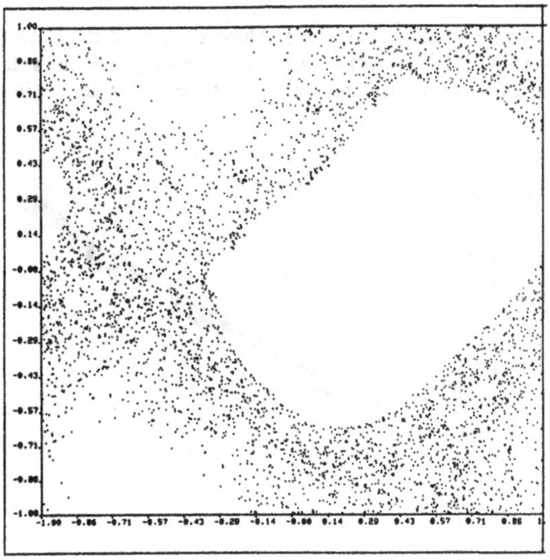

Fig. 2 (b): Four thousand tidal positions. The same velocity field as in (a) but with a different initial position.

4.2. Chaotic trajectories with properties similar to those of a random walk

As seen in fig. 2b, some paths tend to cover a two-dimensional area. These paths have another surprising characteristic: they are chaotic. This means that the separation between two particles that initially have been (infinitely) close to each other, grows exponentially in time. To be more precise, for these chaotic paths one has that

$$\lim_{t\to\infty} t^{-1} \ln \lim_{D(0)\to 0} [D(t)/D(0)] \equiv \mu > 0 \tag{3}$$

where D(t) is the separation between the two particles at time t. The rate μ is called the Lyapunov exponent (or Lyapunov number); when μ is positive we say that the trajectory is chaotic. Notice that μ may depend upon the direction of D(0).

By changing the parameters R_1, A_1, R_o and A_o, a situation may be reached in which the chaotic paths cover all the available space and no patches are observed. Under these circumstances another striking manifestation of the chaotic character of these paths manifests itself: the average squared size of a cloud of particles grows linearly in time, i.e., just like in the case of a cloud of Brownian particles performing random walks. This behaviour is shown in fig. 3.

4.3. Effective shear

When the time-independent homogeneous component of the velocity field vanishes, i.e., $R_o = 0$, there is no long-term preferred direction, so that no long-term drift is possible. When R_o does not vanish, however, the symmetry of the system is broken and a long-term drift along the preferred direction of R_o is possible. Since the velocity field is not homogeneous, different positions in space are not equivalent, and one may observe different long-term drifts associated with different initial positions. This means that in such cases one has an effective (or Lagrangian) periodic shear that leads to stretching of the cloud along the direction of R_o; the average squared cloud size along the preferred direction grows quadratically in time. This behaviour is illustrated in fig. 4.

It should be mentioned that all paths in an island have the same drift velocity; similarly all paths in a connected chaotic region have the same drift velocity. Consequently, when all space is occupied by chaotic trajectories and no regular ones are present, no shear is possible; this is the case in fig. 3.

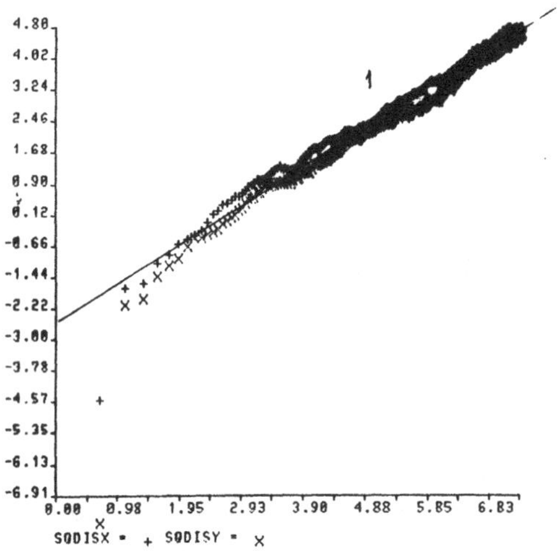

Fig. 3: Logarithm of the average squared size of a cloud of thousand two hundred particles as a function of the logarithm of time. The parameters characterizing the velocity field are $R_1 = (8, 8)$, $R_o = (0.425, 0)$ and $A_1 = A_o = (0.9, 0.9)$. For these values the chaotic region covers all space. The slope of the straight line is very close to 1, like in Brownian-type diffusion.

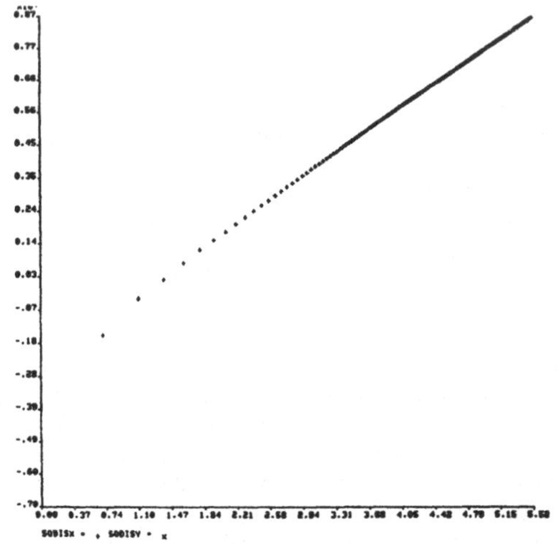

Fig. 4: Logarithm of the average squared size of a cloud of particles as a function of the logarithm of time for a velocity field with parameters $R_o = (0.108, 0)$, $A_o = (0.9, 0.9)$ and $R_1 = A_1 = 0$, i.e., a time-independent velocity field. The slope of the straight line is very close to 2.

4.4. Anomalous diffusion and anomalous stretching

When both regular and chaotic paths are present, many interesting phenomena may take place. In particular, a particle on a chaotic trajectory may come very close to a regular path and "stick" to it, i.e., remain close to it for very long periods of time. One could say that smooth regular paths act like traps for particles on chaotic trajectories.

If the smooth path and the adjacent chaotic area have the same long-term drift, then the diffusion process is effectively blocked and the average squared size of a cloud of particles in the chaotic region grows slower than linearly with time. On the other hand, if the regular path and the paths in the adjacent chaotic area have different drift velocities, then the cloud of particles in the chaotic area tends to stretch and the average squared size of a cloud of such particles grows faster than linearly with time (but slower than summarized as they would in the case of real stretching). These characteristics can be seen in fig. 5.

These results can be summarized in the following expression for $\sigma^2(t)$ the average squared size of a cloud of particles

$$\sigma^2(t) = \text{const } t^\delta \tag{4}$$

where the exponent δ may vary continuously between 0 and 1 when R_o=0 or between 0 and 2 when R_o does not vanish.

5. EFFECTS OF TURBULENT DIFFUSION

Since some of the properties discussed in the previous section are due to a certain instability in the dynamics, one may wonder what happens to them when random perturbations are present, e.g. when turbulent fluctuations in the velocity field are taken into account. In order to study this question, we add a white-noise component to the velocity field eq. 1 and compute the corresponding paths. The results of the previous section are modified as follows:

5.1. Patches

Particles inside a patch can now move out and reach the external chaotic area (or another patch), i.e., all patches become leaky. While the patches disappear in the long run, they may do so on a much longer time-scale than the rest (chaotic part) of the cloud.

5.2. Chaotic paths and diffusion

In some cases, chaotic paths and the diffusion process introduced by them are only slightly affected by the added noise. For example, if to the field of fig. 3 we add random perturbations, the only change is that the diffusion coefficient increases; the increment being essentially the diffusion coefficient associated with the external noise. On the other hand, when chaotic and regular areas coexist, particles initially diffusing in the chaotic area can diffuse into the patches and remain there for a long while until they leak back into the chaotic area. If the adjacent chaotic and regular areas have the same drift, this may reduce the value of the effective diffusion coefficient originally associated with the chaotic area.

5.3. Effective shear

Due to the random perturbations, a particle on a drifting path will move into another path with a different drift, the overall effect being that the particles perform a random walk on the length-scale of the spatial inhomogeneities of the field. Thus, just like the usual (Eulerian) shear, the effective shear becomes shear diffusion. This occurs, however, only after a transient; the time-scale of the transient being the time required by the random noise to diffuse across the inhomogeneities of the field. This is illustrated in fig. 6.

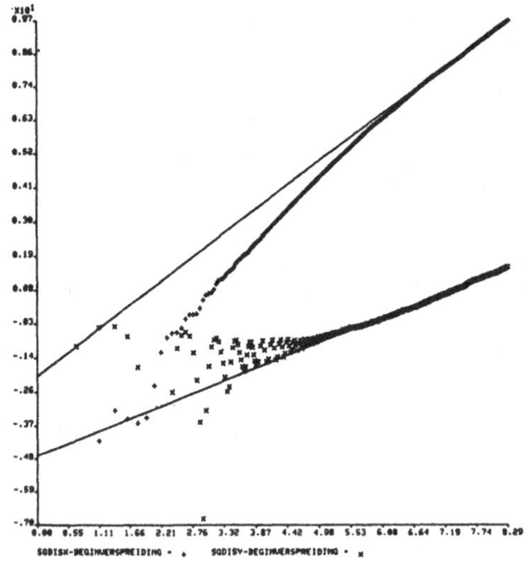

Fig. 5: Logarithm of the average squared size of a cloud of thousand two hundred particles as a function of the logarithm of time. Same parameter values as in fig. 2. Cloud of 1500 particles in the chaotic region of fig. 2b, 4000 tidal periods. The slope of the straight lines are approximately 1.4 (squared size along the direction of R_o) and 0.75 (squared size along the direction perpendicular to R_o).

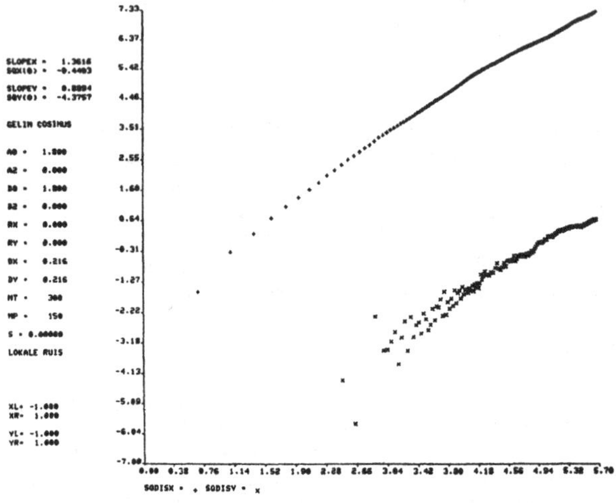

Fig. 6: As in fig. 4, but with weak random noise added to the advection eq. 2. The slopes approach the value 1.

5.4. Anomalous diffusion

After a (long) transient both anomalous diffusion and anomalous shear go over to Brownian-type diffusion and the average squared size of the cloud grows linearly in time.

From the statements above, it follows that as soon as turbulent diffusion is present one may observe extremely complex transient behaviour until the final asymptotic regime is reached. For example, imagine that the cloud of particles is released in a non-chaotic area and slowly leaks out into a chaotic area with a drift velocity equal to that of the regular area. As this process goes on, one will observe a large increase of the "diffusion coefficient" from the value due exclusively to turbulent diffusion to the value due to deterministic (chaotic) diffusion. Subsequently, the particles may reach an area with a different drift velocity. Again, this will lead to an increase of the observed "diffusion coefficient", this time to the value associated with shear diffusion.

6. IMPROVEMENTS

One of the extreme simplifications of the model discussed in the previous sections is that only one component of the bottom irregularities has been included. One may wonder whether the interesting phenomena discussed above are still present when more Fourier components of the topography (and of the velocity field) are taken into account. For this reason we added another Fourier component to eq. 1, performed the numerical integration of the corresponding advection equations, studied the general characteristics of the paths and the time dependence of the average squared size of clouds. No appreciable qualitative or quantitative differences were found.

Another simplification of the model is its two-dimensional character. Should one expect drastic changes when dealing with more realistic three-dimensional flows? From the general theory of non-linear dynamical systems we know that the answer is no: we should observe coexisting patches and chaotic regions, chaotic diffusion, trapping leading into anomalus diffusion, etc. On the other hand, some differences may appear; in the three-dimensional case there is more topological "freedom" (see, e.g., Guckenheimer and Holmes, 1983 and Schuster, 1984) and therefore chaotic regions may exist even when the two-dimensional approximation of the velocity field does not predict them.

7. DISCUSSION AND CONCLUSIONS

What are the observations, either in nature or in more realistic numerical models, that support the predictions of the model? The occurrence in nature of patchiness, i.e., of a strong dependence of the dispersion coefficient upon position in space, lends support to the model, especially when one takes into account that other models cannot predict such phenomenon. Similarly, the variability observed in measurements of the growth of clouds, fig. 1, cannot be predicted by other models. Moreover, numerical calculations of particle paths in realistic velocity fields in the Western Scheldt estuary computed by G.v.Dam (1985) and in the Wadden Sea by Ridderinkhof (private communication) show the characteristic signature of chaotic paths: oscillations on length-scales much smaller than the smallest scale of the velocity field.

Consequently, one has to conclude that chaotic behaviour is present in the trajectories of particles in shallow tidal seas like the Wadden Sea and estuaries like the Western Scheldt, that patches and the anomalous diffusion of fig. 1 can be understood in terms of the fundamental difference between regular and chaotic paths and the trapping of the last ones by the first ones.

La morale de cette histoire: that extremely simple velocity fields can account for essentially all the surprising characteristics of dispersion in the sea. The model has quantitative predictive power if all the relevant mechanisms are included. For example, the very large diffusion coefficient computed from the stationary salt distribution in the Wadden Sea (Zimmerman, 1976) could originate in shear diffusion due to a permanent current as explained above; however, the most probable explanation is the one recently given by Zimmerman (1986), i.e., the inhomogeneity of the velocity across the basin which leads to shear diffusion and was not included in the model discussed in the previous sections.

8. REFERENCES

Chaiken, J., R. Chevray, M. Tabor and Q. M. Tan, 1986: Experimental study of Lagrangian turbulence in a Stokes flow. Proc. Roy. Soc., A408, 165.

Chien, W. L., H. Rising and J. M. Ottina, 1986: Laminar mixing and chaotic advection in several cavity flows. J. Fluid Mech., 170, 355-378.

van Dam, G., 1985: Computations of particle paths and distributions in two and three dimensional velocity fields: Fysische Afd. Colloquium day. Ed. G. v. Dam (in Dutch) RWS/WL/KNMI. The Hague, 19 June 1985.

Guckenheimer J., and P. Holmes, 1983: Nonlinear oscillations, dynamical systems and bifurcation of vector fields. Springer Verlag, New York.

Merlo, V., M. Pettini, aand A. Vulpiani, 1985: Anomalous diffusion of clumps in nonlinear dynamical systems: Lettere Nuovo Cimento, 44, 163-171.

Schuster, H. G., 1984: Deterministic chaos: an introduction. Physik Verlag, Weinheim.

Talbot, J. S., 1974: Interpretation of diffusion data. Proceedings of the International Symposium on Discharge of Sewage from Sea Outfalls, London.

Zimmerman, J. T. F., 1976: Mixing and flushing of tidal embayments in the western Dutch Wadden Sea II: Analysis of mixing processes. Neth. J. Sea Res., 10, 397-439.

Zimmerman, J. T. F., 1978: Topographic generation of residual circulation by oscillatory (tidal) currents. Geophys. Astrophys. Fluid Dyn., 11, 35-47.

Zimmerman, J. T. F., 1986: The tidal whirlpool: A review of horizontal dispersion by tidal and residual currents. Neth. J. Sea Res., 20, 133-154.

NONLINEAR GEODYNAMICAL VARIABILITY: MULTIPLE SINGULARITIES, UNIVERSALITY AND OBSERVABLES

Daniel Schertzer, Shaun Lovejoy*
EERM/CRMD, Météorologie Nationale,
Paris, France.

ABSTRACT. We enlarge on theoretical insights concerning the multiple scaling/multifractal behaviour of geodynamical fields in the space-time domain and the very singular behaviour of their observables which are usually obtained by averaging over scales much greater than that of the homogeneity. We render more direct the link between statistical singularities (divergence of high order statistical moments) and hierarchies of singularities per realization (small scale divergence of densities). In the case of "hard multifractals" (having not only singular realizations but also - contrary to "soft multifractals" - singular statistics), we insist on the importance of the existence of "wild" singularities which although extremely rare, create the statistical divergences, as well as on the need to distinguish between "bare" and "dressed" properties of these multifractal fields.

The bare properties are the properties of a (rather theoretical) process in which nonlinear interactions between scales smaller than the observation scale are filtered out. Conversely, the dressed properties are those of the observables and result from the full hierarchy of nonlinear interactions down to an infinitesimal scale followed by integration over the scale of observation. Both properties involve multiple scaling and hierarchies of dimensions, but the latter introduce statistical divergences, "pseudo-scaling", etc. Observations obtained by averaging over a given dimension therefore "dress" in a drastic manner the "bare" properties of a process. We also underline the fact that in general, multifractals are non-local and hence - contrary to simplistic local multifractal notions - both the scaling exponents and orders of singularities must be understood as statistical exponents, not as point values.

We show that the infinite hierarchies of critical exponents in multifractals may well be very simply determined due to the existence of three-parameter (H, C_1, α) universality classes of the generic multifractal processes. These three fundamental exponents characterize: the degree of non-conservation of flux (H), the deviation of the mean field from homogeneity (C_1), and the deviation of the process from monofractality ($0 \leq \alpha \leq 2$). We discuss other associated fundamental properties. The five main subclasses of these (H, C_1, α) universal canonical multifractals are outlined with their important theoretical and practical consequences.

A quite different aspect of scaling symmetry is that the scale transformations involved can be strongly anisotropic, nonlinear and even stochastic. This leads us to generalize the idea of scale invariance far beyond the familiar self-similar (or even self-affine) notions. We sharpen the ideas of this nonlinear/stochastic Generalized Scale Invariance, thus introducing an enormous diversity of scaling behaviour.

Beyond the many important theoretical and practical consequences of these findings, we are lead to explore a hidden and unexpected face of multifractal chaos: bare universality under dressed Pandemonium.

* Physics dept., McGill University, Montréal, Canada.

D. Schertzer and S. Lovejoy (eds.), Non-Linear Variability in Geophysics, 41–82.

1. INTRODUCTION

1.1. The unification of geophysics ?

An emerging and powerful unifying *problematic* of geophysics is being increasingly recognized: the extreme variability of geophysical phenomena and processes over wide ranges of spatial/temporal scales, which easily cover nine orders of magnitude (earth radius scale/ centimeter scale or e.g. 30 years/second or 10 days/millisecond). Indeed, what has been felt to be a growing and ubiquitous difficulty in geophysics, is more and more perceived as a *fundamental symmetry*: a common behaviour at different scales (*scaling* behaviour). Indeed, this corresponds to *the simplest but also the only symmetry assumption acceptable in the absence of more information or knowledge*. Indeed, we cannot consider the breaking of this symmetry without first exploring its possible manifestations in the largest sense. For instance, the symmetries we will consider are statistical symmetries, *each realization* corresponds in fact to a violation or *a breaking of these symmetries*[1]. The corresponding exponents (dimensions, singularities, ...) are also statistical, not point values.

Since this symmetry is the result of nonlinear interactions -nonlinear (i.e. non proportional) response to a given excitation- between different scales (and/or processes), we are addressing the question of *scaling nonlinear variability*. A general consequence of such variability is that the notion of observables (roughly speaking: what we can observe or measure from a process) is far from trivial, since the details of the process may be overwhelmingly important (due to small scale or high frequency "ultraviolet" divergences or singularities). Unfortunately our observations and measurements are nearly always restricted to resolutions much higher than the scale of the smallest detail, "inner scale" or "scale of homogeneity" which in geophysics is typically of the order of millimeters or less. Full knowledge down to this inner scale is usually out of our scope due to the large number of degrees of freedom involved which can be of the order of *physicists' infinity* such as the Avogadro's number (10^{23}): indeed the number of mm^3 (the number of degrees of freedom) in the atmosphere is of order[2] $10^{10}x10^{10}x10^7= 10^{27}$.

This type of unifying problematic is urgently needed in geophysics, since under the heading of "Global Change Research", the geophysical community is tending more and more to address global questions, particularly those pertaining the climate. Unfortunately, up until now we have faced a rather distressing situation: gigabytes of computer codes which are unable to cope with terabyte flows of (often remotely sensed) data, obtained at finer and finer resolution, all because the numerical models work at far larger scales[3]. It would seem to be of doubtful value to try to answer any of the questions raised by Global Change Research without being able to simultaneously think of the global as well as the detailed characteristics of the variability of geophysical fields. Indeed, it would seem fruitless to design sophisticated integrated data acquisition and processing facilities without having a conceptual framework for handling massive high resolution data sets. Indeed, we desperately need to cast order in geophysical chaos, more properly to discover new order in what according to current knowledge is apparently disorder. In other words to master how simplicity can beget complexity.

---

[1] As in the the widespread image of a marble rolling on a symmetric bottom of a bottle: each of the experiments will violate the rotational symmetry, yet on the average this symmetry is still respected!

[2] Considering the scale of homogeneity of the order of the millimeter, and the (outer) vertical scale of the order of ten kilometers and the horizontal scale of the order of ten thousand kilometers. In a similar manner the Reynolds number (which is the ratio of the nonlinear to dissipation terms and hence characterize the strenght of nonlinearity) of atmospheric turbulence is usually estimated as $\approx 10^{12}$, taking the ratio of injection (1000 km)/dissipation (1mm) (horizontal) scales as 10^9, since it is the 4/3 power of this ratio.

[3] There is even a growing tendancy of evaluating the performance of models by comparing models with models rather than models with data!

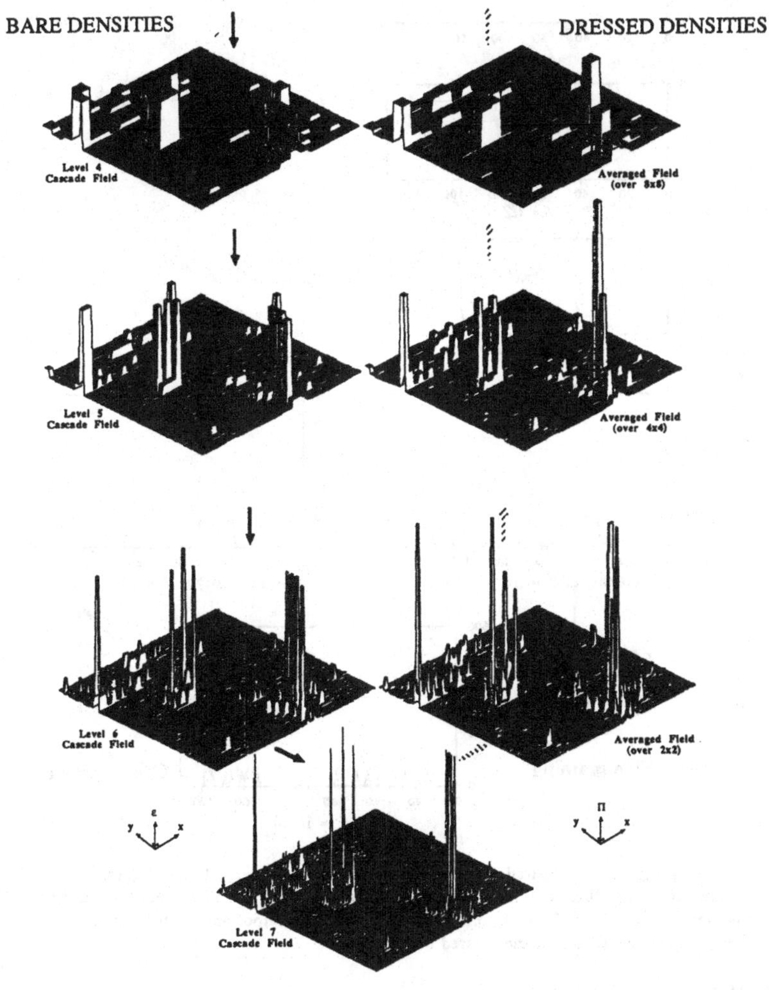

Fig. 1a : Illustration of the "bare"and "dressed" energy flux densities. The left hand side shows the construction step by step of the bare field produced by a multifractal cascade process (the α-model, discussed below) starting with an initially uniform unit density. At each step the homogeneity scale is divided by a constant ratio $\lambda=2$ From top to bottom, the number of cascade steps takes the following values n = 0, 1, 2, 3 and 7, with the corresponding length scale values l = 1, 1/2, 1/4, 1/8, 1/128. When the number of steps n increases, some rare regions of high intensities ("singularities") appear, most of the space becomes inactive. At l =1/8, n=3, one may compare the rather more intense dressed density with the bare density. The sharp contrast arise from the smaller scales singularities, as seen on step n=7, which contribute to high fluctuations of the dressed density.

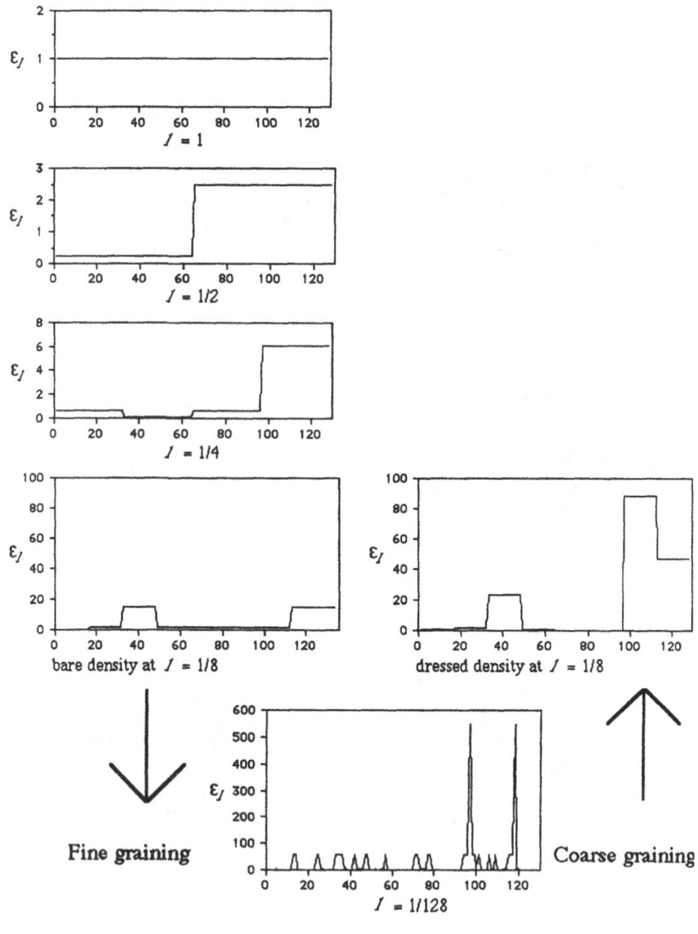

Fig. 1b : as in Figure 1a , illustration of the "bare"and "dressed" energy flux densities, but on a 2 dimensional space. The dressed energy flux densities, obtained by averaging, are presented on the right hand side of the figure. At intermediate scales, level 3 or 4, one may still note the important contributions from smaller scales singularities to high fluctuations of the dressed density.

1.2 The importance of the details

A cornerstone in the early recognition of the importance of the "details" and their appropriate represention seems to be the beginning of our century. For instance, Perrin, in the introduction of his thesis (Perrin, 1913), already pointed out that tangentless curves are the rule rather than the exception, contrary to the academic teaching which tries to render "obvious" a continuous perception of the world. Among the various examples he discussed were flakes and brownian motion. However the example that he stressed was the coastline of Brittany -a question further explored by Richardson (1960) and popularized by Mandelbrot (1982)- even in spite of the fact that our scale dependant representations (maps) are overwhelmingly smooth! He also underscored the conceptual contributions of contemporary mathematicians such as Borel, who by extending in a discontinuous manner the mathematical notion of measures from the volume-like ones (Lebesgue measures) rendered them at the same time more abstract yet nearer to the discontinuous "real world". It is indeed surprising to discover how contemporary Perrin's discussion remains! Ever since then, in order to deal with these "ultra-violet" divergences mentioned many attempts have been made to define smooth macroscopic "effective" fields from irregular microscopic ones using various techniques ("homogeneization", "renormalization"), the permanent question of "coarse graining" vs. "fine graining".

Concerning fluid dynamics, the question of the small scale singularities became more precise with the work of Leray (1934), and in Von Neuman's review on turbulence (Von Neuman, 1963), but also in the debate between Richardson[1] and Bjkernes on the rather fundamental question: is the characterization of a few large scale singularities (the meteorological fronts) sufficient to forecast the evolution of the weather? More recently, under the theme of the "butterfly effect" Lorenz (1963) gave a stunning image and now popular metaphor for the absolute unpredicibility resulting from the small scale singularities or *sensitive dependence on the initial conditions*: as time passes away, the single (small scale) flutter of a butterfly will introduce large scale disturbances in atmospheric dynamics. The current day debate could be much more precise by dealing with the characterization of hierarchies of scaling singularities. In the following we hope to give clearer insights into this fundamental question with the help of apparently (at first glance) simple models (phenomenological models or "mock geophysics"), which nevertheless possess surprising properties which turn out to be quite general.

The exploration of nonlinear variability was maintained in the restrictive frontiers of geometry for too long a period. This period created some unfortunate attempts to bypass various fundamental problems (among which we may cite several abusive uses of adhoc additive processes). Indeed, the development of concrete analytical methods has tended to show that geometrical frameworks can often be misleading and fractal notions have been most fruitful when divorced from geometry. In particular, the abandonment of the dogma of the uniqueness of fractal dimension (Grassberger, 1983; Hentschel and Proccacia, 1983; Schertzer and Lovejoy, 1983,1984; Parisi and Frisch, 1985; Halsey et al., 1986; Pietronero and Siebesma, 1986; Bialas and Peschanski, 1986; Stanley and Meakin, 1988; Levich and Shtilman, 1989; ...) in favour of hierarchies of dimensions and singularities with their non-geometric generators has been one of the most important recent advances. It is now rather obvious that multiple dimensions and singularities are the rule rather than the exception for fields, hence we are now used to "multiple scaling" or "multifractality" associated with highly intermittent processes in which the weak and intense regions have different scaling behaviour. However, as we will discuss after having left this uniqueness for infinity, the important question of the existence of universality classes gives credence to returning to only few fundamental (dynamic, not geometric) parameters!

The scaling symmetries are rather special when compared to the rotational or parity[2] (i.e. mirror reflection) symmetries, since they are not compact[3] contrary to the latter. One may note also that in its simplest, but very restrictive form (the only one explored by fractal geometry[4] up to 1986!), it is not only space/time invariant, hence global, but also isotropic. These two assumptions are obviously unacceptable in geophysics, since we have to deal with anisotropy in the space/time domain, with rotation, stratification[5] (Schertzer and Lovejoy, 1983, 1984, 1985a) or "texture". This is the reason why we developed some new elements (Schertzer and Lovejoy, 1985b, 1987a, b) for a Generalized Scale Invariance

1 Recall that Richardson (1926) didn't hesitate to raise the (sacrilegious?) question "Does the wind have a velocity?" (i.e. are the time derivatives regular?). Indeed, he pointed out the very irregular (fractal) Weierstrass function as a counter-example.

2 The role of symmetry breaking in parity invariance (helicity) for creating large scale disturbances has been emphasized by Levich and Tzvetkov (1985), Levich and Shtilman (this volume) and Moisseev et al. (1988).

3 Since unlike angles the scale ratio is unbounded. This lack of compactness has even lead some to doubt wheter the scaling symmetry is respected (Frisch, 1985).

4 This single fact helps to explain the strangely bitter comments (Mandelbrot, 1986), against some of our earlier papers on scaling anisotropy ...

5 We proposed to add the second stanza: "Flatter whorls have rounder whorls that feed on instability, and roundish whorls have rounder whorls, and so on to spherocity -in the statistical sense" to Richarsdson's celebrated poem : "Big whorls have little whorls that feed on their velocity, and little whorls have smaller whorls and so on to viscosity -in the molecular sense" (Richarsdson, 1922).

(see section 6). In general, we will need both a local[1] and anisotropic scale transformation and we even consider stochastic scale transformations. This leads us to generalize the notion of scaling: a system may be said to be scaling (or scale invariant) over a range if the small and large scale structures/behaviours are related by a scale changing operation involving only the scale ratio. Hence, scale invariance is not restricted to the familiar self-similar (or even self-affine) notions. It is important to distinguish this idea of local scale transformations from the simplistic multifractal notion of local exponents.

1.3 Geophysical observables

The breathetaking pictures of (geometric) fractal objects often inclined us not to explore the rather immediate question: how will we perceive them with the limited resolution of our eyes if the computing process goes down to a much smaller scale? Contrary to what happens with (geometric) monofractal objects, a drastic *symmetry breaking is caused by the observation* not only by the scale of observation, but also by its dimension. This is the reason why we will insist on the fundamental difference between *"bare" and "dressed" properties* at a given (non-zero) scale i.e. the important differences between a process with a cut-off of small scale interactions and one with these interactions restored (cf. fig. 1a-b for illustrations).

The bare properties are related to fine graining (e.g. the development a cascade) and are the properties of the process with the nonlinear interactions at scales smaller than the observation scale filtered out (i.e. the process is truncated at the scale of observation). The dressed properties are coarse grained, they are the observed properties at a given scale of resolution (i.e. obtained by linear or nonlinear averaging over the smaller details of the same process at the observation scale and with all interactions: the process fully developed down to the smallest scale). In other words, only half the problem has been explored (and even a smaller fraction of the real problem): the "dressed" truth is the one which counts! The terms "bare" and "dressed" are borrowed from renormalization jargon, but here due to the extreme variability, their differences become even more important; not only do they involve a renormalizing factor but also quite different statistical behaviour. This raises immediately the overwhelmingly important question of "wild", singular statistics (divergences of statistical moments (Schertzer and Lovejoy, 1987a,b)) linked to multiple ultraviolet divergences.

2. HOW DOES GOD PLAY DICE?

2.1. Scaling nonlinear variability and "Mock Geophysics"

In a very general manner (Noether's theorem), for every (continuous) symmetry we can associate a conservative quantity. For instance in physics: conservation of energy and momentum for time and space translational invariance, angular momentum for rotational invariance... Here however we are investigating dissipative systems, far from equilibrium. As in turbulence theories, the conserved quantities should therefore rather be the rate of dissipation of energy - more properly speaking the *flux of energy* towards smaller scales, not the energy itself (hence the notion of "quasi-equilibrium" with a constant rate of dissipation or flux of energy). We can already anticipate that the fundamental exponents (H, C_1, α) -that we will show to be sufficient to characterizing universal processes of nonlinear variability- are related to various possible deviations from the simplest hypothesis of conservation of the flux, i.e. homogeneous conservation. Indeed each parameter quantifies a distance from homogeneity, H for the degree of non-conservation of the flux, C_1 for the mean deviation from homogeneity, and α (the Lévy index, $0\leq\alpha\leq2$) which indicates how far the process is from monofractality ($\alpha=0$).

The problematic of nonlinear variability over wide ranges of time/space scales, has been considered for a long time with respect to the mysterious turbulent behaviour in fluid dynamics, especially their asymptotic (and universal) behaviour when the dissipation length goes to zero (fully developed turbulence). Conceptual advances occurred using apparently simple models of self-similar cascades, as opposed to the

[1] One may note localization of scale symmetry has been considered by Weyl (Weyl, 1923) under the name of "local gauge symmetry" in the context of (relativistic) electromagnetism, in the spirit of the (already localized symmetry) of Einstein's theory of gravity.

frustratingly tedious developments of renormalization techniques... which still fail to grasp the intermittency problem. From very general considerations (going back to the famous poem of Richardson (1922)), the phenomenological models of turbulence have become more and more explicit although sometimes in an overly restrictive manner (to quote a few: Novikov and Stewart (1964), Yaglom (1966), Mandelbrot (1974), Frisch et al. (1978), ...; for a review see Monin and Yaglom (1975)). However their common theme of how the energy flux is spread into smaller scales in successive cascade steps (while respecting a scale invariant conservation principle) is far from being restricted to turbulence since spreading into small scales is a general theme in geophysics (from concentration of (passive) substances/scalar fields to the spreading of points on strange attractors). Note that the notion of flux at a given scale or through a scale, can be more precisely understood in Fourier space as the flux through the surface of a sphere of constant wave number radius (the inverse of this wave number being proportional to the corresponding scale). In this sense we can speak of a probability flux of points on a strange attractor, e.g. the flux of points flowing to smaller scales on this strange attractor, hence the "flux dynamics" we will discuss is quite general, paralleling classical thermodynamics, but with very strong divergences .... We will also discuss the related fields which are not constrained to such scale conservation (such as the passive scalar concentration and velocity field ...).

2.2. Pixel worlds and weak measurable properties

Geophysical phenomena (especially when remotely sensed) are more and more often represented with the help of digitized "images", pixel sets. In spite of this the "theoretical" representations of the phenomena are still believed to be of a certain continuous type. Such continuous representations are thought to be rather obvious limits of the pixel representation when the resolution (scale of observation) goes to zero. In particular one would usually associate a function with such an image - a "density" - and consider the digitized field as corresponding to averages of this density over a pixel. Hence from a very rough knowledge of the pixel values, one "naturally" tries to associate a hypothetical function. Such a "natural" hypothesis (already critized by Perrin) is far from being physically obvious: it requires ample (mathematical) regularity constraints which are contrary to the observed strong variability down to smaller scales. Mathematically, it corresponds to very particular measurable properties: one considers only regular measures with respect to the usual line, surface and volume measures, i.e. Lebesgue measures. Indeed, the simplest example of scaling and scale invariance is to consider the (apparently "metric" in fact "measure") idea of dimension of a set of points as it often occurs in geophysics. The intuitive (and essentially correct) definition is that the measure of the "size" of the set n(L) at scale L is given by:

$$n(L) \propto L^{D} \qquad (1)$$

where D is the dimension (e.g the length of a line $\propto L$, the area of a plane, $\propto L^2$... or the number of in situ meteorological measuring stations on the earth in a circle radius $L \propto L^{1.75}$ (Lovejoy et al., 1986a, b), the distribution of raindrops on a piece of blotting paper $\propto L^{1.83}$ (Lovejoy and Schertzer, 1990) and the occurrence of rain during a time period $T \propto T^{0.8}$ (Hubert and Carbonnel, 1988, this volume; Tessier et al., 1989) .... The "volume" (actually the measure of the set) is therefore a simple scaling (power law) function, and the dimension is important precisely because it is scale invariant (independent of L). We recall that the Hausdorff dimension D(A) of a (compact) set A may be defined by the generalization to non-integer D of the divergence rule "the length of a surface is infinite, its volume is zero ..." with the rather straightforward extensions (to non-integer D) of the d-Lebesgue measure (defined for integer d) to the D-dimensional Hausdorff measure. Thus we use the notation $\int d^{D}x$ for the D-dimensional Hausdorff measure of a (compact) set A and the Hausdorff dimension D of A is hence defined by the divergence rule[1] (see fig. 2 for illustration and Appendix B for more discussion):

[1] It is easy to check that eq. 1 is consistent with this divergence rule. Indeed, interpreting eq. 1 as the fact that the number of cubes of size $l = L/\lambda$ needed to cover the fractal set will be of the order λ^{D} and since the D'-volume of an elementary cube is $l^{D'}$, it follows that the sum of their D'-volumes -of the order of the D'-Hausdorff measure- will follow the indicated divergence rule.

$$\int_A d^{D'}x = \infty, \quad \text{for } D'<D; \qquad \int_A d^{D'}x = 0, \quad \text{for } D'>D \tag{2}$$

One may note that the D-measure of A is not necessarily finite and non-zero: some logarithmic corrections (exponents Δ_i on the i-th iterate of the logarithm are "sub-dimensions", c.f. Appendix B) may be needed to obtain finiteness and a precise determination of the Hausdorff dimension (they may give rise to the appearance of 'lacunarity', e.g., Smith et al. (1986)).

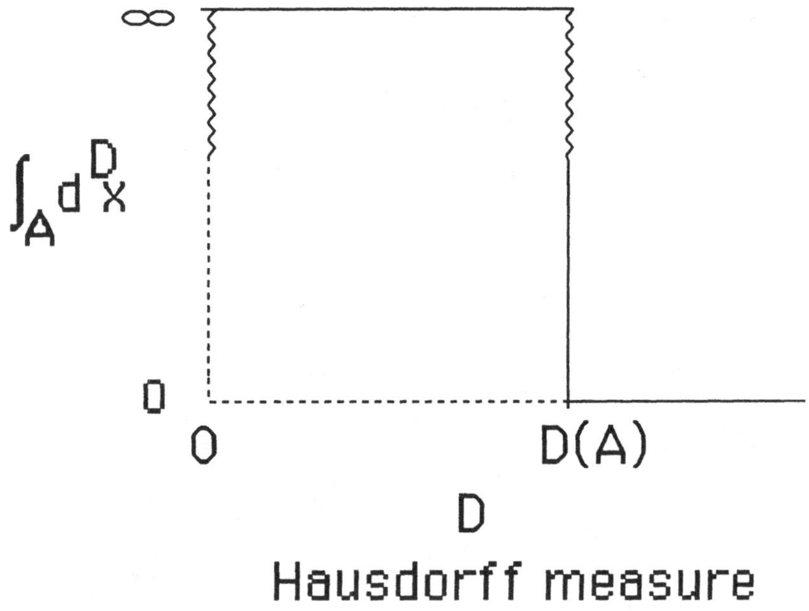

Fig. 2: Illustration of the divergence rule for Hausdorff measures, generalizing the divergence rule "the length of a surface is infinite, its volume is zero...". The transition at D=D(A), from infinity to zero, defines the Hausdorff dimension of the set A.

In other words, the "natural" framework for fields is not functional analysis (nor geometry ...!), but (mathematical) measures. Indeed, the use of functions rather than the (more general) measures is often a purely mathematical artifact. It is unnecessarily restrictive since really what we can empirically measure or describe is not in fact a value at a (geometric) point, but rather a value on "nearly any" (small) set surrounding this point. Such considerations are at the basis of (mathematical) measure theory which renders quite precise the notion of "nearly any" set[1] as small as we wish. Thus geophysics already seems to be more and more associated with singular measures with respect to Lebesgue measures[2]. Going a step further we will be interested in (random) *linear operators acting on measures*, as fundamental tools to study nonlinear variability. Such apparently abstract questions can be concretely addressed by apparently simple-minded geophysical models, but with rather general and non-trivial consequences and properties corresponding to the more abstract tools mentioned above.

---

1 It needs to be member of the "tribe", usually the borelian tribe...

2 Regular (respectively singular) measures with respect to Lebesgue measures means that (almost everywhere) they correspond to a product of a density (a function) and a Lebesgue measure (resp. they don't).

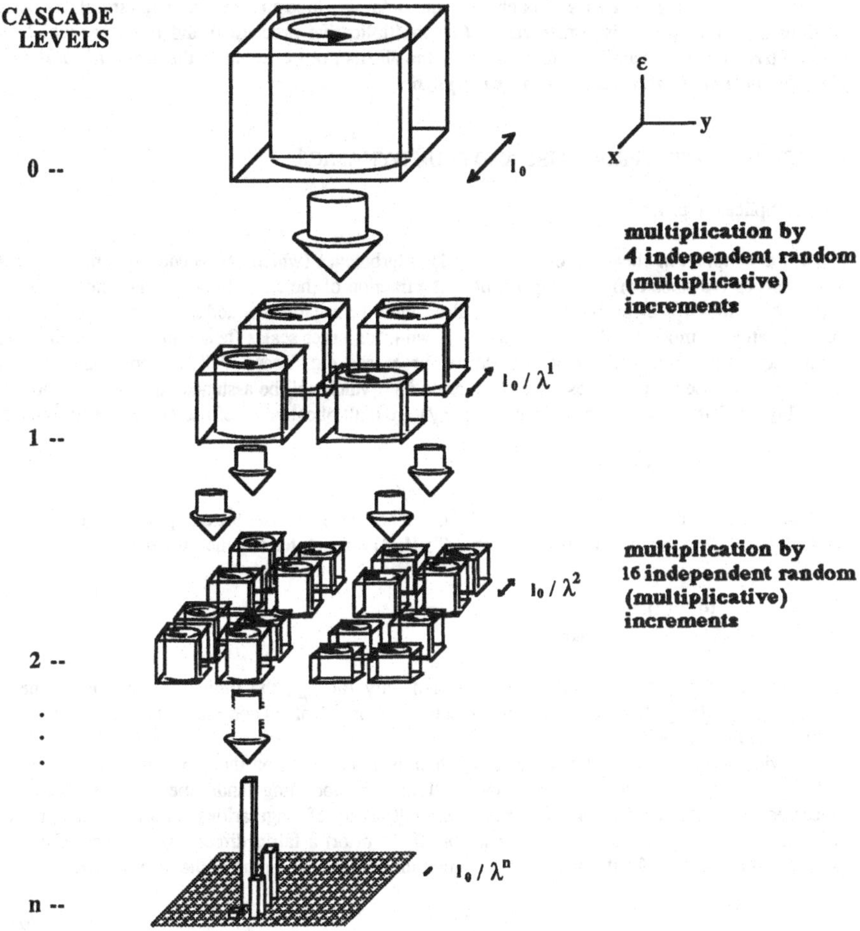

Fig. 3: A schematic diagram showing few steps of a discrete multiplicative cascade.process, here the "α-model" with two pure singularity orders γ^- (>-∞) and γ^+ (corresponding to the two values taken by the independent random increments, $\lambda^{\gamma^-}<1$ and $\lambda^{\gamma^+}>1$) leading to the appearance of mixed singularity orders γ ($\gamma^- \leq \gamma \leq \gamma^+$).

On the contrary, failure to adequately recognize some of these fundamental properties (such as the bare/dressed distinction) has lead to a simplistic notion of multifractals as mathematical functions (rather than measures) involving properties such as the order of singularities and even dimensions at mathematical points rather than on neighbourhoods of points. This "local" multifractal notion - and its"soft" consequences as discussed in section 4.1 - has been influenced, if not inspired, by an exaggerated emphasis of the geometry of multifractals rather than statistics, and has lead to various frustrating attempts to calculate various local multifractal properties. The difficulties and apparent contradictions which result from local multifractal notions include the apparent existence of negative fractal dimensions, the difficulty of obtaining converging estimates of local orders of singularity. Examples include the use of (Arneodo et al., 1988; Farge and Rabreau, 1988) wavelet analysis (Grossman and Morlet, 1987; Meyer, 1987).

We will try to give two approaches to the weakly[1] measurable scale invariant properties of geodynamics: one which is constructivist (the multiplicative processes) and the other which is non-constructivist ("flux dynamics"). In both cases, the abstract object remains the same: *to study how God plays dice in creating a stochastic chaos in Geophysics!*

3. MULTIPLICATIVE PROCESSES AND FLUX DYNAMICS

3.1 Multiplicative processes

The key assumption in phenomenological models of turbulence (which has recently become more explicit) is that successive steps define (independently) the fraction of the flux of energy distributed over smaller scales. Note that it is clear that the small scales cannot be regarded as adding some energy but can only (multiplicatively) modulate the energy passed down from larger scales (hence in spite of their occasional visual success the lack of relevance of additive processes, e.g. Voss (1983)). Hence bare densities ε_λ, resulting from cascade processes from outer scale l_0 (which will be assumed equal to 1, without loss of generality) to l (the homogeneity scale of ε_λ)$=l_0/\lambda$ are multiplicatively defined (see fig. 3 for illustration):

$$\varepsilon_{\lambda\lambda'} = T_\lambda(\varepsilon_{\lambda'})\ \varepsilon_\lambda \tag{3}$$

T_λ denotes a spatial contraction of ratio λ (>1). In the isotropic case, for any point x; $T_\lambda x = x/\lambda$; for any set A: $T_\lambda(A) = \{T_\lambda x\ /x \in A\}$; for any function f :$T_\lambda\,[f(x)] = f(\lambda x)$; for any measure μ and any set A:

$$\int_A d[T_\lambda(\mu)] = \int_{T_\lambda(A)} d\mu$$

and more generally for any function f (i.e. not only for 1_A, the indicator function of the set A) $\int f d[T_\lambda(\mu)) = \int T_\lambda(f) d\mu$. In case of (scaling) anisotropy, more involved contractions of space are required as discussed in the section 6.

Leaving additive (stochastic) processes (which had been used on the purely geometrical grounds of fractal geometry, e.g. fractional brownian motions, for modelling landscapes, etc.) to multiplicative processes, one encounters surprising properties: multiplicity of singularities and dimensions, rather than uniqueness. Let us discuss these properties briefly: a priori a fairly direct consequence of eq. 3 is the existence of a generator for the one parameter multiplicative (semi-) group of the bare densities:

$$\varepsilon_\lambda = e^{\Gamma_\lambda} \tag{4}$$

where Γ_λ is its generator, still with the homogeneity scale $l = l_0/\lambda$. Γ_λ is a certain operator whose main properties (especially its asymptotic behavior, $l \to 0$ or $\lambda \to \infty$) we will analyze. Γ_λ should in *some sense* (see below) become independent of λ, i.e. approach its limit Γ as the homogeneity scale approaches zero. For positive values γ of Γ_λ, divergence of ε_λ occurs as λ tends to ∞, hence such values correspond to (algebraic) orders (γ) of singularity. Conversely negative values correspond rather to (algebraic) orders of regularity. Nevertheless for brevity, we will frequently keep the expression "singularity" (instead of "regularity") for both cases. Note we are studying a whole family of measures defined by just one density, this is the reason why our notation doesn't reduce to the very specialized notation[2] (α, $f(\alpha)$) introduced by Halsey et al. (1986), where they refer to a single dimension (the dimension d of the embedding space) and the corresponding specialized measure. Hence, $\alpha = d-\gamma$ is the order of singularity of the d-dimensional Lebesgue measure, whereas γ is the order of singularity of the corresponding density of the measure, and $f(\alpha)=d-c(\gamma)$. As soon as this generator does not reduce (Schertzer and Lovejoy, 1983, 1984) to only two values $\gamma^+>0$ and $\gamma^- = -\infty$ (the once celebrated "β-model" (Novikov and Stewart, 1964; Mandelbrot,

[1] "Weak" refers to the type of convergence of the process and of statistical estimators of the process.
[2] Do not confuse this α with Lévy index α used below

1974; Frisch et al., 1978) corresponding to the alternative of dead (λ^{γ^-}=0) or alive (and λ^{γ^+}>1) sub-eddies, the *pure singularity orders* γ^- and γ^+ lead to the appearance of *mixed singularity orders*. In particular, as soon as $\gamma^- > -\infty$ (the "α-model"), mixed singularities of different orders γ, are built up step by step (cf. fig. 3) and are bounded by γ^- and γ^+ ($\gamma^- \leq \gamma \leq \gamma^+$, γ^- and γ^+ corresponding then to the alternative of weak ($1>\lambda^{\gamma^-}>0$) or strong ($\lambda^{\gamma^+}>1$) sub-eddies). In other words, as pointed out by Schertzer and Lovejoy (1983), leaving the simplistic alternative dead or alive ("β-model") for the alternative weak or strong ("α-model") leads to the appearance of a full hierarchy of levels of survival, hence the possibility of a hierarchy of dimensions of the set of survivors for these different levels. In this α-model (as in more elaborate ones) the different orders of singularities (or survival levels) define the multiple scaling of the (one-point) probability distribution :

$$\Pr(\varepsilon_\lambda \geq \lambda^\gamma) \approx N_\lambda(\gamma)/N_\lambda \approx \lambda^{-c(\gamma)} \qquad (5)$$

where $N_\lambda(\gamma)$ is the number of occurrences of singularities with order greater than γ, N_λ is the total number of events examined. We temporarily postpone discussion on the accuracy of the approximations indicated in eq. 5 -e.g. the sub-multiplicity problem[1] already discussed by Schertzer and Lovejoy (1987a, b) and point out the convenient empirical analysis technique to measure the the probability distribution multiple scaling (PDMS, introduced by Lavallée et al. in this volume) in order to estimate $c(\gamma)$:

$$c(\gamma) \approx -\mathrm{Log}_\lambda \Pr(\,(\mathrm{Log}_\lambda(\varepsilon_\lambda) \geq \gamma\,) \qquad (6)$$

Multiple scaling is obviously not restricted to one-point distributions (the latter being incomplete statistical descriptors of a field), indeed we need to know the behaviour of the joint n-point probability distribution (for the n-position vectors $(x_1, x_2, ..x_n)=\underline{x}$). It suffices to consider a n-dimensional vector $\underline{\gamma} =(\gamma_1, \gamma_2, ..\gamma_n)$ -instead of a scalar γ– with corresponding codimension $c_n(\underline{\gamma})$, and we should have the n-dimensional multiple scaling of the (n-dimensional) probability distribution:

$$\Pr(\{\varepsilon_\lambda(x_i) \geq \lambda^{\gamma_i}\}_{i=1,n}) \approx \lambda^{-c_n(\underline{\gamma})} \qquad (7)$$

The hierarchy ($n \to \infty$)of the codimensions $c_n(\underline{\gamma})$ would be sufficient to assess the statistical behaviour of the field and we will discuss the interpretation of the $\underline{\gamma}$ as extensions of phase portraits in section 4.3. However, the behaviour of $c_n(\underline{\gamma})$ is very sensisitive to the distances between the n points x_i and the scale ratio λ. It will be often easier to consider the characteristic functional, as in Schertzer and Lovejoy (1987a, b), which is even more general than the n-dimensional extension considered here (see also Appendix C).

3.2. Wild singularities and the sampling dimension:

We would like to insist on the interest of the above formulae (eq.5-7) in gaining insights into different fundamental aspects of multiple scaling. Obviously singularities will prevent convergence in the usual sense, i.e. even if the ε_λ are smooth functions (for a given λ), they do not admit a function as their limit. Indeed, their limit will rather be defined by the limit of the fluxes (i.e. integrals of the density) over different sets. One may note also that N_λ will be proportional to λ^d -the number of boxes, size l_0/λ, required to cover the relevant region of the embedding space (which can be fractal) of dimension d (integer or not)- multiplied by the number (N_i) of realizations (e.g. images) examined. Hence, when $c(\gamma)$ is smaller than d it has a rather immediate meaning of a codimension[2] = $d-d(\gamma)$; $N_\lambda(\gamma) \approx \lambda^{d(\gamma)}$, where $d(\gamma)$ (>0) is the dimension of the fraction of the space occupied by the singularities of order greater than γ on "nearly" each realization.

Larger values of $c(\gamma)$ (>d, i.e. "negative dimensions": $d(\gamma)=d-c(\gamma)<0$), which have often been disregarded, correspond to more rare events: singularities of orders which "nearly" never appear on a realization, but nevertheless give overwhelmingly important statistical contributions since they prevent

[1] This leads to log corrections ignored in Eq. 5, hence the sign ≈.

[2] In particular, in the case of the β-model there is a unique codimension c, characterizing the fraction of the space occupied by the alive sub-eddies. The parameter β is λ^{-c}.

convergence of (statistical) moments as shown below! This is the reason we call them "wild singularities". At first glance they seem to correspond to negative dimensions, sometimes mysteriously called "latent dimensions". However, there is no mystery at all, since $c(\gamma)$ still has a meaning of a codimension: no longer in an individual realization, but in the subspace of the (infinite dimensional) probability/phase space that our finite sample size enables us to explore as a finite dimensional cut. Indeed the dimension of this subspace can be estimated as $d + d_s$, where d_s is called the "sampling dimension" (at scale l_0/λ)- and is estimated by writing the number of images (or realizations) N_i as λ^{d_s}. Indeed when $c(\gamma)$ is smaller than $d+d_s$, γ occupies a fraction of the accessible subspace having dimension $d(\gamma) = d + d_s - c(\gamma)$. Of course, increasing the number of images, hence the sampling dimension, allows us to more readily encounter higher order singularities occupying a fraction of the accessible subspace, with well defined dimension ($d(\gamma) = d + d_s - c(\gamma) > 0$). The corresponding mathematical subtlety underlying the important difference between cases $c(\gamma)<d$ and $c(\gamma)>d$, is the "almost surely" or not properties, the latter do correspond to extremely rare events.

Although extremely rare, the wild singularities will be of overwhelming importance since they will prevent convergence of all moments of (high enough) orders. Indeed, the smoothing introduced by integrating the density over a set A with dimension D (to obtain the flux through A) may be sufficient to ensure the convergence for low order statistics, but not for orders higher than a critical order h_D of divergence. Let us point out this rather immediate consequence of eq. 5, by introducing first the trace (paralleling the definition of the trace of the density operator in Quantum Statistical Mechanics, see below) of the hth power of the flux Π_λ over an (averaging) set A of dimension D (with integration performed with resolution l_0/λ on A_λ, which denotes the set A measured with the same resolution):

$$\mathrm{tr}_{A_\lambda}\, \varepsilon_\lambda^h = \int_{A_\lambda} \varepsilon_\lambda^h\, d^{hD}x \approx \sum_{A_\lambda} \varepsilon_\lambda^h\, \lambda^{-hD} \qquad (8)$$

a priori any singularity of order higher than D, may create divergences of the trace but are extremely rare (since their frequency of occurrence tends to zero as $\lambda^{-c(\gamma)}$). One may evaluate the importance of these by considering their statistics (the trace-moments introduced by Schertzer and Lovejoy (1987 a, b)) for an arbitrary singularity of order γ:

$$\mathrm{Tr}_{A_\lambda}\, \varepsilon_\lambda^h = <\mathrm{tr}_{A_\lambda}\, \varepsilon_\lambda^h> \geq N_\lambda(\gamma)\, \lambda^{h\gamma}\, \lambda^{-hD} = \lambda^{[h\gamma - c(\gamma)]-(h-1)D} \qquad (9)$$

which diverges, for some orders of singularity, as soon as:

$$K(h) \geq (h-1)D \text{ or } K_D(h) \geq 0; \quad K_D(h) \equiv K(h)-(h-1)D \qquad (10)$$

where:

$$K(h) \equiv \sup_\gamma[h\gamma - c(\gamma)] \quad \{\text{hence: } c(\gamma) \equiv \sup_h[h\gamma - K(h)]\}$$

or:

$$h = dc(\gamma)/d\gamma,\ K(h) = h\gamma - c(\gamma) \quad \{\gamma = dK(h)/dh,\ c(\gamma) = h\gamma - K(h)\} \qquad (11)$$

On the one hand, eq. 11 corresponds to the Legendre transform of $c(\gamma)$ as pointed out by Parisi and Frisch (1985), Halsey et al. (1986) and as the resulting K(h) does correspond -by the method of steepest descent- to the exponent of the moment of the density of the flux (at least to first order, i.e. omitting logarithmic corrections):

$$<\varepsilon_\lambda^h> = \lambda^{K(h)} <\varepsilon_1^h> = e^{K(h)\mathrm{Log}\lambda} <\varepsilon_1^h> \qquad (12)$$

The Legendre transform establishes a well defined relation between orders of singularities and orders of moments. It is worth noting that it is straightforward to obtain the n-point statistics for $\mathbf{x} = (x_1, x_2, ..., x_n)$ by replacing the scalar h by the vector $\mathbf{h} = (h_1, h_2, ..., h_n)$ and K(h) by $K_n(\mathbf{h})$:

$$<\prod_{i=1}^{n}\varepsilon_\lambda(x_i)^{h_i}> = \lambda^{Kn(h)}<\prod_{i=1}^{n}\varepsilon_1(x_i)^{h_i}> \quad (13)$$

As for $c_n(\gamma, K_n(h)$ will depend sensitively on the distances betweeen the n points x_i and λ. It will far more general (and easier) to consider the characteristic functional K(f) of the generator (see following subsection and appendix C).

Note that conservation in ensemble average of the flux requires conservation of densities ($<\varepsilon_\lambda>=<\varepsilon_1>$) thus K(1)=0. On the other hand, as pointed out by Schertzer and Lovejoy (1983), the divergence rule, eq. 10, introduces a hierarchy of critical codimensions C(h), simply defined as:

$$C(h)\,(h-1) = K(h) \quad (14)$$

since the former divergence rule (eq. 8) can be rewritten ($\lambda\rightarrow\infty$, $\varepsilon_\lambda\rightarrow\varepsilon$):

$$Tr_A\,\varepsilon^h = \infty, D<C(h), \text{ i. e. } h>h_D, C(h_D)=D \;\{\gamma_D=dK(h)/dh|_{h_D}\} \quad (15)$$

where h_D is the critical moment order, and γ_D the critical singularity order (the wild singularities correspond to $\gamma>\gamma_D$) at which divergence of trace moments of the flux on the set A of dimension D occurs. For $h_D>1$ it implies the divergence of the usual statistical moments, since:

$$Tr_{A\lambda}\,\varepsilon_\lambda{}^h \leq <\Pi_\lambda{}^h(A)> \text{ any } h\geq 1 \quad (16)$$

Conversely, as discussed more thoroughly by Schertzer and Lovejoy (1987b), convergence of statistical moments of order h (h>1) is assured by the convergence of the hth trace moment; for h<1 divergence of the trace moment implies degeneracy of the flux (the set A has a so small dimension ($D<C_1\equiv C(1)$) that almost surely the flux is null). We thus obtain a twin divergence rule for the trace moments (represented in fig. 4) implying non-degeneracy of the flux ($h\leq 1$) and divergence of the flux ($h\geq h_D>1$). Note that non-degeneracy of the flux implies conservation of the ensemble average flux[1]:

$$<\varepsilon_\lambda> = <\varepsilon_1> = 1 \text{ and } D > C_1\,(\equiv C(1)) \Rightarrow <\Pi_\lambda(A)> = <\Pi_1(A)>\;(\equiv \int_A d^Dx) \quad (17)$$

Note that C_1 ($\equiv C(1) = K'(1)$, due to eq. 14) is at the same time the codimension of singularities contributing to the average (h=1) and the order of these singularities, since by virtue of Legendre transform it is the fixed point of $c(\gamma)$:

$$c(\gamma) = \gamma \Rightarrow \gamma = C_1\;(\equiv C(1) \equiv K'(1)) \quad (18)$$

---

1 As it corresponds to a "martingale "property, it assures a "weak measurable" convergence of the process (see Schertzer and Lovejoy 1987b for discussion).

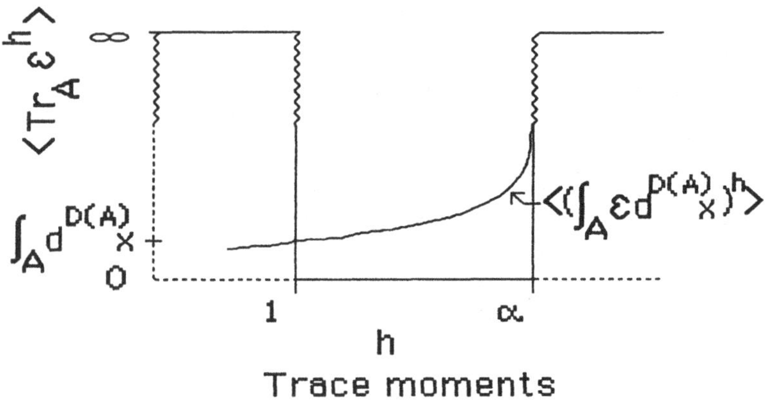

Fig. 4: the twin divergence of trace-moments.

3.2. Characteristic functionals and Fluxdynamics

Multiple scaling (for the statistical moments) corresponds to the fact that unlike the β-model K(h) is no longer linear (= C_1(h-1)) and depends on a whole hierarchy of codimensions C(h) (≠C_1, for h≠1). Since the first (Laplacian) characteristic function (or moment generating function) $Z_\lambda(h)$ and second characteristic function (or cumulant generating function) $K_\lambda(h)$ of the generator Γ_λ, are by definition:

$$Z_\lambda(h) = e^{K_\lambda(h)} = \langle e^{h\Gamma_\lambda}\rangle \quad (\equiv \langle \varepsilon_\lambda{}^h\rangle) \tag{19}$$

multiple scaling corresponds to algebraic divergence (λ→∞) of Z(h) and thus to logarithmic divergence of $K_\lambda(h)$ (≈ K(h) logλ, see eq. 12), a fundamental property which we will exploit below. This property is in fact far more general if we consider not only the n-point characteristic function $K_{n,\lambda}$ (eq. 13), but the characteristic functional $\mathbf{K}_\lambda(f)$ on "any test" function f of the generator (in eq. 13, we have considered the particular case: $f(x) = \Sigma_{i=1,n}\, h_i\, \delta_{x_I}$):

$$Z_\lambda(h) = e^{K_\lambda(f)} = \langle \exp(\int_A h(x)\Gamma_\lambda(x) d^D x) \rangle \tag{20}$$

Note here, we are dealing with characteristic functions or functionals in the Laplace sense, since $Z_\lambda(h)$ or $\mathbf{Z}_\lambda$ are obtained by Laplace transform (instead of Fourier transform) of the probability distribution. In order to make connections with statistical physics, $-\Gamma_\lambda$ can formally be considered as an Hamiltonian[1] (H_λ) and h as the inverse of temperature (h=1/T, the Boltzmann constant being set equal to 1), Z_λ is called a partition function and the "free-energy" (F_λ) would correspond to $K_\lambda(h)/h$. More generally (in Statistical Quantum Mechanics), the "density operators" $\rho_\lambda = e^{-H_\lambda/T}$ (corresponding to $\varepsilon_\lambda{}^h = e^{h\Gamma_\lambda}$) are considered along with their trace over different sub-spaces of states, each trace corresponding to a partition function. The densities ρ_λ and ε_λ are both defined on a fairly abstract space (e.g., in quantum mechanics the space of wave functions). The trace moments correspond to the average ensemble of the trace of the density operator integrated over an (ordinary or fractal) set A, this integration corresponds to the linear action of an operator generated by N_λ. This is the analogue of the operator for the number of particles; here it is rather the

[1] As done in random energy models (Derrida and Gardner, 1986; Gardner and Derrida, 1989)

generator of the fractal set A, seen at resolution l_0/λ, i.e. over which the "boxes" where we integrate the flux:

$$\varepsilon_\lambda{}^h\ (= e^{h\Gamma_\lambda}) = \rho_\lambda\ (= e^{-H_\lambda/T}); \qquad 1_{A_\lambda} d^{D_x} \approx e^{-N_\lambda};$$

$$Z_\lambda(h) = Tr_A(e^{-N_\lambda/T}e^{-H_\lambda/T}) = Tr_A(\varepsilon_\lambda{}^h) \tag{21}$$

on the space of measures of (compact) supports (for the different sets A, 1_{A_λ} is the indicator function for the set a at resolution λ).

If we consider now the ensemble average of the fluctuations of the operator N_λ itself (i.e. we are averaging over a certain subspace of the random measures of (compact) supports) we define a "grand ensemble" partition function $Z_{G,\lambda}(h)$:

$$Z_{G,\lambda}(h) = Tr(e^{-N_\lambda/T}e^{-H_\lambda/T}) = \lambda^{G_\lambda(T)} \tag{22}$$

which makes explicit in a rather formal manner the crucial problem of observations obtained by integration on a scale ($l'=l_0/\lambda'$) much larger than the homogeneity scale of the process ($l=l_0/\lambda$): the possible non-commutation between $N_{\lambda'}$ and H_λ (especially when $\lambda>>\lambda'$), thus the possible divergence of moments for dressed fluxes ("dressed" by the observation). We may also understand the divergence of moments as a phase transition, i.e. "solidification" by extreme localization in phase space of the contributions (of wild singularities) to high order moments (low temperatures 1/h) The second characteristic function of the trace moment $K_D(h)/h$ is rather the equivalent of a chemical potential and is simply related to K(h) by:

$$K_D(h) = K(h)-(h-1)D = (C(h)-D)\ (h-1) \tag{23}$$

One may note that if the observation sets A are multifractal sets their own nonlinear characteristic function $K_A(h)$ will intervene instead of (h-1)D ($\equiv K_A(h)$ in the monofractal case). On the other hand, since $c(\gamma)$ characterizes the logarithm of the probability distribution of Γ_λ, it corresponds to the entropy (S_λ) (of the state γ), and indeed the Legendre duality between $K_\lambda(h)$ and $c(\gamma)$ corresponds to the same (and more familiar) duality between $F_\lambda(T)/T$ and $S_\lambda(E)$ (the conjugate variables being 1/T and the energy E). Let us emphasize that in both cases, this property simply results from the fact that the Laplace transform of the probability distribution[1] (or conversely of the partition function) reduces to a Legendre transform of the exponents. In order to develop a nonconstructivist approach, which we call "fluxdynamics", we consider ε *per se* (the limit ε of the ε_λ, at zero homogeneity scale, λ going to infinity) as a linear operator on the measures (converting the D-volume, D being the dimension of A, integer or not, into the flux over the set A). However, we need to investigate some basic properties of this limit and its generator.

4. SINGULAR STATISTICS, TYPES OF CONSERVATION AND CHAOS

4.1 "Hard" (wild) vs. "soft" multifractality

The divergence of moments is a wild statistical behaviour very far from "soft" statistics (e.g. Gaussianity, quasi-Gaussianity...), and corresponds to "hyperbolic" (algebraic) fall-off of the probability distribution:

$$Pr(|X|\geq s) \approx s^{-\alpha}\ (s>>1) \Rightarrow \text{any } h\geq\alpha : \langle|X|^h\rangle = \infty \tag{24}$$

Among these "hyperbolic" random variables some are rather well defined, since they are mostly (but surprisingly!) generalizations of Gaussian laws. These are the Lévy stable random variables ($0<\alpha<2$) satisfying "generalized central limit theorems", hence involved in additive processes as discussed in subsequent sections and especially with the help of Appendix A which deals with a particular class of them. We used the expression "hyperbolic intermittency" (Schertzer and Lovejoy, 1985) to describe the effect of

[1] It also implies the convexity of K(h) (or F((T)/T), hence of $c(\gamma)$ (or S(E)).

this strong variability for a wider range of α (i.e. $\alpha \geq 2$) and we pointed out that this divergence is a general consequence of multiplicative processes and that the corresponding critical order of divergence $\alpha = h_D$ (theoretically, determined by eq. 15) has no absolute bound. Waymire and Gupta (1985) have used the expression "fat-tailed" for such (asymptotically algebraic) distributions, and "long-tailed" for the log-normal law, to distinguish these distributions from standard exponential "thin-tailed" distributions. In the preceding section we showed that hyperbolic behaviour is expected from averaging a multifractal field over a set with too small a dimension D. Its value has been empirically estimated in a variety of meteorological fields: $h_D \approx 5$, for temperature (Lovejoy and Schertzer, 1986a, b; Ladoy et al., 1986), $h_D \approx 1.66$ for changes in storm integrated rainrates (Lovejoy, 1981), $h_D \approx 1.06$ in radar reflectivity factors of rain (Schertzer and Lovejoy, 1987), and respectively $h_D \approx 5$ and 3.33 for wind speed and potential temperatures, $h_D \approx 1$ for the Richardson number (Schertzer and Lovejoy, 1985)...

An important consequence of the divergence of statistical moments is that the usual estimation procedures no longer work efficiently, but rather exhibit spurious (or pseudo-) scaling exponents. Indeed, these methods rely heavily on the law of large numbers which blows up due to the statistical divergences. Fig. 5 shows how the appearance of spurious scaling can be quite misleading. The classical estimation leads to bounded codimensions C(h)... even though the (well understood) simulated field has a linear C(h)! Conversely, clear understanding of spurious scaling can be used to explain most of the behaviour of certain data. For instance we argue (Schertzer and Lovejoy 1983, 1984, 1985; Lovejoy and Schertzer, 1986a) that the presumed critical moment order $h_D \approx 5$ for wind speed, may well explain the overall behaviour of the observed scaling exponents of the structure functions of the velocity field collected by Anselmet et al. (1984)!

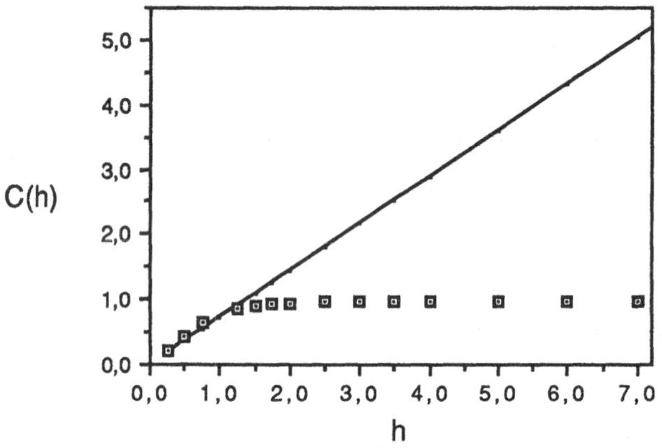

Fig. 5 : Illustration of the consequences of spurious scaling of the "dressed" quantities for the estimated C(h), that stay bounded for large values of h, in disagreement with the linear behaviour of the theoretical C(h) given by the continuous curve The estimated C(h) is obtained by trace moment analysis of 2000 independent samples of density fields induced by log-normal ($\alpha = 2$) multiplicative cascade process, the scale ratio $l = 2^{10}$.

A direct consequence of the hyperbolic behaviour of the dressed densities $\varepsilon_{\lambda,D}$ (obtained by D-dimensional averaging, at scale l/λ) is that their singularity codimensions $c_D(\gamma)$ are quite different from their bare counterparts $c(\gamma)$, since they become linear for orders greater than the critical singularity order γ_D:

$$c_D(\gamma) \approx c(\gamma_D) + h_D(\gamma - \gamma_D),\ \gamma \geq \gamma_D \quad (25)$$

this is an immediate consequence of hyperbolic behaviour as described by eq. 24, as well as from the corresponding divergence of the characteristic function ($K_D(h) = \infty$, $h \geq h_D$) and the fact that the Legendre transform breaks down[1] for linear functions. Conversely, for the same reasons, K(h) becomes linear as soon as there is an upper bound (γ_0) of the singularity order:

$$c(\gamma)\rightarrow\infty,\ \text{when}\ \gamma\rightarrow\gamma_0 \Leftrightarrow K(h)\approx\gamma_0 h\ (h>>1),\ \text{hence}\ C_\infty\equiv \lim_{h\rightarrow\infty}\{K(h)/(h-1)\} = \gamma_0. \quad (26)$$

4.2 Canonical vs. microcanonical conservation

Hyperbolic behaviour is expected only for singularities of order greater than the dimension of the averaging set A. It obviously can't occur if we are imposing a much more strict conservation than conservation of ensemble average such as a strict conservation on A of the flux in each realization. This follows from the fact that in the latter case we have:

$$\varepsilon_\lambda\, \lambda^{-D}\leq\Pi_\lambda(A)=\Pi_1(A) \quad (27)$$

Paralleling, once again classical thermodynamics, one can speak respectively of canonical conservation (or cascade) in the former case, and micro-canonical conservation (or cascade) in the latter (see for instance[2] Benzi et al. (1984), Pietronero and Siebesma (1986), Sreenivasan and Meneveau (1988)). The micro-canonical conservation assumption has many defects: not only are we usually dealing with open systems (as in thermodynamics), but this assumption turns out to be quite demanding and restrictive. In the framework of scale invariance, it requires *strict conservation at every scale*, so we can even speak of a "*pico-canonical*" assumption: strict conservation is implied not only on the largest scale of A, but on the smallest scale due to scaling behaviour of the process! Hence, we have rather sharp distinctions between *hard (wild) multifractality* allowed by canonical conservation, and *soft multifractality* implied by micro- (in fact pico-) canonical conservation. Furthermore, one may note that micro-canonical conservation refers to a given dimension: it no longer holds on sets with dimensions smaller than the characteristic dimension of the micro-canonical conservation. Hence, micro-canonical conservation is at the same time too precise and too vague. In contrast, realizations of a canonical process can be understood as low dimensional cuts of very high dimensional processes (the strict canonical case corresponds to phase spaces of infinite dimension).

4.3 Stochastic chaos vs. deterministic chaos

Considering the n-points statistics at points $(x_1, x_1,..., x_n)=\mathbf{x}$, via the n-dimensional singularity vectors $\underline{\gamma}=(\gamma_1, \gamma_2,..., \gamma_n)$ with conjugate vectors of moments order $\mathbf{h}=(h_1, h_2,..., h_n)$, we are in fact exploring the (stochastic) phase space by using "phase portraits". Such protraits have been often considered in the very particular framework of deterministic chaos (e.g., Grassberger and Procaccia (1983)) for time series $\{\varepsilon(t), \varepsilon(t+\tau), \varepsilon(t+2\tau),..., \varepsilon(t+n\tau)\}$ where the time lag τ is of no fundamental significance if scaling is observed. In this approach, finite dimensionality D of the attractor is infered if:

$$n>>1 \quad n-c_n(\underline{\gamma})\approx D \quad (28)$$

Obviously the condition that n should be large is very demanding because we need an enormous number N of points (N>>n) to obtain a reasonable estimate of $c_n(\gamma)$, hence it may easily require a prohibitive number of samples (see, for instance, the theoretical estimates of Essex, in this volume). Indeed, numerous doubts have been raised (Grassberger (1986), see also Nerenberg et al. and Viswanathan et al. in this volume) concerning some preliminary results (with n of the order 10, and N of the order of several hundred) which reported very low dimensionality ($D\approx 3$) of climatic and other geophysical attractors (e.g., Nicolis and

[1] This is a direct consequence of the geometrical interpretation of the Legendre transform as the envelope of the tangents.

[2] In fact a micro-canonical version of the α–model is often called a "random β–model" (Benzi et al., 1984) or "p model" (Sreenivasan et al., 1988). The former expression (which refers to the fact that the fraction of the space occupied by active sub-eddies is randomly chosen) is somewhat misleading, since the "β–model" is already a random model...

Nicolis (1983)). These results could be profitably reexamined with a straightforward generalization of sampling dimension for singularity vectors.

It is important to call attention to two basic facts. On the one hand, low values of D do not imply deterministic chaos (even if the converse is by definition true!). Indeed, an obvious counter-example is the brownian motion which in any space (of dimension d≥2) will be concentrated on a stochastic attractor of dimension 2. On the other hand, and more fundamentally, deterministic chaos (as the result of a set of coupled ordinary differential equations) was originally studied as an oversimplified version of the chaos generated by partial differential equations (Lorenz 1963), which a priori has an infinite dimensional phase space. It is then not logically consistant to take too literally what the simplified model tells us, otherwise we may be lead to simplistic conclusions. Indeed, this image and the "real world" share the common property that only a very small fraction of the phase space is active, thus corresponding to nonzero codimensions $c_n(\underline{\gamma})$ for any n. In this respect, eq. 28 is too strong a requirement. Let us briefly point out one of the major differences between the deterministic image and the stochastic one presented by multiplicative processes: in the latter, time is treated on an equal footing with space (even if we introduce scaling anisotropy in the space-time domain, hence scaling anisotropic space/time transformations) in the sense that the variability in time and space will be of the same type, whereas in the former case the process will be regular with respect to time (i.e. the time variable is very smooth, contrary to the space variables).

In our opinion, the little empirical evidence collected to date indicates no more than that only a small fraction of the phase space is active. And due to the preceding arguments, it seems more natural to interpret time series as low dimensional cuts of a very high dimensional process (in the canonical case discussed in the preceding section), i.e. stochastic chaos.

5. (H, C_1, α) UNIVERSAL MULTIFRACTALS

5.1 Characterization of the generator

Since a multiplicative process is a one-parameter group, the characterization of its generator is fundamental. Corresponding to the definition of the Γ_λ (eq. 4), we have - at least formally - for the limit Γ of the Γ_λ ($\lambda \to \infty$):

$$\varepsilon \equiv \text{limit}_{\lambda\to\infty}(\varepsilon_\lambda) = e^{\Gamma} \tag{29}$$

One may also note that we have (corresponding to eq. 4) the following "dynamical" (and somewhat formal) equation for the ε_λ:

$$\partial\varepsilon_\lambda/\partial\lambda = \gamma_\lambda\, \varepsilon_\lambda \;\; ; \gamma_\lambda = \partial\Gamma_\lambda/\partial\lambda \tag{30}$$

where γ_λ is the infinitesimal generator. This equation gives the cascade dynamical sense when we are studying a cascade on a space-time domain[1]. As discussed by Schertzer and Lovejoy (1987a, b) and expanded here in Appendix C, the generator must satisfy four main properties:

i) Γ is a random noise, with infinite band-width $[1, \infty]$. The bare or (finite resolution) generator Γ_λ should rather be understood as the corresponding filtered noise restricted to the wave-number band $[1, \lambda]$ of Γ.

ii) the second characteristic function (or cumulant generating function) $K_\lambda(h)$ of the (bare) generator Γ_λ has a logarithmic divergence ($\lambda \to 0$) in order to assure multiple scaling, i.e.:

$$K_\lambda(h) \approx \text{Log}(\lambda)\, K(h) \tag{31}$$

---

[1] On the space-time domain, the scaling is usually strongly anisotropic as we will discuss more thoroughly in section 6.

iii) in order to obtain some finite moments of positive order ($Z_\lambda(h)$ and $K_\lambda(h)<\infty$ for $h>0$), the probability distribution of positive fluctuations of the bare generators Γ_λ must fall off more quickly than exponentially.

iv) the generator needs to be normalized ($K_\lambda(1)=0$) in order to assure (canonical) conservation of the flux.

It turns out that properties i) and ii) correspond to the fact that the (generalized) spectrum $E_\Gamma(k)$ of the generator should be then proportional to the inverse of the wave-number:

$$E_\Gamma(k) \propto k^{-1} \tag{32}$$

since the characteristic function will correspond to its integral and we will have the appropriate log divergence. Such noises are often called "1/f noises" or "pink noises". Usually, one considers only Gaussian noises or quasi-Gaussian noises. We have already indicated that there is no fundamental reason to restrict our attention to quasi-Gaussianity, and thus we consider hyperbolic noises. Indeed, among the hyperbolic noises, Lévy stable noises ($0<\alpha<2$) are particularly important, since they define a family of universal generators as we will discuss latter. However, the third property, which is due to the fact the moments of ε_λ are Laplace transforms of the probability density of Γ_λ, lead us to restrict our attention to extremely unsymmetric hyperbolic noises, since we can accept a hyperbolic fall-off of the probability distribution only for the negative fluctuations of Γ_λ. Considering Lévy-stable noises (or hyperbolic noises with $0<\alpha<2$), one has to generalize the notion of spectrum (the usual spectrum diverges, since it is a second order moment) as discussed by Schertzer and Lovejoy (1987a, b). The fourth property is easy to satisfy since if Γ'_λ is not yet normalized, we can obtain a normalized generator Γ_λ by:

$$e^{\Gamma_\lambda} = e^{\Gamma'_\lambda} / \langle e^{\Gamma'_\lambda} \rangle \tag{33}$$

however, as we will later insist this method of normalization is not unique, since a fractional integration may achieve the same result.

Note that the properties of the generators stressed above are on the one hand quite different from the usual properties of Hamiltonians, since they have a 1/f spectrum and the equivalent of negative energies (the positive singularities). On the other hand, they give a precise definition of multiple scaling, especially by specifying the necessary and sufficient properties of the second characteristic functional, which might be called the "free flux" since it is the analogue of the free energy (as outlined in section 3.2).

5.2 The conservative $(0, C_1, \alpha)$ universal canonical multifractals

5.2.1. *The universal generators obtained by densification of scales.* In this subsection we will concentrate our attention on conservative fluxes, and we will show that they indeed depend only on two fundamental parameters (C_1, α). In the following section we will consider generalizations to nonconservative fields (such as the temperature and velocity fields, ...).

Just as in additive processes, one may look for universality classes in the sense that whatever generator is used (here the flux generator, the infinitesimal increment in additive processes) under repeated iteration - through (renormalized) multiplication or addition - it may converge to a well defined limit which depends on relatively few of its characteristics. Appendix A first recalls the classical (but not well enough known) results for additive processes associated with the generalized central limit theorem, here the classes and "basins of attraction" are primarily[1] defined by the Lévy index α, the critical order of moments of the increments (moments of order $h>\alpha$ diverge).

One has to be careful about the definitions of convergence and universality, since it has been obscured by some misplaced claims (Mandelbrot, 1989) that such universality cannot exist in multifractal processes. Indeed, it is easy to check that repeated multiplication corresponding to a process with fixed discretization

[1] There are two subsidiary Lévy parameters which are fixed in our case: the 'location parameter' (fixed by the normalization constraint) and the 'skewness' (set to its extremal value -1 by the condition iii, as explained below). The third subsidiary parameter, the 'scale parameter' is defined by C_1.

(i.e. a fixed elementary ratio of scale $\lambda_0>1$) fails to create a simplifying convergence to universal generators (e.g. the α-model remains an α-model), and it seems that this is the reason why Kolmogorov (1962) postulated a lognormal behaviour, without claiming convergence[1] to it. However, if we are discussing continuous cascade processes, i.e. processes which have an infinite number of cascade steps over any *finite* range of scales (i.e. the elementary ratio of scales $\lambda_0 \rightarrow 1_+$), we are facing quite a different limiting procedure, namely the *densification of scales at fixed overall scale range*, instead of *increasing the range at fixed density of scales*. Indeed, the continuous processes are obtained from a discrete model (finite number of discrete steps over the given ratio of scales) by introducing more and more steps up to an infinity of infinitesimal ones and while fixing some properties (e.g. the variance of the generator on this given scale ratio). This densification of scales can be done explicitly either with Fourier techniques (Schertzer and Lovejoy, 1987a, 1988, Wilson et al. , this volume) or with wavelets[2]. Obviously while such properties are mathematically best studied directly on the generator, we should also establish the physical relevance of doing so. Indeed, -generalizing the test field method introduced in homogeneous turbulence by Kraichnan (1971)- we may introduce new intermediate scales first as rather passive components, advected by the others, and then include them in the whole set of "active" scales. In this respect, the passive scalar example (studied by Schertzer and Lovejoy (1987a,1988), Wilson et al., (this volume)), is illustrative: the density of the flux (φ) controlling the passive scalar diffusion is a product of powers of densities of the energy flux (ε) and the scalar scalar variance flux (χ) -mainly from dimensional arguments, we have: $\varphi=\chi^{3/2}\varepsilon^{-1/2}$. In the first step, χ (corresponding to ε on the new intermediate scales) and ε can be considered as independent but of the same type. In the second step we identify φ as a more complete ε. Hence, we are multiplying densities by densities, or simply adding generators to generators...

Now, we have to investigate which classes of generator are stable and attractive under addition and such that for the corresponding density ε_λ will at least converge for some positive order moments (i. e. the probability density of the generator admits a Laplace transform as already discussed). Either we examine those Lévy stables -usually studied in a Fourier framework (e.g. Lévy (1924, 1925, 1954), Gnedenko and Kolmogorov (1954), Gnedenko (1969), Feller (1971), Zolotarev (1986), Gupta and Waymire (1990)- which also satisfy a Laplace transform or we directly study (as done in Appendix[3] A) the generalized central limit theorem in the Laplace framework which is much more immediate and natural.

In any case, it is clear that the restriction imposed by Laplace transforms is that we require (as condition (iii) already discussed) a steeper than an algebraic fall-off of the probability distribution for the (positive) orders of singularities, hence with the exception of the Gaussian case ($\alpha=2$), we have to employ strongly asymmetric, "extremal" Lévy laws, as explicitly emphasized by Schertzer et al. (1988). In our case, we are not considering random variables but noises, however the same characterization is relevant (characteristic functionals are involved rather than characteristic functions).

Let us examine the classes of universal generators (ranging from $\alpha=2$ down to $\alpha=0$), recalling that the corresponding characteristic functional K(h) and codimension functions c(γ) estimated by Legendre transform, are (Schertzer and Lovejoy, 1987a,b), since h^α/α and $\gamma^{\alpha'}/\alpha'$ are Legendre dual with $1/\alpha+1/\alpha'=1$, $0\leq\alpha\leq2$, $-\infty<\alpha'\leq0$ or $2\leq\alpha'<\infty$):

$$\alpha\neq1:\ K(h)=\frac{C_1\alpha'}{\alpha}(h^\alpha-h)\qquad \text{(only for } h\geq0 \text{ when } \alpha<2;\ =\infty \text{ for } h<0);$$
$$\alpha=1:\ K(h)=C_1\,h\,\mathrm{Log}(h)\qquad(34)$$

and (restricted to increasing branches when $\alpha<2$, since $dc/d\gamma=h$):

$$\alpha\neq1:\ c(\gamma)=C_1\left(\frac{\gamma}{C_1\alpha'}+\frac{1}{\alpha}\right)^{\alpha'}\qquad (dc/d\gamma>0 \text{ when } \alpha<2)$$
$$\alpha=1:\ c(\gamma)=C_1\exp\left(\frac{\gamma}{C_1}-1\right)\qquad(35)$$

---

[1] Yaglom (1966) seems to be less cautious on this point.

[2] Work is in progress on this, in collaboration with P. Brenier.

[3] One may note that classically only the case $0<\alpha<1$ is treated by Laplace transforms, Appendix A extends the result for $1\leq\alpha\leq2$. See also Schertzer and Lovejoy (1990) for discussion of this point.

We recall that C_1 ($\equiv C(1) = K'(1)$) is the fixed point of $c(\gamma)$, being simultaneously the codimension of singularities contributing to the average and the order of these singularities (see eq. 18). We may introduce another convenient characteristic order of singularity :

$$\gamma_0 = -\frac{C_1\alpha'}{\alpha} \tag{36}$$

γ_0 is[1] either the lower bound ($\alpha>1$, $\alpha'>2$) of fractal singularities ($c(\gamma)=0$, i.e. singularities occupying all the space) or ($\alpha<1$, $\alpha'<0$) the upper bound of singularities ($c(\gamma)=\infty$, i.e. unreachable singularities). It is then also the slope of the tangent at the origin of K(h) ($\gamma_0=K'(0)$; $\alpha>1$) or of the asymptote ($\gamma_0=K'(\infty)$; $\alpha<1$). We may then rewrite eq. 35 ($\gamma\geq\gamma_0$ when $2\geq\alpha>1$; $\gamma<\gamma_0$ when $\alpha<1$) as:

$$\alpha\neq 1,\ \gamma_0 = -\frac{C_1\alpha'}{\alpha},\ c_0=c(0):\ c(\gamma) = c_0\left(1-\frac{\gamma}{\gamma_0}\right)^{\alpha'} \tag{37}$$

One may note that the $c(\gamma)$ introduced here corresponds rather to the probability density, instead of the probability distribution. Both are equal when $c(\gamma)$ is increasing (e.g. for extreme singularities: $\gamma>>0$), but obviously decreasing $c(\gamma)$ (e.g. for extreme regularities: $\gamma<<0$) of the probability is offset for the probability distribution by its minimum value (see below the role of γ_0 for the Gaussian case). On the other hand, the $c(\gamma)$ don't coincide with the log of the probability density due to (at least!) some logarithmic terms (corresponding to sub-codimensions) which are missed by the Legendre transform, but are of no fundamental importance (as easily seen by considering the exact log of the probability density).

5.2.2. *The five main universality classes.* Let us review briefly the properties of the five main classes (α going from 2 to 0, hence α' going from 2 to ∞, then from $-\infty$ to 0), ranging from the Gaussian generator to the β-model, crossing three Lévy cases (see the corresponding fig. 6a and fig. 6b):

i) $\alpha=\alpha'=2$: the Gaussian generator is almost everywhere (almost surely) continuous. K(h) and $c(\gamma)$ are parabolas, $c(\gamma)$ is tangent on the γ axis at $\gamma_0=-C_1$, C(h) is linear ($= C_1$ h).

ii) $2>\alpha>1$ ($2<\alpha'<\infty$): the Lévy generator is almost everywhere (almost surely) discontinuous and is extremely asymmetric. The lower bound γ_0 ($=-C_1\,\alpha'/\alpha$) of fractal singularities decreases from $-C_1$ to $-\infty$, as α decreases from 2 to 1. The corresponding $c(\gamma)$ of the probability distribution, will remain on the γ axis for $\gamma\leq\gamma_0$, these singularities are space-filling (distributed on "fat fractals" which involve only sub-codimensions). It is also strongly asymmetric (even for the probability density, since $K_\lambda(h)=\infty$ for h<0). The wild singularities ($\gamma>>1$) give rise to a steeper algebraic branch than before ($c(\gamma)\propto\gamma^{\alpha'}$, $\alpha'>2$).

iii) $1>\alpha>0$ ($-\infty<\alpha'<0$) the generator is everywhere (almost surely) discontinuous, and is obtained in fact by a one-sided unnormalized generator hence the orders of singularities are bounded by γ_0 (thus decreasing, with α, from $-\infty$ to C_1), which thus defines a vertical asymptote, and now the algebraic asymptote intervenes for the large orders of regularity ($\gamma\to-\infty$, $c(\gamma)\propto(\gamma-\gamma_0)^{-|\alpha'|}$). As the singularities are bounded, the same occurs for the hierarchy of critical codimensions C(h) of the different moments, since we can now always smooth out the highest singularity on a set A of high enough dimension D. Indeed γ_0 bounds also C(h) (see eq. 26), hence to obtain convergence of every (positive order) moment of the flux it suffices that: $D>C_\infty=\gamma_0=C_1|\alpha'|/\alpha$. We thus leave hard (wild) multifractality for soft multifractality.

iv) $\alpha=1$ ($\alpha'=\pm\infty$!): it is the special in-between case ("extremal Cauchy[2]"), associated with the ambiguity on α' (note that γ_0 has opposite limits: $\gamma_0=\pm-\infty$), this corresponds in fact to a special case of quasi-stability (or not strict stability) briefly outlined in Appendix A. Note that the curves K(h) and $c(\gamma)$ are nevertheless the limits ($\alpha\to 1\pm$, $\alpha'\to\pm-\infty$) of the two preceding cases, especially the former algebraic asymptotes of $c(\gamma)$ tend to exponential behaviour since: $(x/\alpha'+1)^{\alpha'}\to e^x$ when $\alpha'\to\infty$.

v) $\alpha=0+$ ($\alpha'=0-$) : this limiting case corresponds to divergence of every statistical moment of the generator and seems at first glance very strange, but one of its representations is none other than the once

[1] This is the negative of the former γ_0 introduced by Schertzer and Lovejoy (1987a, b). The change of sign is required to obtain directly the bounds of singularities/regularities as explained.
[2] The usual Cauchy variable is the symmetric stable Levy variable, with $\alpha=1$.

celebrated β-model (γ^-=-∞, γ^+=C_1 =limit $_{\alpha\to 0+}(\gamma_0)$)! This fact in turn, shows clearly the peculiarities of the β-model, once thought to be a more or less crude approximation of intermittency ...

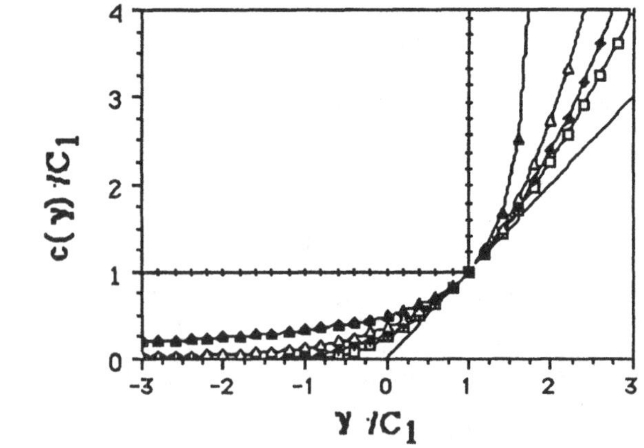

α=2 (–□–), α=1.5 (–♦–), α=1 (–△–), α= .5(–▲–), α=0 (–+–)

Fig. 6a: universal (bare) singularities codimension $c(\gamma)/C_1$ corresponding to the five classes; here α=2, 1.5, 1, .5, 0.

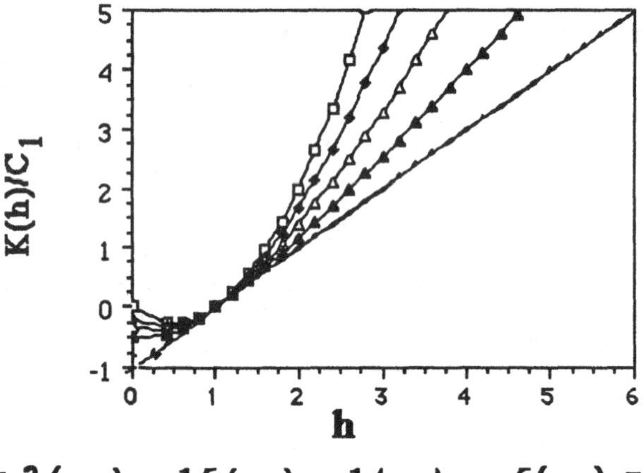

α=2 (–□–), α=1.5 (–♦–), α=1 (–△–), α= .5(–▲–), α=0 (–+–)

Fig. 6b: universal (bare) second characteristic function $K(h)/C_1$ (≡ $h.F(h)/C_1$, F(h) being the "free energy"), corresponding to the five classes; here α=2, 1.5, 1, .5, 0.

Let us point out briefly some consequences:

- the Lévy cases fill the gap between the two more or less classical cases the so-called lognormal (α=2) and the β-model, which now represent just two extremes of the whole spectrum of universal generators. The corresponding bare processes are log-Lévy, but not the dressed ones. The Lévy index α indicates how far we are from monofractality (the β-model) in the precise sense that it measures the convexity of K(h) for h≈1 ($K''(1) = C_1\alpha$).

- the very symmetric Gaussian case is the exception which assures the existence of negative order moments, on the contrary the asymmetric "extremal" character of the Lévy cases corresponds to the fact that we are "digging" wild regularities ("holes") with the algebraic extremes of the Lévy generator which preventing convergence of any negative order moment[1] (see the relevant fig. 7a, b).

- the links between regularity of the generator and the resulting flux are at first glance somewhat paradoxical since as α decreases the generator is more and more wild but the resulting fluxes more easily have finite positive order moments since the upper bound C_∞ of C(h) ($\equiv\infty$ for $\alpha\geq 1$; $\equiv\gamma_0<\infty$ for $\alpha<1$) decreases with α, leading to the finiteness of all moments for set A of dimension $D>C_\infty$. However, it is merely due to the fact that the wild behaviour of the generator is restricted to regularities (hence the particular problem of negative order moments since they interchange regularities with singularities and conversely). The β-model yields the extreme regularity since: $C_\infty = C_1 = C(h)$, any set A where the process is not degenerate, will have regular flux at all positive orders (but still none at negative orders!).

One may note that exact mathematical results have been obtained on the case α=2 (Kahane, 1985, 1987) and α<1 (Fan, 1989).

Fig. 7a : a Gaussian white-noise (α=2), fluctuations are symmetric.

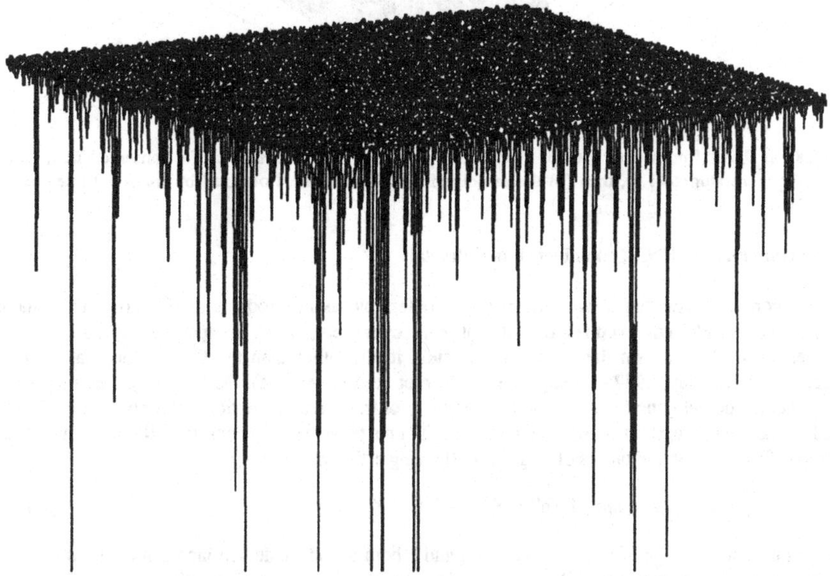

Fig. 7b: an extremal Lévy white-noise (α=1.5), fluctuations are extremely asymmetric: only *negative* hyperbolic jumps are allowed for the generator, "digging" wild regularities ("holes")

[1] Note that this may explain the many difficulties discussed in the physics literature connected with the estimation of moments of negative order of multifractal fields... since in general these moments don't exist!

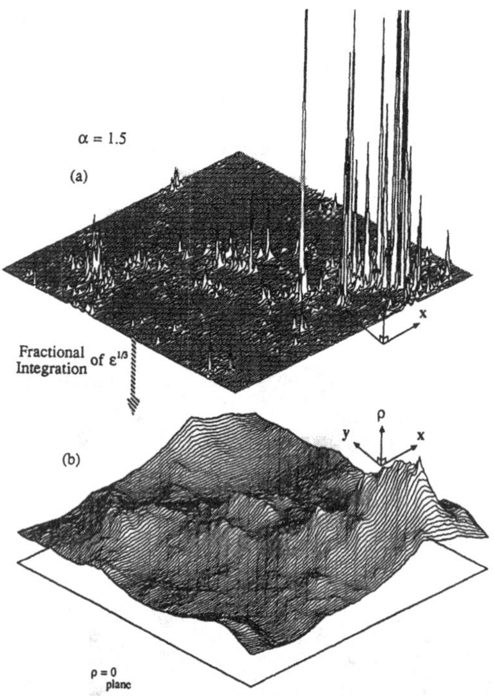

Fig. 8 : It shows from top to bottom, over $2^8 x 2^8$ pixel grid, (a) the density ε of a conserved flux, obtained with a Lévy generator (α=1.5), then (b) the associated concentration field obtained by fractional integration (of order 1/3).

5.3 Nonconservative (H, C_1, α) universal multifractals

From the conservatives fluxes, we may build up others by taking products of them or raising them to different powers. We may even fractionally integrate over them, which is especially desirable when for example, we want to obtain the concentration field itself, rather than the flux of the scalar variance (Schertzer and Lovejoy, 1987a, 1988, 1989; Wilson et al., this volume). However, by doing so, we will fundamentally add only one extra parameter[1] (the order of fractional integration) to our two basic C_1 and α. Indeed, a fractional integration/derivation of order[2] (b) on a power (a) of a conserved flux corresponds to an affine transformation on the orders of singularity (leaving c(γ) invariant):

$$\gamma' = a\gamma - b \; ; \; h' = h/a \; ; \; K'(h') = K(h) - bh' \; ; \; c'(\gamma') = c(\gamma) \tag{38}$$

staying in the same α-type of universality, but a and b both introduce deviations from conservation of the flux, hence the new parameter H. We can restrict our attention to transformations with a=1, since the

[1] At least considering bare properties. Things are more involved when considering dressed properties: apparently equivalent processes may create quite different wild singularities.
[2] Integration when b is positive, derivation when negative. This explains the change of sign we made in comparaison with Schertzer and Lovejoy (1987a, 1990).

deviation (K(a)) resulting from raising to the power a the density (ε) of a conservative flux, corresponds at least formally[1] , to changing the order b of fractional integration to H = b-K(a) on the conserved flux (obtained by fractional integration/derivation of order -K(a) on the density ε^a).

Fig. 8 shows the two main steps needed to obtain a concentration field over a $2^8 x 2^8$ pixel grid simulated (a=b=1/3) on a personal computer, fig. 8a shows the corresponding conserved flux (identified to ε), then the resulting concentration field after fractional integration (fig. 8b), see also Wilson et al. (this volume) for more discussion.

6. GENERALIZED SCALE INVARIANCE

6.1 Scaling anisotropy

We have already pointed out that the time-space domain is usually strongly (scaling) and anisotropic. This remark can be rendered a bit more obvious when considering formal scaling transformation on the velocity field (as done in Schertzer and Lovejoy (1987a, Appendix D2)) since (on purely dimensional grounds):

$$x \to x/\lambda,\ v \to v/\lambda^H \Rightarrow t \to t/\lambda^{1-H} \tag{39}$$

thus as soon as H≠1, we have space/time anisotropy (in case of homogeneous turbulence we would have H=1/3[2]).

However, strong anisotropy is already present in the space domain with oriented forces such as buoyancy (due to gravity) as well as the Coriolis force (due to the earth's rotation). These forces, which may introduce anisotropic differential operators, e.g., a fractional differential operator with the order of differentiation depending on the direction instead of (isotropic) gradients, are responsible for the (fractional) differential stratification and rotation of the atmosphere respectively. For instance, in order to avoid the classical but untenable 2D/3D dichotomy between large and small scale atmospheric dynamics, we have proposed an anisotropic scaling model of atmospheric dynamics (Schertzer and Lovejoy (1983, 1984, 1985a,b, 1987a,b); Lovejoy and Schertzer (1985); Levich and Tzvetkov[3] (1985)). In this model, the anisotropy introduced by gravity via the buoyancy force results in a differential stratification and a consequent modification of the effective dimension of space, involving a new "elliptical" dimension (d_{el}, see below), with resulting anisotropic shears. In isotropy, d_{el}=3, while in completely flat (stratified) flows, d_{el}=2. Empirical and theoretical evidence is given indicating that for the horizontal components of the velocity filed d_{el} is rather the intermediate value $d_{el} \approx 23/9 = 2.5555...$ (and $d_{el} \approx 2.22...$ for the rainfield (Lovejoy et al., 1987)).

Indeed, the requisite scale changes $\mathbf{T}_\lambda$ can be far more general than simple magnifications or reductions. It turns out that practically the only restrictions on $\mathbf{T}_\lambda$ are that it has group properties, viz: $\mathbf{T}_\lambda = \lambda^{-\mathbf{G}}$ where G is a the generator of the group of scale changing operations, and that the balls $E_\lambda = \mathbf{T}_\lambda(S_1)$ (S_1 being the unit sphere) decrease with λ. In the simplest case of "Generalized Scale Invariance" ("GSI"), G is a matrix -"linear GSI" (Schertzer and Lovejoy, 1983, 1984) E_λ are self-affine ellipsoids (see fig. 9) rather than the self-similar spheres of the isotropic case (G=1= identity), it already allows a tremendous variety of behaviour, since the only constraint on G turns out to be that every (generalized) eigenvalue of G has a non-negative real part (Schertzer and Lovejoy, 1985b), we can speak more concisely of positive (generalized) spectrum (σ(G)):

$$\sigma(G)\geq 0 : \inf_\mu \left[\mathrm{Re}\, \sigma(G)\right] \geq 0\ ;\ \sigma(G) = \{\mu \in C \mid G - \mu \mathbf{1} \text{ non-invertible on } CX\Re^d\}. \tag{40}$$

anisotropic ("elliptical") scale ϕ_{el} is then defined by the volume of the E_λ (hence is a measurable property, rather than a metric property)

---

[1] Indeed there is a priori no equivalence between the different ways of maintaining conservation of fluxes.
[2] This is indeed consistent with the value empirically measured in rain according to preliminary results (Lovejoy and Schertzer, this volume), and work in progress.
[3] They also pointed out the possible breaking of mirror symmetry for atmospheric dynamics, hence the importance of the associated helicity.

$$\phi_{el}{}^{d_{el}}(E_\lambda) = \phi^d(E_\lambda) = \lambda^{d_{el}}\,\phi^d(S_1) = \lambda^{d_{el}}\,\phi_{el}{}^{d_{el}}(S_1) \tag{41}$$

where the effective dimension of the space, the "elliptical" dimension d_{el} of the space, is simply[1] the trace of G: $d_{el}=Tr(G)$. This anisotropic framework allows rather straightforward extensions of Hausdorff measures and dimensions, still respecting the divergence rule (eq. 2):

$$\int_A d^{D_{el}}x = \lim_{\delta\to 0}\ \inf_{\substack{A\supset\cup E_i\\ \phi_{el}(E_i)<\delta}}\ \sum_{E_i} \phi_{el}{}^{D_{el}}(E_i) \tag{42}$$

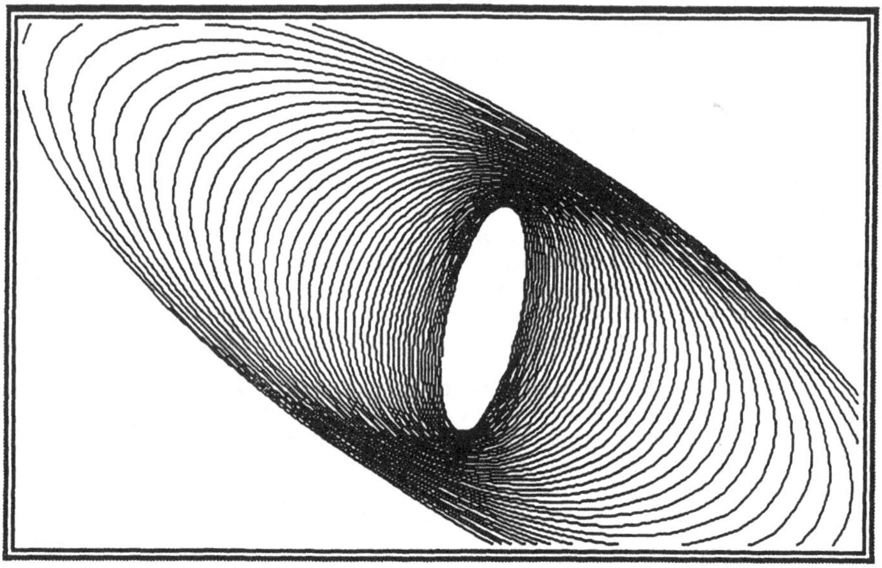

Fig. 9: Family of "ellipsoids" E_λ obtained by linear GSI. Due to linearity the E_λ remain convex.

In two dimensions it is rather convenient to use a representation of quaternions (Schertzer and Lovejoy, 1985b; Lovejoy and Schertzer, 1985) for the generator:

[1] However on any Euclidean subspace, G needs to be appropriately normalized (as discussed by Schertzer and Lovejoy (1987b)).

$$\mathbf{1}=\begin{bmatrix}1 & 0\\ 0 & 1\end{bmatrix};\ \mathbf{I}=\begin{bmatrix}0 & -1\\ 1 & 0\end{bmatrix};\ \mathbf{J}=\begin{bmatrix}0 & 1\\ 1 & 0\end{bmatrix};\ \mathbf{K}=\begin{bmatrix}1 & 0\\ 0 & -1\end{bmatrix} \quad (43)$$

with the following commutation rules:

$$\mathbf{IJ} = -\mathbf{JI} = -\mathbf{K};\ \mathbf{JK} = -\mathbf{KJ} = \mathbf{I};\ \mathbf{KI} = -\mathbf{IK} = -\mathbf{J}\ ;\ \mathbf{1} = -\mathbf{I}^2 = \mathbf{J}^2 = \mathbf{K}^2 \quad (44)$$

in this representation **G** is a linear combination of **1, I, J, K:**

$$\mathbf{G} = d\mathbf{1}+c\mathbf{K}+ e\mathbf{I}+f\mathbf{J}\ ;\ \mathrm{Tr}(\mathbf{G})=2d=d_{el} \quad (45)$$

I is a rotation generator, **J** and **K** are stratification generators (as seen by the commutation rules they corresponds to each other via a rotation **I**), and **1** remains of course the generator of isotropic contraction. Indeed, the effect of rotation dominates as soon as a is complex (negative a^2), otherwise stratification dominates (positive a^2):

$$a^2 = c^2 + f^2 - e^2\ ;\quad u = \log(\lambda)\ ;\ \det(\mathbf{G}) = d^2 - a^2$$

$$\mathbf{T}_\lambda = \lambda^{-\mathbf{G}} = \lambda^{-d}\,[\mathbf{1}\cosh(au) - (\mathbf{G}-d\mathbf{1})\sinh(au)/a] \quad (46)$$

in the stratification dominated case, the axis of the ellipsoids rotates only by $\tan^{-1}(e/a)$ when λ goes from 0 (infinite outer scale) to ∞ (zero inner scale), and interchange of minor and major axis occur at $\lambda=1$ (the sphero-scale, which corresponds to isotropy), yielding a total rotation of $\pi/2 + \tan^{-1}(e/a)$ for each of the axes). The condition of positivity for the spectrum is simply:

$$\sigma(\mathbf{G})\geq 0 \Leftrightarrow \det(\mathbf{G}) \text{ and } \mathrm{Tr}(\mathbf{G})\geq 0 \Leftrightarrow d\geq 0 \text{ and } d^2\geq a^2 \quad (47)$$

thus always satisfied in the case of rotation dominance. Note we use the following fundamental identity for the commutators ($[\mathbf{X},\mathbf{Y}] = \mathbf{XY}-\mathbf{YX}$) :

$$[\lambda^{\mathbf{A}},\lambda^{\mathbf{B}}]= \lambda^{[\mathbf{A},\mathbf{B}]} \quad (48)$$

One may also note that reducing the scale by a factor λ via $\mathbf{T}_\lambda = \lambda^{-\mathbf{G}}$ in the physical space corresponds to magnification by the same scale ratio of wave-vector **k** in Fourier space by the transposed $\tilde{\mathbf{T}}_\lambda{}^{-1}$ of $\mathbf{T}_\lambda{}^{-1}$ since wave vector are conjugate via the scalar product (denoted $(\cdot,\cdot)$) of physical coordinates, indeed:

$$(\tilde{\mathbf{T}}_\lambda{}^{-1}\,\mathbf{k}, \mathbf{T}_\lambda\,\mathbf{x}) = (\mathbf{k}\ ,\ \mathbf{T}_\lambda{}^{-1}\mathbf{T}_\lambda\mathbf{x}) = (\mathbf{k}, \mathbf{x}) \quad (49)$$

hence the generator of anisotropy in Fourier space is the transposed $\tilde{\mathbf{G}}$ of **G**, in the quaternion representation (with obvious notation):

$$(\tilde{d}, \tilde{c}, \tilde{e}, \tilde{f}) = (d, c, -e, f) \quad (50)$$

6.2. Local scale transformations, nonlinear GSI

In the framework of linear GSI, the anisotropic scale transformations remains global, i.e. invariant for any time-space translation The same is true of multifractality, even if overly simplistic presentations of multifractals tend incoherently to speak of "local fractal dimension"[1]. However, we already argued for the indispensable necessity of using local scale transformations, as in the original (Weyl's) local gauge invariance. We then have to consider nonlinear GSI, with nonlinear scale operator $\mathbf{T}_\lambda$. The main new feature is that the generator of the scale transformation becomes local and $\mathbf{T}_\lambda$ is only a semi-group. The general framework is that of differential manifolds, and on each tangent space we will return to consider a

[1] Indeed, they often falsely present fractal dimensions as pointwise and/or scale dependent notions.

linear scale transformation which is the tangent application of T_λ. Hence, we will once again find the condition of a positive (generalized) spectrum (eq. 47), but here in each tangent space at every point of the manifold. As preliminary examples, we may consider nonlinear GSI (still on the plane) with a representation of quaternions (Schertzer and Lovejoy, 1985b, 1989; Lovejoy and Schertzer, 1985), hence with spatially variable coefficients (d,c,e,f). In order to obtain the image point x_λ (belonging to E_λ, the "ellipsoid" of scale reduced by factor λ) of a point x_1 (belonging to the sphero-scale S_1) we now have to solve the nonlinear differential equation:

$$dx_\lambda/d\lambda = -G(x_\lambda)/\lambda \tag{51}$$

this leads to the striking figures, note due to nonlinearity the E_λ, are no longer convex sets (see fig. 10).

Fig. 10 : Family of "ellipsoids" E_λ obtained by nonlinear GSI . Due to nonlinearity the E_λ, are no longer convex sets.

These preceeding considerations are not at all incidental, since the quaternions are nothing but one of the simplest examples of Lie algebras. Indeed, more generally we have to consider scale changing operators T_λ as forming a Lie group acting on a manifold, and their (local) generators in the associated Lie algebra. More specifically we have to consider the subset satisfying the condition of positive spectra, and the scale changing operators are then obtained by their exponentiation on the manifold, i.e. by integration of eq. 51. Doing so we indeed obtain a connection on this manifold (local properties get transportable on the global manifold, or at least on a part of it). The structure of the Lie algebra $\{G | G = \Sigma_i g^i G_i, g^i \in R \text{ or } C\}$ of the generators depends primarily on its structure constants c_{ijk} given by the commutation rule (eq. 44 is a particular case):

$$[G_i, G_j] = c_{ijk}\, G_k \tag{52}$$

The choice of the algebra depends on the symmetries other than scaling which we want to be respected (as a consequence of eq. 48, the generators S of these symmetries must commute with the generator G of the scale changing operator T_λ). In this respect, Lie groups lead to the important Casimir invariant which commutes with every generator of the group and in particular for compact groups they appear as the sum of the squares of the generators (e.g. the square of the angular momentum in orthogonal group O(3) in dimension 3), in the quaternion case, the corresponding invariant of this type is:

$$1^2 + I^2 + J^2 + K^2 = 2\,1 \tag{53}$$

6.3 Stochastic GSI

We can now add a new and important ingredient: the random behaviour of the scale transformation T_λ, i.e. of its generator G which up to now were considered deterministic. This opens up a wide variety of possible behaviour. Indeed, we can understand it as being produced by the infinitesimal (and random) generator (γ, G), more specifically a conservative flux as resulting from the nonlinear integration on the manifold of (eq. 30), coupled with the nonlinear scale transformation (eq. 51):

$$\partial\varepsilon_\lambda(x_\lambda)/\partial\lambda = \gamma_\lambda(x_\lambda)\,\varepsilon_\lambda(x_\lambda)\ ;\ dx_\lambda/d\lambda = -G(x_\lambda)/\lambda \tag{54}$$

But now G is random as γ. Of course the statistical interrelations between both are of prime importance. Note that the *linear stochastic* case is rather simple since $G(x_\lambda)$ corresponds simply to the action of a random matrix (still denoted G) on the vector x_λ. The constraints will be on the one hand those already discussed on the monotony of the ellipsoids E_λ (i.e., eq. 40) and on the other hand the statistical behaviour of G will be essentially the same[1] type as those of the singularities γ, except we have to consider a matrix pink noise. Once again, the decomposition into elementary components (such as quaternions) can be quite helpful (e.g. we have only to consider the random behaviour of (d, c, e, f) subject to $d^2 \geq a^2$ (eq. 47).

The extension to the nonlinear case is more or less straightforward still using decomposition of the Lie groups (for instance fig. 11 gives an example nonlinear stochastic GSI using quaternions). Note that the 1/f noise may also be obtained by the same kind of integration on the manifold of a white-noise, but of fractional order. In this respect, Fourier techniques[2] seem to be manageable with the use of the transposed generator $\tilde{G}$.

Understanding such a symmetry group raises highly important theoretical and empirical questions. On the one hand, it renders much more precise and abstract the question of the generation of nonlinear variability: we need no longer care much about more or less complex sets of coupled partial differential equations, but rather to find their generators (γ, G), the analog of the "action" to use the consecrated expression in physics. Furthermore, it may also be important to realize that such an approach is also very concrete, and may indeed extract valuable information through the large geophysical data sets which are more and more frequently available. Indeed, empirical determination of the generators may be sought in generalizing the elliptical dimension sampling, designed up to now to explore GSI (as done by Lovejoy et al. (1987)), since the theoretical arguments (Schertzer and Lovejoy, 1987a (Appendix B2)) leading to this method seems to have rather straightforward generalizations for nonlinear and/or stochastic GSI. Nevertheless, their concrete exploration will require in certain cases gargantuan amounts of data, particularly since geophysical multifractals are often hard (with wild singularities), not soft.

---

[1] Deterministic GSI corresponds to a unique and homogeneous singularity for the generator of scale transformation (e.g. a unique matrice in the linear deterministic GSI.

[2] Although presumably under their localized version, i.e., the "wavelet" techniques.

Fig. 11 : Family of "ellipsoids" E_λ obtained by stochastic nonlinear GSI . Due to nonlinearity the E_λ are no longer convex sets and their shapes randomly fluctuate scale by scale.

7. CONCLUSIONS

We sharpened the theoretical foundations of the singular statistics of multifractal fields. We discussed in a rather general manner the conditions of their appearance. We clarified the fundamental difference between "bare" and "dressed" properties at a given (non-zero) scale, i.e. the important differences between a process with small scale interactions cut-off and one with the full range of interactions. We thus emphasized the nontrivial behaviour of geophysical observables depending on the type of the process, as well as on the observation (both its scale and dimension). We also pointed out general properties of the generators of multifractal fields and their links with notions of classical statistical physics, emphasizing their particularities.

We demonstrated the existence of three-parameter (H, C_1, α) universality classes of the generic multifractal processes. These three fundamental exponents characterize the degree of flux non conservation flux (H), the deviation of the mean field from homogeneity (C_1), and the deviation of the process from monofractality ($0\leq\alpha\leq2$). These three exponents correspond to fundamental properties of the process, for instance: to the fractional order of integration over a conservative flux (H), the sparseness of average singularities measured by the codimension (C_1), and the type and regularity of the generator (2-α indicates the deviation from Gaussianity). The five main sub-classes of these (H, C_1, α) universal generators are:

Gaussian generator (α=2), unbounded Lévy generator (2>α>1), bounded Lévy generator (1>α>0), a very special in-between case Cauchy generator (α=1) ... as well as the once celebrated β-model (α=0)!

We also investigated the anisotropic and/or local scaling symmetry. This leads us to generalize the idea of scale invariance far beyond the familiar self-similar (or even self-affine) notions. We pointed out some of the basic ingredients of the nonlinear Generalized Scale Invariance (or anisotropic/local/stochastic scale transformations).

8. ACKNOWLEDGMENTS

We thank especially D. Lavallée and J. Wilson, for stimulating discussions and help preparing figures. We also acknowledge discussions with G. L. Austin, P. Brenier, J. P. Carbonnel, A. Davis, A. H. Fan, U. Frisch, P. Hubert, J. P. Kahane, P. Ladoy, E. Levich, A. Seed, G. Sèze, G. Sarma, L. Smith, Y. Tessier and R. Viswanathan.

APPENDIX A: GENERALIZED CENTRAL THEOREM, EXTREMAL LEVY STABLE GENERATORS (in collaboration with R. Viswanathan[1])

A.1. Fixed points for sums of independent and identical distributed (i.i.d.) random variables and central limit theorems

In this sub-section we review briefly the classical features of Lévy stable variables, stressing that these variables emerge as generalizations of Gaussian variables, which then are seen to be a very particular case of Lévy stable variable. Indeed, we are interested in the universal stable and attractive fixed points of renormalized sum of independent and identical distributed (i.i.d.) variables, consider first the stable fixed points of renormalized sum ($\stackrel{d}{=}$ means equality in probability[2]):

$$X_i \stackrel{d}{=} X_1 \quad i=1,n \text{ are stable points under renormalized sum iff} \tag{A1}$$
for any (integer) n (≥2), ,there exists a (positive) b_n and a (real) a_n
$$\sum_{i=1,n} X_i \stackrel{d}{=} b_n X_1 + a_n$$

The well-known Gaussian case corresponds to:

$$<X_1^2> \leq \infty \Rightarrow b_n = n^{1/2},\ a_n = (n-1)<X_i> \tag{A2}$$

hence the assumption of finite variance which has been considered as so "natural" that it has become a kind of dogma. The usual central limit theorem corresponds simply to the limit $n \to \infty$ in eq. A1:

$$X = \lim_{n\to\infty} [(\sum_{i=1,n} X_i) - a_n]/b_n \tag{A3}$$

even though the X_i on the r.h.s. are not necessarily assumed to be Gaussian, the X will be, hence the Gaussian law is attractive :

$$<X^2> = <X_i^2> \leq \infty \Rightarrow b_n = n^{1/2},\ a_n = n <X_i> - <X> \tag{A4}$$

one may note that the average $<X>$ can be arbitrarily set to 0 (as usual) due to the expression of the a_n.

Lévy (1925,1954) generalized the Gaussian case by relaxing the hypothesis of finite variance for the X_i (which implies finiteness of every statistical moment for the limit itself) introducing on the contrary an

---

[1] Now at Mitsubishi, London, England.

[2] Note in order to be consistent, the use of this symbol requires that the corresponding variables should be mutually independent.

order of divergence (α, $0<\alpha<2$) for the moments of the X_i (α is often called the Lévy index) which satisfies either A1 or A3:

$$h<\alpha \Rightarrow <|X_i|^h> < \infty \text{ and } h'\geq\alpha \Rightarrow <|X_i|^{h'}> = \infty \quad \text{i. e. : } Pr(|X_i|\geq s) \sim s^{-\alpha} \qquad (A5)$$
$$\text{A1 or A3} \Leftrightarrow b_n = n^{1/\alpha} \quad (\text{and, if } \alpha>1;\ a_n= n<X_i> - <X>)$$

the variables X_i are very often termed "hyperbolic variables" (or even "hyperbolics") due the algebraic fall-off of their probability distribution tails, which are themselves sometimes termed "fat tail" due to their (unusually) important contribution. Hence, the Lévy stable variables are the stable and fixed points of (renormalized) sums of i. i. d. hyperbolic variables. Note that for $\alpha\leq1$, as the mathematical expectations of the X_i and their sums are divergent, the required recentring is a bit more involved than that indicated for $1<\alpha\leq2$ (subtracting out the averages) and will be only discussed later. The very special Gaussian case appears as the (extreme) regular case $\alpha=2$, after a highly critical transition since for any $\alpha=2-\varepsilon$ (ε arbitrarily small) we have divergence of all orders greater than α whereas all divergences are suppressed for $\alpha=2$. One may note that the stable variables were introduced in a slightly different form (Lévy 1925, 1954) who addressed the stability under any linear combination:

$X_i \stackrel{d}{=} X_2$ are said stable under linear combination iff (A6)
for any (positive) b_1 and b_2, there exists (real) a and (positive) b, such that:
$b_1X_1+ b_2 X_2 \stackrel{d}{=} b X_1+a$

It is rather easy to check (by induction) that eq. A1 and eq. A6 are equivalent. One may furthermore note that "any n" in A1 can be equivalently reduced to "n=2,3" due essentially to the density of numbers $2^j 3^k$ among positive numbers, j and k being relative integers (Zolotarev, 1986).

Note that there exists a sub-class of stable variables which do not require recentring (i.e. a=0, -it is rather obvious in the cases $1<\alpha\leq2$). These special cases (to which, Lévy (1925) restricted his study) are frequently called "strictly stable" (Feller 1971, Zolotarev 1986), more rarely the complementary cases (i.e., $a\neq0$) are called (Lévy, 1954) "quasi-stable".

A.2. Characteristic functions of Lévy laws

With a few notable exceptions (α=2, 1, 1/2) the probability distributions of Lévy stable variables are not expressible in a closed form. However, the second (Fourier or Laplace) characteristic function is easily expressible due to the basic properties of stability. K(h) is the logarithm of the first characteristic function Z(h), i.e. the (Fourier or Laplace) transform of the probability distribution dP(x)) and the argument h is purely imaginary in case of Fourier (h=ih'), real in case of Laplace (we will discuss later the restrictive conditions under which such a transform is possible) and a complex number (h+ih') in the case of Fourier-Laplace (or two-sided Laplace) transform:

$$e^{K(h)} = Z(h) = <e^{hX}> = \int e^{hx}\, dP(x) \qquad (A7)$$

the fundamental property of the fixed point (eq. A1) or the equivalent form (eq. A6) are easily transposed for the characteristic functions:

$X_i \stackrel{d}{=} X_1$ (i=1,n) of second characteristic function K(h) (A1')

are stable points under renormalized sum iff. for any (integer) n ($\geq$2), there exists a (positive) b_n and a (real) such that:

$$a_n nK(h)= K(b_nh)+ a_nh;$$

and:

$$X_i \stackrel{d}{=} X_2 \text{ of second characteristic function } K(h) \tag{A6'}$$

are said stable under linear combination iff. for any (positive) b_1 and b_2 there exists a (real) a and a (positive) b such that:

$$K(b_1h)+K(b_2h) = K(bh) + ah$$

the limit theorems correspond to (K_i second characteristic function of the X_i, K second characteristic function of X):

$$K(h) = \lim_{n\to\infty} n[K_i(h/b_n) - ha_n/nb_n] \tag{A3'}$$

We may infer, especially from A6', that these characteristic functions, up to the recentring term, should be of power law form whose exponent α is bounded above by 2 (the extreme regular Gaussian case) and must of course be positive (to avoid divergence at h=0). It is obvious that the case α=1 is very special since this hyperbolic exponent becomes equal to the (linear) recentring term exponent, we may guess that conflict and compensation between the two terms will introduce logarithm corrections. For other values of α the (linear) recentring term has no importance and we can restrict our attention to strictly stable cases (a_n=0). As h^α (or h Log(h) for α= 1) is not analytic (except once again in the Gaussian case α=2) in the complex plane, we clearly expect on the one hand divergence of moments of order greater or equal to α, on the other hand that the inferred "power law form" may be rendered more precise in order to obtain a second characteristic function (e.g., Z(h) must be positive definite in case of Fourier transform (Bochner's theorem) or absolutely monotone in case of Laplace transform). Indeed, considering the symmetric (or symmetrized) probability distributions lead to the following law (partially) known ... since Cauchy 1853, but essentially obtained by Lévy (1925) of Fourier characteristic functions, since K(h) must be also symmetric:

$$K(ih') = -\lambda_\alpha |h'|^\alpha \tag{A8}$$

with the obvious Gaussian case when α=2, and Cauchy case when α=1. The λ_α characterizes the width of the probability distribution as in the Gaussian case ($\lambda_2 = \sigma^2/2$) but doesn't correspond to the evaluation of an α-moment, since it diverges, but rather the rate of divergence of this moment.

A.3. Particular properties of extremal Lévy laws

The symmetric case corresponds to limit of sums of symmetric hyperbolics or mixing with equal probability (p=q=1/2) positive (with probability p) and negative (with probability q) one-sided hyperbolic distributions (concretely: just multiplying by a random sign positive one-sided hyperbolics). Asymmetric cases correspond to $p \neq q$. It is time to stress that if we want to have a Laplace transform. we can only consider extremal (asymmetric) hyperbolics, simply because algebraic fall-off could not tame an exponential divergence, hence we restrict here our attention to negative hyperbolics (p=0, q=1). However, note that the corresponding limits, the extremal Lévy stables, are not always one sided (in the sense of having only positive or negative values) -precisely one sided probability distributions only occur for $0<\alpha<1$.

In order to assess different statements, it is interesting to consider the characteristic functions under their "canonical form" i. e:

$$K(h) = \int (e^{hx} - 1 + hx)\, dF(x) \approx Z(h) - 1 + a'h \tag{A9}$$

dF is called the Lévy canonical measure or the spectral measure, which needs not be a probability measure. Indeed $\int dF(x)$ is not bounded, with the exception of the Gaussian case, and we introduced a recentring term a' just to cancel the second term of the exponential development (near h=0) when needed (for $1<\alpha<2$). This form corresponds on the one hand to a first order term development of log[1+(Z(h)-1)], which is the only term kept in the limit theorem besides recentring and normalization of K(h) (i.e. K(0)=0 corresponding to $\int dP(x)=1$). On the other hand, it corresponds to the limit of (Poisson) random (renormalized) sums of i. i.

d. variables, instead of uniform sums as discussed up to now. Indeed considering the characteristic function of the limit ($n \to \infty$) of the Poisson compound probability distribution generated by renormalized sum of n i. d. d. hyperbolic variables (as given by A1) lead us to a new version of the limit theorem (earlier stated under the forms A3 and A3') keeping in mind that the second characteristic function K(h) of the Poisson compound probability distribution is c(Z(h)-1), where c is the parameter of the Poisson process, Z(h) the first characteristic function of the generating probability distribution):

$$K(h) = \lim_{n \to \infty} K_n(h)$$

$$K_n(h) = n[Z_n(h) - 1]$$

$$Z_n(h) = Z_i\,(h/b_n)\exp(-\,h a_n / n b_n) \qquad \text{(A3'')}$$

the "canonical form" (eq. A9) is obtained by slightly recasting this equation to take directly into account arbitrary centering directly on K (no longer on Z_n or Z).

Let us consider the negative hyperbolic generation of extremal Lévy stable by negative hyperbolic, it suffices to put $dF(x) \propto 1_{x<0} x^{-\alpha}\, dx/x$ ($1_{x<0}$ being the indicator function of the negative x) and with repeated use of the identity:

$$\Gamma(\beta) = z^{-\beta} \int_0^{\infty} e^{-zt} t^{\beta} dt/t \ ; \ \ Re(z) \geq 0 \qquad \text{(A10)}$$

and integrations by parts, we obtain easily for $dF(x) = 1_{x<0} C(2-\alpha)\, x^{-\alpha}\, dx/x$:

$$\alpha \neq 1:\ K(h) = C\, h^{\alpha}\, \Gamma(3-\alpha)/\alpha(\alpha-1);\ \alpha \neq 1$$
$$\alpha = 1:\ K(h) = C\, h\log(h) \qquad \text{(A11)}$$

one may note that the expressions for the corresponding Fourier transforms are a bit more complex (i.e. Fourier transforms, so convenient for symmetric laws, are inconvenient for extremal (and more generally for asymmetric laws), on the contrary Laplace is only fitted for the extremal, useless for the others):

$$\alpha \neq 1:$$
$$K(h) = |h|^{\alpha}\ C\, [\Gamma(3-\alpha)/\alpha(\alpha-1)]\, [\cos(\pi\alpha/2) + -i(\mathrm{sgn}(h)\,(p-q)\,\sin(\pi\alpha/2)]$$
$$\alpha = 1:$$
$$K(h) = -|h|\ C\, [\pi/2 + i\,(\mathrm{sgn}(h)(p-q)\log|h|] \qquad \text{(A12)}$$

As a last general remark, one may note (from eq. A11 or eq. A12) it is only in the case of extremal stable distribution ($p-q=\pm 1$) that an analytic extension on the whole complex plane of K is possible (but with a cut along the ray $\arg(h) = -3\pi/4$), as it is for $\alpha=2$, or that the double-sided Laplace transform applies only to extremal stable variables.

APPENDIX B: HAUSDORFF MEASURES, FRACTIONAL INTEGRALS AND DERIVATIVES

B.1. Hausdorff measures

We first recall the geometric definition of Hausdorff measures and dimensions for a compact set A. The Hausdorff measure relative to a convex function g (m_g) of A is defined as:

$$m_g(A) = \lim_{\lambda \to \infty} m_{\lambda,\, g}(A)$$

$$m_{\lambda,\,g}(A) = \inf_{\substack{A \supset \cup B_i. \\ \mathrm{diam}(B_i)<\delta=l_0/\lambda}} \sum_{\{B_i\}} g(\mathrm{diam}(B_i)) \tag{B1}$$

where $m_{\lambda,\,g}(A)$ can be understood as the Hausdorff measure (relative to the function g) with resolution λ, and corresponds to the infinimum over all possible coverings by balls B_i such that the diameter ($\mathrm{diam}(B_i)$) is smaller than the resolution scale $\delta=l_0/\lambda$.

The D-dimensional Hausdorff measure of A, $m_D(A)=\int_A d^Dx$ is obtained in the particular case $g(t)=t^D$ and the Hausdorff dimension (D) of the set A is obtained by the divergence rule:

$$\int_A d^{D'}x = \infty, \ \text{for } D'<D \ ; \qquad \int_A d^{D'}x = 0, \quad \text{for } D'>D \tag{B2}$$

One may note that the D-measure of A is not necessarily finite and non-zero. In order to obtain a finite and non-zero value of the D-like measure of A, one may have to change slightly of the basic function g of the Hausdorff measure introducing on iterates of the logarithm:

$$\begin{aligned} &g \rightarrow g(t)=t^{D(A)}(\mathrm{Log}_1\, t)^{\Delta 1}(\mathrm{Log}_2)^{\Delta 2}\ldots \Rightarrow 0<m_g(A)<\infty \\ &\mathrm{Log}_i=\mathrm{Log}(\mathrm{Log}_{i-1});\ \mathrm{Log}_1=\mathrm{Log} \end{aligned} \tag{B3}$$

the logarithmic correction exponents Δ_i on the i-th iterate of the logarithm, are called sub-dimensions and correpond to the fact that the volume of an elementary ball (l^D) is now 'corrected' by factors of the type $[\mathrm{Log}_i\,(l/l_0]^{\Delta i}$. For instance, Mauldin and William (1986) have shown that such log-corrections arise in evaluating the dimension of the support of the β-model with the help of a simple box-counting algorithm, and Lovejoy and Schertzer (this volume) give indications of the presence of log corrections in experimental analysis using the same algorithm.

On the other hand, it is worthwhile to note that in fact that the D-dimensional Hausdorff measure of A can be defined direcrtly in a measure sense (for balls of topological dimension d), i.e.:

$$\phi^d(B_i) = \int_{B_i} d^dx$$

$$\int_A d^Dx = \lim_{\lambda\rightarrow\infty} \ \inf_{\substack{A \supset \cup B_i. \\ \phi(B_i)<\delta=l_0/\lambda}} \sum_{\{B_i\}} \phi^D(B_i) \tag{B4}$$

This measure definition allows us to deal with more complex integrands (such as h powers of the flux "density"[1] in the trace moments, Schertzer and Lovejoy (1987a)) or anisotropic scaling (sect. 6.1) by replacing balls B_i by their anisotropic analogues and henceforth introducing "elliptical Hausdorff dimensions" instead of the isotropic ones.

B.2. Fractional integrals and derivatives

These correspond to extensions to non-integer orders (H) of integrations (I^H) or differentiations (D^H). These extensions are rather straightforward in Fourier space for 1-dimensional (scalar) functions, since integrations -up to a constant of integration discussed below- or differentiations of integer order n, correspond respectively to division or multiplication by $(ik)^n$, where k is the wave number (Fourier transforms of physical space quantities will be denoted by a circonflex (^):

$$\hat{I}^{-H}(f) = \hat{D}^H(f) = (ik)^H\,\hat{f}(k)) \tag{B6}$$

[1] which is no longer a function in the limit of zero homogeneity-scale length.

In the usual physical space for non-integer H, we obtain an ordinary (i.e. of integer order n) derivation (D^n, positive n) or integration (I^{-n}, negative n) of a convolution:

$$I^{-H}f = D^{H}f = 1/\Gamma(n-H))\, D^n[f*x^{n-H-1}] \qquad \text{(B7)}$$

here Γ is the Euler gamma function and intervenes as in Appendix A where we already encounter integration of the same type. Eq. B7 is more general than B6, since directly written in physical space, but introduces ambiguities in the definition on non-integer integration or differentiation[1] because they will clearly depend on the domain of defintion of the convolutions (cf. e.g. Ross (1975)). The same techniques can be extended to functions defined on R^d, however the analysis becomes more complex because various combinations of partial derivatives are now possible. Nevertheless, one can consider the following strongly isotropic extension:

$$I_d^{-H}f = D_d^{H}f = 1/\Gamma(n-H))\, D_d^{n}[f*|x|^{n-H-d}];$$
$$\hat{I}_d^{-H}(f) = \hat{D}_d^{H}(f) = |k|^H \hat{f}(k)) \qquad \text{(B8)}$$

which in fact, corresponds to fractional powers of the Laplacian (or of the Poisson solver):

$$I_d^{-H}f = D_d^{H}f = (-\Delta)^{H/2} \qquad \text{(B9)}$$

Extensions for $R^{d'}$-valued (vectorial or even tensorial) fields can be also considered, but the variety of possible differential operators still increases, although this variety can be reduced by considering certain symmetries as previously.

As a final and important remark for applications, one must take into account the modification of the average of the integrand by the fractional integration, i.e. consider closely the role of the constant of integration. When working in the Fourier space (on a periodic box of size L), it corresponds to the modification of the Fourier component at wave number k=0. Indeed, splitting f in its average ($\overline{f}$) and fluctuating parts (f') we obtain:

$$f = \overline{f} + f':$$
$$I^m(f) = I^m(\overline{f}) + I^m(f'); \qquad I^m(\overline{f}) = \overline{f}\, L^m/m \qquad \text{[B10)}$$

These considerations are especially important when we must preserve the sign of a field after fractional integration, e.g. the sign of the extreme fluctuations of an extremal "white" Lévy noise, in order to obtain extremal "pink" Lévy noise by fractional integration (as needed for a Lévy generator of a multiplicative cascade process, see also Wilson et al., this volume).

APPENDIX C: CHARACTERISTIC FUNCTIONALS AND UNIVERSAL GENERATORS

C.1. Characteristic functionals of generators

We consider the finite exponential increments Γ_λ as noises concentrated in the wave number band $[1/l_0, \lambda/l_0]$ (filtered out or strongly damped for other wave numbers) obtained by filtering their limit Γ:

$$\Gamma_\lambda = \Gamma * F_\lambda \qquad \text{(C1)}$$

where F_λ is the filter and * denotes the convolution product corresponding to a product in Fourier space. This is an explicit definition of the scale of homogeneity $l = l_0/\lambda$, although at a more sophisticated

[1] It is important to note that fractional derivations are obtained with the help of integrations, thus are depending in fact on "constants of integration" (contrary ot their integer counter parts)!

mathematical level Γ needs to be defined as a limit (Eq. 27) and not the Γ_λ as restrictions of Γ. Fourier transforms of physical space quantities (e.g. $\Gamma_\lambda(\mathbf{x})$) will be denoted by a circonflex (e.g., $\hat{\Gamma}_\lambda(\mathbf{k})$), $\mathbf{k}$ indicating a wave vector and k (=|k|) the corresponding wave vector, and we will take for sake of notational simplicity, $l_0=1$. Hence, we have in the Fourier space:

$$\hat{\Gamma}_\lambda(\mathbf{k}) = \hat{\Gamma}(\mathbf{k})\, \hat{F}_\lambda(\mathbf{k}) \qquad \text{(C1')}$$

If we strictly filter out any wave number ouside the range $[1,\lambda]$, $\hat{F}_\lambda$ is simply the indicator function $\hat{1}_{S\lambda}$ of the spherical (hyper) volume S_λ delimited by the spheres of radius 1 and λ, both centered at the origin of Fourier space ($\hat{1}_{S\lambda}(\mathbf{k}) = 1$ if $1 \leq k \leq \lambda$, 0, otherwise). However, its Fourier transform F_λ corresponds to Bessel functions. The second characteristic functional (or cumulant generating functional) K_λ of the generator Γ_λ is defined by the scalar product with any "test function f", as:

$$K_\lambda(f) = \log(\langle \exp \int \Gamma_\lambda(x)\, f(x)\, d^d x \rangle) \qquad \text{(C2)}$$

We have not only a similar definition (with the hermitian product) in (complex) Fourier space for the characteristic functional of $\hat{\Gamma}_\lambda(\mathbf{k})$, denoting by $\hat{f}^*$ the conjugate of any complex function $\hat{f}$ (and the same for noises, e.g. $\hat{\Gamma}_\lambda^*$):

$$\hat{K}_\lambda(f) = \log(\langle \exp \int \hat{\Gamma}_\lambda(\boldsymbol{k})\, \hat{f}^*(\boldsymbol{k})\, d^d\boldsymbol{k} \rangle) = \log(\langle \exp \int \hat{\Gamma}_\lambda^*(\boldsymbol{k})\, \hat{f}(\boldsymbol{k})\, d^d\boldsymbol{k} \rangle) \qquad \text{(C3)}$$

but, as the scalar product (on L^2 space, and more generaly the duality product[1] between the dual spaces L^α and $L^{\alpha'}$, $1/\alpha + 1/\alpha' = 1$) is conserved by Fourier transforms, we have also the equality between these two characteristic functionals:

$$K_\lambda(f) = \hat{K}_\lambda(\hat{f}) \qquad \text{(C4)}$$

In order to have multiple scaling, the characteristic functionals K_λ and $\hat{K}_\lambda$ of Γ_λ and $\hat{\Gamma}_\lambda$, respectively, must be logarithmically divergent, namely:

$$K_\lambda(f) = \hat{K}_\lambda(\hat{f}) = (\mathrm{Log}\,\lambda)\, K(f) \qquad \text{(C5)}$$

at least for the test functions corresponding to n-points statistics, i.e. :

$$f(\mathbf{x}) = \Sigma_{i=1,n}\, h_i \delta_{x_i}; \qquad \hat{f}(\mathbf{k}) = \Sigma_{i=1,n}\, h_i e^{i(\mathbf{k}.\mathbf{x}_i)}; \qquad \text{(C6)}$$

C.2. Universal generators

In order to satisfy eq. C5 , it is rather obvious that Γ_λ should be a coloured noise. Indeed, as we have shown that universal generators should be either Gaussian ($\alpha=2$) or extremal Lévy-stable ($0<\alpha<2$), let $\Gamma_{\alpha\lambda}$ be defined as

$$\Gamma_{\alpha\lambda} = s * \gamma_\alpha * F_\lambda; \qquad \hat{\Gamma}_{\alpha\lambda}(\mathbf{k}) = \hat{s}(\mathbf{k})\, \hat{\gamma}_\alpha(\mathbf{k})\, \hat{F}_\lambda(\mathbf{k}) \qquad \text{(C7)}$$

γ_α being a (unit) white noise, either Gaussian ($\alpha=2$) or extremal Lévy-stable of index α ($0<\alpha<2$), $\hat{\gamma}_\alpha$ its Fourier transform, and s a non-random weighting function determined below, corresponding in fact to fractional integration (see Appendix B), hence we consider fractional Lévy noises. Loosely speaking, the (unit) white noise noises may be understood as $\gamma_\alpha(x)$ $(=\int \gamma_\alpha \delta_x dx)$ for the different x, are independently

[1] This is in fact a convenient way to define Fourier transforms of distributions or "generalized functions". In the L^2case, this property is known as Parseval's theorem.

identically distributed according to either (symmetric) Gaussian law (α=2) or (extremal) Lévy-stable law (of index α). Hence, their characteristic functionals generalizes what we obtain for the extremal Lévy characteristic funtion (eq. A11), in the sense that for any function f(x):

$$K_{\gamma_\alpha}(f) = \mathrm{Log}(<\exp(\int f(x)\gamma_\alpha d^d x)>) = \int f(x)^\alpha d^d x \tag{C8}$$

As the same relation holds for the convolution product, the desired log divergence (eq. C5) of $\Gamma_{\alpha\lambda}$ is obtained for fractional intergration of order d/α' (Appendix B, with as usual $1/\alpha+1/\alpha'=1$)), i.e.:

$$s(\mathbf{x}) \propto |\mathbf{x}|^{-d/\alpha};\ \hat{s}(\mathbf{k}) \propto |\mathbf{k}|^{-d/\alpha'} \tag{C9}$$

One may note that the fact that the exponents are not the same are not the same in the physical space (d/α) and the Fourier space (d/α'), with the notable exception of the Gaussian generator ($\alpha=\alpha'=2$), corresponds to the introduction of interrelations between the componenents of the noise by the Fourier transform, i.e. the Fourier transform of a white Lévy noise is a coloured Lévy noise (the Fourier transform of a white Gaussian noise remains white). Nevertheless, the Fourier transform of a white Lévy noise can be "whitened" down by dividing its components by $|\mathbf{k}|^{-d}$ in order to obtain a flat (generalised) spectrum. We recall that such a generalised spectrum can be defined (Schertzer and Lovejoy 1987a; see also Fan, 1989) for fractional Lévy noises (as defined by eq. C7) as:

$$E_{f_\lambda}(k) = \int_{\partial S_\lambda} |\hat{s}(\mathbf{k})|^\alpha\, d^{d-1}\mathbf{k} \tag{C10}$$

∂S_λ being the surface of the sphere (S_λ) of radius λ), corresponding to the usual definition of the spectrum in the Gaussian case and to its natural extension for Lévy-stable noises. In particular, we still have:

$$\hat{K}_{f_\lambda}(k) = \int_1^\lambda E_{f_\lambda}(k)dk \tag{C11}$$

thus log divergence of $\hat{K}$ requires a k^{-1} spectrum, obtained with eq. C9. More precisely, taking:

$$\hat{s}(\mathbf{k}) = \left[\frac{|\mathbf{k}|^{-d}\, C_1}{\alpha-1}\right]^{1/\alpha'} \tag{C12}$$

leads to the (non normalized) universal characteristic function ($\alpha\neq1$):

$$K(h) = \frac{C_1\alpha'}{\alpha} h^\alpha \tag{C13}$$

hence the corresponding (normalized) universal characteristic function and singularity codimension function (eq. 34-35).

REFERENCES

Anselmet, F., Y. Gagne, E.J. Hopfinger, and R. A. Antonia, 1984: High order velocity structure functions in turbulent shear flows, J. Fluid Mech., 140, 63-75.

Arneodo, A., G. Grasseau, M. Holschneider, 1988: Wavelet transform of multifractals. Phys. Rev. Lett., 61, 2281-2284.

Benzi, R., G. Paladin, G. Parisi, A. Vulpiani, 1984: J. Phys. A, 17, 3521.

Bialas, A., R. Peschanski, 1986: Moments of rapidity distributions as a measure of short-range fluctuations in high-energy collisions, Nucl. Phys. B, B 273, 703-718.

Brax, Ph., R. Peschanski, 1990: Multifractal analysis of intermittency in random cascading models. Nuclear Physics B, (submitted).

Cahalan, R. F., 1990: Landsat Observations of Fractal Cloud Structure, this volume.

Cho, H. R., 1990: Energy spectrum and intermittency of semi-geostrophic flow, this volume.

Corrsin, S., 1951: On the Spectrum of Isotropic Temperature Fluctuations in an isotropic Turbulence, J. Appl. Phys., 22, 469-473 .

Davis, A., S. Lovejoy and D. Schertzer, 1990: Radiative transfer in multifractal clouds, this volume.

Derrida, B., E. Gardner, 1986: Magnetic properties and the function q(x) of the generalised random-energy model. J. Phys. C: Solid State phys., 19, 5783-5798.

Essex, C., 1990: Correlation dimension and data sample size , this volume.

Fan, A H., 1989a: Chaos additif et multiplicatif de Lévy. C. R. Acad. Sci. Paris, I, 308,151-154.

Fan, A H., 1989b: Décomposition des mesures et recouvrements aléatoires. Ph.D. Thesis, Paris-Sud U., Orsay.

Farge, M., G. Rabreau, 1988: Transformée en ondelettes pour détecter et analyser les structures cohérentes dans les écoulements turbulents bidimensionnels. C. R. Acad. Sci. Paris, II, 1479 1486.

Feller, W. 1971: An introduction to probability theory and its applications, vol.2. Wiley, New-York.

Frisch, U., P. L. Sulem, and M. Nelkin,.1978: A simple dynamical model of intermittency in fully developed turbulence. J. Fluid Mech., 87, 719-724.

Gabriel, P, S. Lovejoy, D. Schertzer, 1988a: Multifractal analysis of satellite resolution dependence. J.Geophys. Res. Lett., 1373-1376.

Gardner, E, B. Derrida, 1989: Magnetic properties and the function q(x) of the generalised random-energy model. J. Phys. A: Math Gen., 22, 1975-1981

Gnedenko, B. and A. N. Kolmogorov, 1954: Limit distribution for sums of independent random variables. Addison-Wesley, Cambridge, Massachusetts.

Gnedenko, B. , 1969: The theory of probability. MIR, Moscow

Grassberger, P., 1983: Generalized dimensions of strange attractors. Phys. Lett., A 97, 227.

Grassberger, P., 1986: Are there really climate attractors? Nature, 322, 609.

Grassberger, P., I. Procaccia, 1983: On the characterization of strange attractor. Phys. Rev. Lett., 50, 346.

Grossman, A., J. Morlet, 1987: Decomposition of functions into wavelets of constant shape and related transforms. Mathematics and physics lectures on recent results, ed. L. Streit, World Scientific, Singapore.

Gupta, V.K, E. Waymire, 1990: On scaling and log-normality in rainfall?, this volume.

Gurvitch E., M. Yaglom, 1967: Breakdown of eddies and probabilty distributions of small-scale turbulence, Phys. Fluids, 16, Suppl. S, S59-S65.

Halsey, T.C., M.H. Jensen, L.P. Kadanoff, I. Procaccia, B. Shraiman, 1986: Fractal measures and their singularities: the characterization of strange sets. Phys. Rev. A, 33, 1141-1151.

Hentschel, H.G.E., I. Proccacia, 1983: The infinite number of generalized dimensions of fractals and strange attractors, Physica, 8D, 435-444.

Herring, J.R., D. Schertzer, M. Lesieur, G.R. Newman, J.P. Chollet, and M. Larchevêque, 1982: A comparative assessment of sopectral closures as applied to passive scalar diffusion. J. Fluid Mech., 124, 411-420.

Hubert P., J. P. Carbonnel, 1988: Caractérisation fractale de la variabilité et de l'anisotropie des précipitations tropicales. C. R. Acad. Sci. Paris, 2, 307, 909-914.

Hubert P., J. P. Carbonnel, 1990: Fractal caracterisation of intertropical precipitations variability and anisotropy, this volume.

Kahane, J. P., 1985: Sur le Chaos Multiplicatif, Ann. Sci. Math. Que., 9, 435-444.

Kahane, J. P., 1987a: Martingales and Random Measures, Chinese Ann. Math., 8B1, 551-554.

Kahane, J. P., 1987b: Multiplications aléatoires et dimensions de Hausdorff. Ann. Inst. Henri Poincaré, 23, 289-296.

Kahane, J. P., 1988: Désintégration des mesures selon la dimension. C. R. Acad. Sci. Paris, I, 306,107-110.

Kolmogorov, A. N., 1949: Local structure of turbulence in an incompressible liquid for very large Reynolds numbers. Proc. Acad. Sci. USSR., Geochem. Sect., 30, 299-303

Kolmogorov, A. N., 1962: A refinement of previous hypothesis concerning the local structure of turbulence in viscous incompressible fluid at high Reynolds number. J. Fluid Mech., 13, 82-85

Kraichnan, R. H., 1971: An Almost-Markovian Galilean-invariant turbulence model. J. Fluid. Mech., 83, 349-367.

Kraichnan, R. H., 1980: Realizability inequalities and closed moment equations. Nonlinear Dynamics (Ann. N. Y. Acad. Sci., 357), ed. R. H. G. Helleman., New. York. Academy of Sciences.

Ladoy, P., D. Schertzer and S. Lovejoy, 1986: Une étude d'invariance locale-regionale des temperatures, La Météorologie, 7, 23-34..

Ladoy, P., S. Lovejoy and D. Schertzer, 1990: Extreme variability of climatological data: scaling and intermittency, this volume.

Landau, L. D., and E. M. Lifshitz, 1963: Fluid Mechanics, Pergamon, New York.

Lavallée, D,. D. Schertzer and S. Lovejoy, 1990: On the determination of the codimension function, this volume.

Leray, J., 1934: Sur le mouvement d'un liquide visqueux emplisent l'espace. Acta Math., 63, 193-248.

Levich E., and E. Tzvetkov, 1985: Helical inverse cascade in three-dimensional turbulence as a fundamental dominant mechanism in meso-scale atmospheric phenomena, Phys. Rep., 128, 1-37.

Levich, E., I Shtilman, 1990: Helicity fluctuations and coherence in developed turbulence, this volume.

Lévy, P., 1924: Théorie des erreurs, la loi de Gauss et les lois exceptionnelles. Bull. Soc. Math., 52, 49-85.

Lévy, P., 1925: Calcul des Probabilités, Gauthier Villars, Paris

Lévy, P., 1954: Théorie de l'addition des variables aleatoires, Gauthier Villars, Paris

Lilly, D. K., 1983: Meso-scale variability of the atmosphere in Mesoscale Meteorology Theories, Observations and Models, edited by D. K. Lilly and T. Gal'Chen, pp.13-24, D. Reidel, Hingham, Mass.

Lorenz, E. N., 1963: Deterministic nonperiodic flows. J. Atmos. Sci., 20, 130.

Lovejoy, S., D. Schertzer, 1985: Generalized scale invariance in the atmosphere and fractal models of rain. Wat. Resour. Res, 21, 1233-1250.

Lovejoy, S., D. Schertzer, 1986: Scale invariance, symmetries fractals and stochastic simulation of atmospheric phenomena. Bull AMS 67, 21-32.

Lovejoy, S., D., Schertzer, 1989: Comment on "Are Rain Rate Processes Self-Similar?" by B.Kedem and L. S. Chiu. Wat. Resour. Res, 25, 3,577-579.

Lovejoy, S., D. Schertzer,1990a: Multifractals, universality classes and satellite and radar measurements of cloud and rain fields, J. Geophy. Res., 95, 2021-2034.

Lovejoy, S., D., Schertzer, 1990b: Multifractal analysis techniques and the rain and cloud fields from 10^{-3} to 10^{6}m, this volume.

Lovejoy, S., D. Schertzer, A.A. Tsonis, 1987: Functional box-counting and multiple elliptical dimensions in rain. Science, 235, 1036-1038.

Mandelbrot, B., 1974: Intermittent turbulence in self-similar cascades: Divergence of high moments and dimension of the carrier. J. Fluid Mech., 62, 331-350.

Mandelbrot, B, 1982: The Fractal Geometry of Nature, Freeman, 465pp.

Mandelbrot, B, 1984: fractals in Physics: squig clusters, diffusions, fractal meaures and the unicity of fractal dimensionality. J. Stat. Phys., 34, 895-930.

Mandelbrot, B., 1986: Self-affine fractal sets, I, the basic fractal dimensions, Fractals in Physics, edited by L. Pietronero and E. Tosatti, pp. 3-15, Elsevier North-Holland, New York.

Mandelbrot, B., 1989: An introduction to multifractal distribution functions. Fluctuations and Pattern Formation, Eds. H. E. Stanley and N. Ostrowsky, Kluwer, Dordrecht-Boston

Mauldin, R.D., S.C. Williams, 1986: Random recursive constructions: asymptotic geometric and topological properties. Trans. Am. Math. Soc., 295, 325-346.

Meneveau, C. ,K. R. Sreenivasan, 1987: Simple multifractal cascade model for fully developed turbulence. Phys. Rev.Lett., 59, 13, 1424-1427.

Meyer, Y., 1987: Wavelets viewed by a mathematician. Proceedings Ondellettes, méthodes temps-freequences et espaces de phases, CIRM, Luminy.

Moisseev, S. S., P. B. Rtkevitch, A. V. Tur and V. V. Yanovskii, 1988: Vortex dynamo in a convective medium with helical turbulence. Sov. Phys. JETP, 67, 2, 294-299

Monin, A.S., A. M. Yaglom, 1975: Statistical Fluid Mechanics, vol. 2. MIT Press, Boston.

Nerenberg, M. A. H., T. Lookman and C. Essex, 1990: On the existence of low dimensionsal climatic attractors, this volume.

Nicolis, C., G. Nicolis, 1984: Is there a climate attractor. Nature, 311, 529.

Novikov, E. A., R. Stewart, Intermittency of turbulence and spectrum of fluctuations in energy-dissipation, Izv. Akad. Nauk. SSSR, Ser. Geofiz., 3, 408-412, 1964.

Obukhov, A., 1949: Structure of the Temperature Field in a Turbulent Flow, Izv. Akad. Nauk. SSSR Ser. Geogr. I Jeofiz., 13, 55-69.

Obukhov, A., 1962: Some specific features of atmospheric turbulence. J. Geophys. Res, 67, 3011-3014

Parisi, G., U. Frisch, 1985: A multifractal model of intermittency, in Turbulence and predictability in geophysical fluid dynamics and climate dynamics, 84-88, Eds. Ghil, Benzi, Parisi, North-Holland.

Perrin, J., 1913 (and 1948): Les Atomes, NRF-Gallimard, Paris.

Pietronero, L., and A.P. Siebesma 1986: Self-similarity of fluctuations in random multiplicative processes. Phys. Rev. Lett., 57, 1098-1101.

Rényi, A., 1966: Calcul des Probabilités. Dunod, Paris, 620 pp.

Richardson, L. F., 1926: Atmospheric diffusion shown on a distance neighbor graph. Proc. R. Soc., London, Sec. A, 110, 709-722.

Richardson, L. F., 1922, (Republished by Dover, New York,1965.): Weather prediction by numerical process, Cambridge U. Press.

Ross, B.: Fractional Calculus and its applications. Lecture Notes in Mathematics 457, Springer-Verlag, Berlin.

Sarma, G., 1989: Analyses et simulations multifractales des champs de nuages, applications à la télédétection. Engineer's Thesis, Ecole Nationale Supérieure des Techniques Avancées, Paris.

Schertzer, D., S. Lovejoy, 1983: On the dimension of atmospheric motions. Preprint Vol., IUTAM Symp on Turbulence and Chaotic Phenomena in Fluids, 141-144.

Schertzer, D., S. Lovejoy, On the dimension of atmospheric motions, in Turbulence and chaotic phenomena in fluids, edited by T. Tatsumi, pp.505-508, Elsevier North-Holland, New York, 1984.

Schertzer, D., S. Lovejoy, 1985a: The dimension and intermittency of atmospheric dynamics. Turbulent Shear Flow 4, 7-33, B. Launder ed., Springer, NY.

Schertzer, D., S. Lovejoy, 1985b: Generalized scale invariance in turbulent phenomena. P.C.H. Journal, 6, 623-635.

Schertzer, D., S. Lovejoy, 1987a: Physically based rain and cloud modeling by anisotropic, multiplicative turbulent cascades. J. Geophys. Res., 92, 9693-9714.

Schertzer, D., S. Lovejoy, 1987b: Singularités anisotropes, et divergence de moments en cascades multiplicatifs. Annales Math. du Qué., 11, 139-181.

Schertzer, D., S. Lovejoy 1989: Generalized Scale invariance and multiplicative processes in the atmosphere. Pageoph, 130, 57-81.

Schertzer, D., S. Lovejoy, 1990: Nonlinear variability in geophysics: Multifractal simulations and analysis, in Fractals: Physical Origin and Consequences, Ed. L. Pietronero, Plenum, New York, 49-79.

Schertzer, D., S. Lovejoy, R. Visvanathan, D. Lavallée, and J. Wilson, 1988: Universal Multifractals in Turbulence, in Fractal Aspects of Materials: Disordered Systems, Edited by D.A. Weitz, L.M. Sander, B.B. Mandelbrot, 267-269, Materials Research Society, Pittsburg, Pa.

Seed, A., D. Lavallée, S. Lovejoy, D. Schertzer, G.L. Austin, 1990: Multifractal analysis of radar rain measurements. (in preparation).

Smith, L A , 1988: Intrinsic limits on dimension calculations. Phys. Lett. A, 133, 6, 283-288

Smith, L A , J. D. Fournier, E. A. Spiegel , 1986: Lacunarity and intermittency in fluid turbulence. Phys. Lett. A, 8,9.

Stanley, H. E. P. , P. Meakin, 1988: Minireview on multifractals. Nature, 6, 116.

Todoeschuck, J. P. , O.G. Jensen, 1990: !/f geology and seismic deconvolution, this volume.
Tessier, Y., S. Lovejoy, D. Schertzer, 1989: Multifractal analysis of global rainfall from 1 day to 1 year. XXIV th. European Geophysical Society Assembly, Barcelonna.
Viswanathan, R., C. Weber and P. Gibart, 1990: Stochastic coherence and the dynamics of global climate models and data, this volume.
Von Neumann, J., 1963: Recent theories in turbulence, in Collect. Works, 6, 437-450, Pergamon, New York.
Voss, R., 1983: Fourier Synthesis of Gaussian fractals: 1/f noises, landscapes and flakes. Proceedings, Siggraph conf., Detroit, p1-21.
Waymire, E., and V.K. Gupta, 1981: The mathematical structure of rainfall representations, parts 1-3, Water Resour. Res., 17, 1261-1294.
Wilson, J., D. Schertzer, S. Lovejoy, 1990: Physically based cloud modelling by scaling multiplicative cascade processes, this volume.
Yaglom, A. M. 1966: The influence of the fluctuation in energy dissipation on the shape of turbulent characteristics in the inertial interval, Sov. Phys. Dokl., 2, 26-30.
Zolotarev, V. M, 1986: One-dimensional stable distributions. Translations of Mathematical Monographs, Americ. Math. Soc., Providence, Rhode Island.

SCALING AND STRUCTURE FUNCTIONS IN TURBULENT SHEAR FLOWS

C. W. Van Atta and K. Poddar
Department of Applied Mechanics and Engineering Sciences
University of California, San Diego
La Jolla, CA 92093

ABSTRACT. An experimental investigation was made to obtain the statistics of velocity differences measured simultaneously in time at laterally separated spatial locations in a laboratory turbulent boundary layer flow. For large separations the measured probability density function of velocity differences is self-similar and near Gaussian. The probability distribution of squared velocity differences exhibits exponential behavior, rather than hyperbolic as was suggested for atmospheric turbulence by Schertzer and Lovejoy in the limit of very high Reynolds number.

1. INTRODUCTION

Recent applications of scaling and fractal ideas to the behavior of turbulent velocity fields and in particular to structure functions of turbulent flows have drawn the attention of geophysical fluid dynamicists.

From atmospheric boundary layer measurements over the open ocean, Van Atta (1971) found that the measured probability density function of the single point velocity difference at two different times continuously evolves from a strongly non-Gaussian form for smaller separation to a nearly Gaussian form at the largest time separations. Schertzer and Lovejoy (1983) suggested that their atmospheric measurements yielded hyperbolic distribution functions for squared velocity differences, implying the nonexistence of higher order moments (structure functions) of velocity differences. They conjectured that such hyperbolic, strong intermittency is built up as the turbulent cascade concentrates energy flux into smaller and smaller region and suggest that a wide inertial range and hence very high Reynolds numbers should be required.

Our measurements were stimulated by these observations of Schertzer and Lovejoy. Since not much appears to be known about the statistics of velocity differences measured at laterally separated locations in turbulent shear flows (as opposed to a considerable amount of data taken exploying time delays at a fixed longitudinal position), we have carried out a series of measurements in a laboratory turbulent boundary layer. It is hoped that the laboratory results will be useful in applying scaling hypotheses to the interpretation of similar geophysical data, which is necessarily obtained under essentially nonstationary and nonrepeatable conditions. The results of our investigation in a laboratory turbulent boundary layer are discussed in the following sections.

2. EXPERIMENTAL SET-UP AND MEASUREMENTS

Our experiments were carried out as part of an ongoing investigation of the effect of large eddy breakup devices on turbulent boundary layers on axisymmetric bodies. The basic experimental setup is discussed in Poddar and Van Atta (1985). All the experiments were performed at a free stream speed (U_∞) of 9.5 m/sec and at a distance of 2.5 m from the nose of the model. The boundary layer thickness (δ) at this measuring station was about 30 mm, so the Reynolds number based on δ, $U\delta/\nu$, was 2×10^4.

D. Schertzer and S. Lovejoy (eds.), Non-Linear Variability in Geophysics, 83–89.

The longitudinal velocity was measured simultaneously at pairs of spatial locations with the same streamwise location but at different heights in the turbulent boundary layer using two TSI 1218-T1.5 single hot-wire probes. One hot-wire probe was kept fixed, while the height of the other was varied to obtain different spatial separations. Hot-wire sensors were operated at an overheat of about 50% by DISA 55M01 constant temperature anemometers. The output of each anemometer was lowpass filtered and connected to a buck and gain amplified before digitizing on a 14-bit A/D converter at a typical sample rate of 7.5 kHz.

The probability density functions of the velocity difference (Δu) and the squared velocity difference $(\Delta u)^2$ and the probability distribution function of $(\Delta u)^2$ were computed for different separations (Δy) across the boundary layer.

3. EXPERIMENTAL RESULTS

Mean velocity and turbulence intensity profiles in the boundary layer are shown in figures 1 and 2 respectively. Poddar and Van Atta (1985) had found that the flow at this station is fully developed. In fig. 3 the 1-D spectrum of streamwise velocity fluctuations is plotted for different separations (Δy) across the boundary layer. The spectrum does not show any -5/3 slope, suggesting that the Reynolds number is too small for the existence of an inertial subrange.

The probability density functions of the velocity differences (Δu) are plotted in fig. 4a and the same quantities plotted with non-dimensional velocity scale are shown in fig. 4b. We notice from fig. 4a that the peak of the probability density function occurs at higher values of Δu as the separation increases. This is because of the fact that the convection speed increases with height in the boundary layer. From fig. 4b, where the probability density function is plotted in self-similar form, it is clear that for separations (Δy) greater than 4 mm the probability density function is self-similar and nearly Gaussian. This self-similarity and near Gaussian nature of the probability density function for large separations was also noticed by Van Atta (1971) in the atmospheric boundary layer. In fig. 5 the probability density function of $(\Delta u)^2$ is plotted for different Δy across the boundary layer. Here we notice again that the peak of the distribution occurs at higher values of Δu as the separation increases.

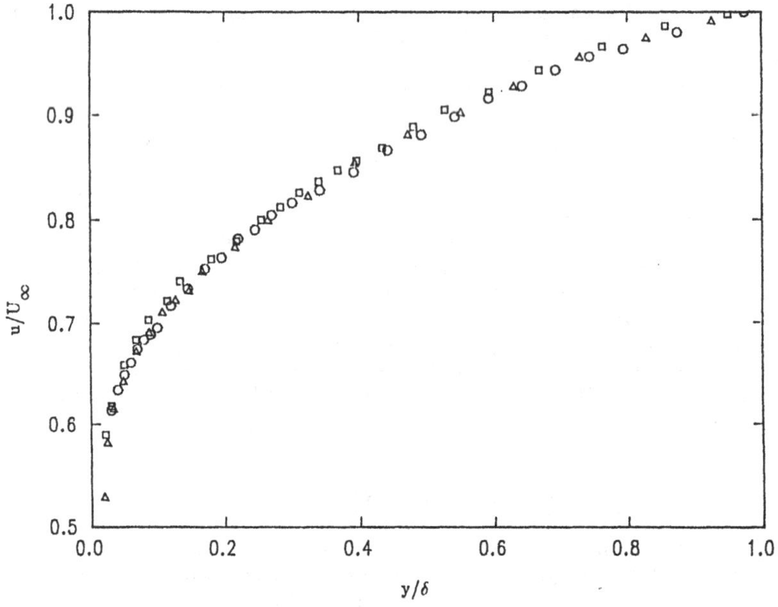

Fig. 1: Boundary layer velocity profile 2.5 m from the nose.

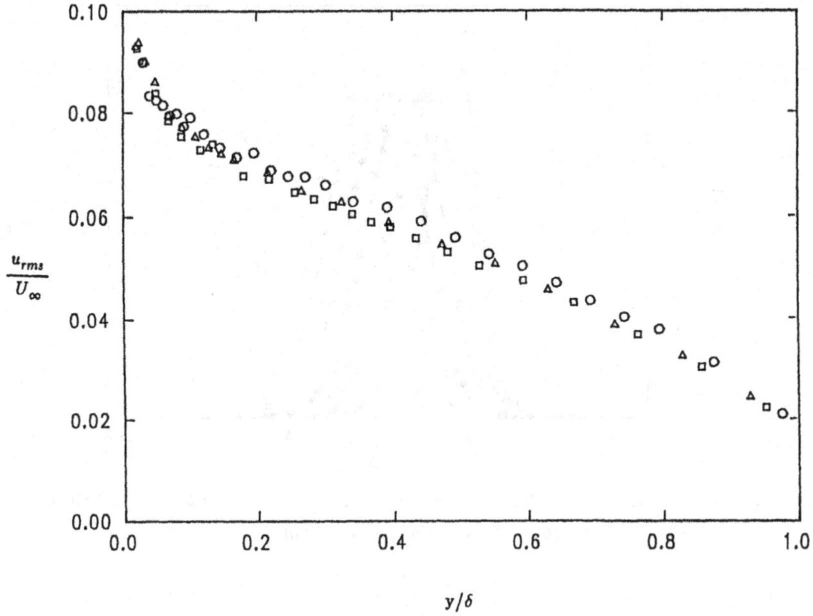

Fig. 2: Streamwise turbulence intensity profile.

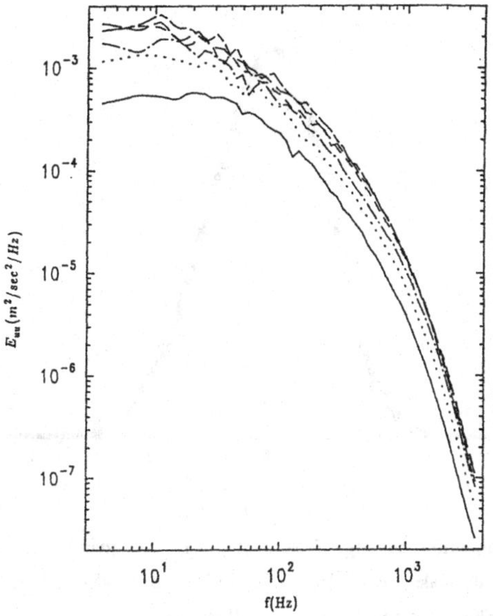

Fig. 3: 1-D spectrum of streamwise velocity fluctuations across the boundary layer. —, y/δ=0.83; ···, y/d=0.66; ——, y/d=0.50; - - , y/δ=0.33; ——, y/δ=0.20; —, y/δ=0.07.

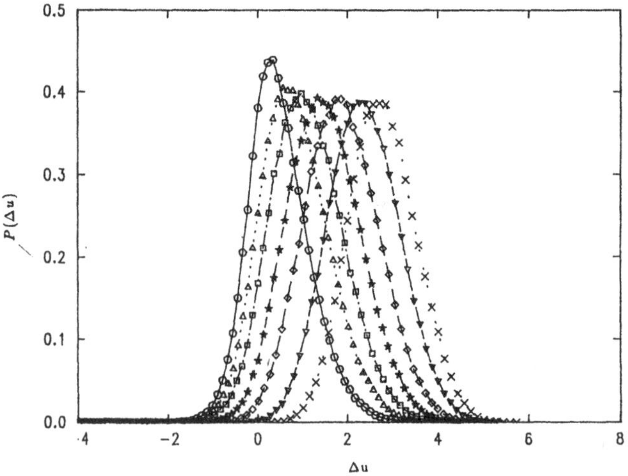

Fig. 4a: The probability density fluctuations in the quantity Δu(Δy). O, Δy=2mm; Δ, Δy=5mm; □, Δy=8mm; *, Δy=12mm; ◊, Δy=16mm; ∇, Δy=20mm; ×, Δy=24mm.

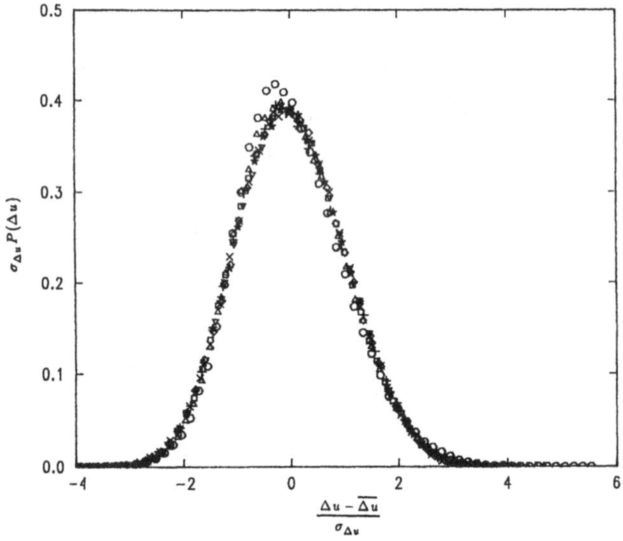

Fig. 4b: The probability density of fluctuations in the quantity Δu(Δy) plotted with non dimensionalized velocity scale $(\Delta u - \bar{\Delta u})/\sigma_{\Delta u}$. O, Δy=2mm; Δ, Δy=5mm; □, Δy=8mm; *, Δy=12mm; ◊, Δy=16mm; ∇, Δy=20mm; ×, Δy=24mm; +, Δy=28mm.

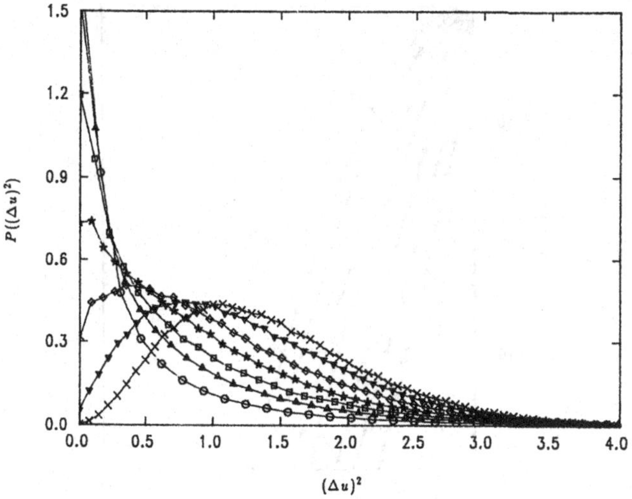

Fig. 5: The probability density fluctuations in the quantity $(\Delta u)^2$ for different.(Δy). O, Δy=2mm; Δ, Δy=5mm; $\square$, Δy=8mm; *, Δy=12mm; $\Diamond$, Δy=16mm; ∇, Δy=20mm; $\times$, Δy=24mm.

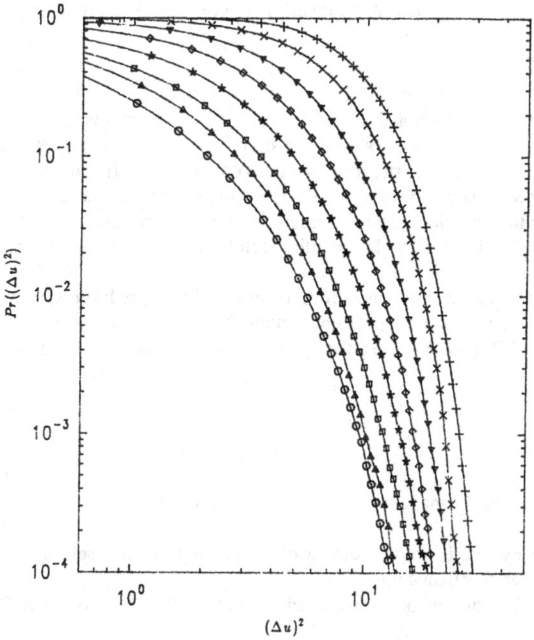

Fig. 6: The probability distribution of fluctuations in the quantity $(\Delta u)^2$ for different.(Δy) plotted with logarithmic horizontal scale. O, Δy=2mm; Δ, Δy=5mm; $\square$, Δy=8mm; *, Δy=12mm; $\Diamond$, Δy=16mm; ∇, Δy=20mm; $\times$, Δy=24mm; +, Δy=28mm.

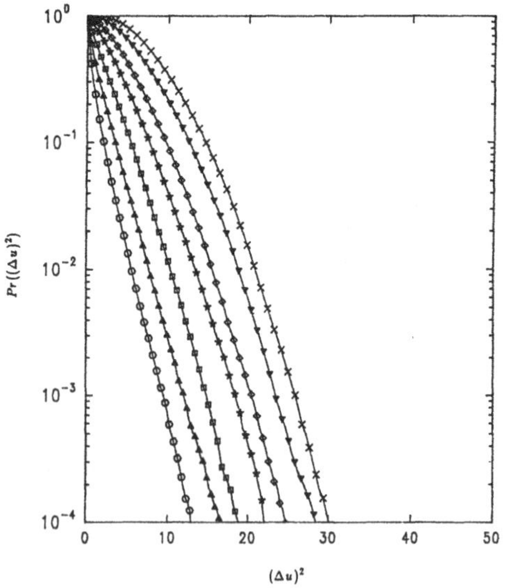

Fig. 7: The probability distribution of fluctuations in the quantity $(\Delta u)^2$ for different.(Δy) plotted with linear horizontal scale. O, Δy=2mm; Δ, Δy=5mm; □, Δy=8mm; *, Δy=12mm; ◊, Δy=16mm; ∇, Δy=20mm; ×, Δy=24mm.

The probability distribution of $(\Delta u)^2$ is plotted in fig. 6 with logarithmic horizontal and vertical scales for different Δy to see if our distribution followed any power law. From this figure it is very clear that the distribution function for any separation Δy does not show hyperbolicity even for large Δu, implying the existence of all higher order moments (structure functions) of velocity differences. This contrasts with the previous observation by Schertzer and Lovejoy (1983) in atmospheric turbulence. In fig. 7 the probability distribution is plotted with a logarithmic vertical scale and a linear horizontal scale to test the existence of exponential behavior in the distribution. It is readily seen from fig. 7 that the distribution does exhibit an exponential behavior.

It could be useful to examine similar data for a laboratory boundary layer with much higher Reynolds number, in which the velocity energy spectrum contained an extensive inertial subrange, to see if the suggested hyperbolicity of $Pr((\Delta u)^2)$ of the Schertzer and Lovejoy atmospheric data would appear at higher Reynolds number.

4. CONCLUSIONS

The following conclusions can be made from the experimental results:

1. The probability density function of the velocity difference appears to be self-similar and near Gaussian for separations (Δy) greater than 4 mm.
2. The probability distribution of the squared velocity difference does not follow a hyperbolic distribution, but rather an exponential distribution, implying the existence of all higher order moments in the flow considered in the present investigation.
3. Similar laboratory measurements at significantly higher Re should be performed to further examine the question of hyperbolicity.

5. ACKNOWLEDGEMENTS

This work was supported under a DARPA University Research Initiative, "Program on Research and Eduction in Chaotic and Turbulent Fluid Dynamics", in the DARPA Applied and Computational Mathematics Program, grant # N00014-86-K-0758.

6. REFERENCES

Poddar, K. and C.W. Van Atta, 1985: Turbulent boundary layer drag reduction on an axisymmetric body using LEBU manipulators. Proceedings of the 5th Symposium of Turbulent Shear Flows, Cornell, Ithica.

Schertzer, D. and S. Lovejoy, 1983: The dimension and intermittency of atmospheric dynamics. Turbulent Shear Flows 4. ed. Bradbury et al., Springer-Verlag.

Van Atta, C.W., 1971: Statistical self-similarity and atmospheric turbulence. Proc. International Symp. on Probability and Statistics in the Atmospheric Sciences, Honolulu.

2 - DATA ANALYSIS TECHNIQUES

CORRELATION DIMENSION AND DATA SAMPLE SIZE

Christopher Essex
Department of Applied Mathematics
University of Western Ontario
London, Ontario, Canada N6A 5B9

ABSTRACT: A general method for estimating the amount of data sufficient to reliably determine the correlation dimension from a time series is presented. The results of this method are discussed in terms of previous attempts to calculate this quantity from climatological data.

1. INTRODUCTION

The theory behind the algorithm that has been used to determine dimensionality, ν, from a time series (Takens, (1981)) assumes an infinite number of values in a data string. Naturally, in practical analyses of this type only finite data sets are relevant. The extent to which one can account for this disparity between theory and practice represents a limit to the validity of this type of analysis, and any broader conclusions, such as those discussed by Nerenberg et al. (1988) that can be drawn from this analysis.

This issue has been discussed by various authors: Takens (1985), Grassberger (1986), Denker and Keller (1986), Holzfuss and Mayer-Kress (1986), and Essex et al. (1987). Takens calculated statistical error estimates for values of ν, as did Denker and Keller. Both estimates considered only a single embedding dimension, for which scaling behaviour in the correlation function, C(r), was assumed to be present. Holzfuss and Mayer-Kress considered the effect of statistical error as a function of embedding dimension for a fixed number of data points. Grassberger demonstrated that false scaling behaviour of a system could arise by inflating the amount of data through interpolation. Essex et al. introduced the idea of a critical embedding dimension, D_c, beyond which an embedding from a data set of a given size cannot exhibit scaling behaviour.

Calculations of this type from solutions of known systems of equations, where one can check the results, have dealt with values of $\nu \cong 1$-3 (e.g., Grassberger and Procaccia (1983a, b), while typically using data sets with $1\text{-}5\times10^4$ members. However, such calculations performed directly on observations, where the results are not easily checked because the appropriate governing equations are not normally known, have typically dealt with larger values of ν, while using data sets with much smaller memberships (Grassberger (1986), Nicolis and Nicolis (1984), Fraedrich (1986), and Hense (1986)). There is a risk in this case that underestimations of the correlation dimension can arise through spurious scaling behaviour, or that scaling behaviour could be missed entirely because data sets are too small for the embedding dimensions examined (Essex et al. (1987)).

The idea of critical embedding dimension, D_c, is put into concrete terms here by introducing a simple procedure for estimating this quantity directly, for any data set, given the number of values in that data set. Estimates can be made of how large a data set must be in order to be confident that scaling behaviour is observable.

2. CRITICAL EMBEDDING DIMENSION

Scaling behaviour is discussed here in terms of the correlation function C(r), which is defined by

D. Schertzer and S. Lovejoy (eds.), Non-Linear Variability in Geophysics, 93–98.

$$1/N^2 \sum_{\substack{i=1 \\ i \neq j}}^{N} \sum_{j=1}^{N} \Theta(r - | X_i - X_j |), \tag{1}$$

where $\Theta(u) = 1$ for $u>0$, and $\Theta(u) = 0$ for $u \leq 0$, and $\{x_i\}$ is the set of N points on the attractor constructed from a time series of $M \geq N$ values (Grassberger and Procaccia (1983a, b)). See Nerenberg et al. (1988) for some other references.

If scaling behaviour is present in C(r), arising from a particular D dimensional embedding in R^D of a data set containing N values, the interval in r over which the scaling appears will be bounded (Essex et al. (1987)). On the large scales, the scaling region is limited by the nature of C(r). As r approaches the set diameter, C(r) becomes saturated and turns toward a constant in r. Such turning is inconsistent with scaling behaviour.

On the small scales, the presence of scaling is limited by the intrinsic sparseness of a finite data set. The number of neighbours closer than a distance r declines as r decreases, until the D dimensional ball of radius r around most points is empty. Over those smallest values of r, C(r) is effectively zero, which is also inconsistent with the scaling behaviour that is sought.

The scaling region is always sandwiched between depopulation and saturation. For a particular D, the scaling region can be squeezed by decreasing N until the points are populated sparsely enough that depopulation and saturation occur on the same scale. In such a case, the scaling region will have entirely collapsed, and C(r) will not exhibit scaling on any interval in r as N is further decreased.

The scaling region can be similarly squeezed by considering a sequence of embeddings of increasing D, while holding N fixed. As D increases, the population density of the points must decrease, implying that points grow further apart. It follows that, as D grows, increasing numbers of points centered in balls of a given r contain empty regions of embedding space. Morever, C(r) can become saturated at even lower values of r as D increases. The first value of D for which the value of r where depopulation occurs equals or exceeds the value of r where saturation occurs is the critical embedding dimension, D_c. A particular data set is insufficient in size for embeddings where $D \geq D_c$.

3. ESTIMATION OF D_c

D_c should, in general, depend upon the specific nature of the reconstructed object, as well as upon N. However, if the scaling dimension, ν, exceeds the embedding dimension, then all compact objects share, with random numbers, the property of filling space. That is, these compact objects will have dimensions equal to that of the embedding space. This fact ensures the value of using N random numbers as a control sequence in such computations. Moreover, it provides a practical framework for presenting an explicit estimate of the critical embedding dimension with some general validity.

Suppose that a time series reveals that a system's dynamical variable is nearly uniformly distributed over the interval $-l$ to l, that $D< \nu$, and that the constructed vectors are largely uncorrelated. Then the points in any particular embedding dimension, D, will fall uniformly into a hyper-cube of volume,

$$(2l)^D. \tag{2}$$

The volume of a hyper-cube with edges of length 2x, and the same center and orientation inside that one, is given by

$$(2x)^D \tag{3}$$

where $0 \leq x \leq l$, and V is the hyper-volume of this smaller hyper-cube. If v_o is defined as the volume per point for some D and N, we have from expression (2),

$$v_o = \frac{(2l)^D}{V} \tag{4}$$

The number of points dN in an interval dx will be given by

$$dN = \frac{dV}{v_o} = \frac{DN(x/l)^D}{x}\,dx\,. \tag{5}$$

The mean distance from the boundary at $x = l$, r_s, will be given by

$$r_s = \frac{\int (l-x)dN}{\int dN} = \frac{\int (l-x)dN}{N} = \frac{l}{D+1}\,, \tag{6}$$

where the integrations are carried out over the entire hyper-cube through equation (5).

The characteristic nearest neighbour spacing, r_n, can be extracted from v_o and equation (4)

$$r_s = (v_o)^{1/D} = 2l(1/N)^{1/D}\,. \tag{7}$$

When $r > r_s$, significant saturation in C(r) is expected, because the average point is itself saturated. That is, D dimensional balls centered at an average point must contain elements of volume outside of the hyper-cube when $r > r_s$, and therefore cannot represent a population with scaling behaviour in r. However, when $r < r_n$ one expects to observe the effects of depopulation, because D dimensional balls centered at some average point will be empty. Thus if $r_n > r_s$, there can be no scaling region, because any non-empty D dimensional ball centered on an average point must contain volume outside the hyper-cube.

It follows that the transition from $r_n < r_s$ to $r_n > r_s$ as D increases represents the collapse of the scaling region. If D_r is taken as the value of D when $r_n = r_s$, then the next higher integer value of D is D_c. Thus from eqs. (6) and (7)

$$N^{1/D_r} = 2(D_r + 1)\,, \tag{8}$$

and

$$D_c = G(D_r)\,, \tag{9}$$

where G(u) is the greatest-integer step function.

TABLE 1

N	D_c	D_r	ν
184	3	2.63	–
500	3	2.99	3.1 (Nicolis and Nicolis 1984)
750	4	3.13	3.5-6 (Hense 1987)
1008	4	3.24	2.5-4.5 (Hense 1987)
5475	4	3.80	4.4 (Fraedrich 1986)
7100	4	3.89	>10 (Grassberger 1986)
12084	5	4.06	–
108756	5	4.75	~6.5 (Essex et al. 1987)

These equations indicate that D_c does not depend upon l, but upon N alone. This supports the position of Holzfuss and Mayer-Kress (1986), and Essex et al. (1987) that the number of data points is the primary limitation on the validity of computations of ν. Moreover, it agrees with the prediction of Holzfuss and

Mayer-Kress that the necessary number of points would need to grow rapidly, even exponentially, with embedding dimension. Table 1 provides a list of values of D_c and D_r calculated from eqs. (7) and (8) for selected values of N, as well as other columns which are discussed below.

4. COMPARISON TO COMPUTATIONS OF ν

The technique of computing ν from a single dynamical variable was introduced by Grassberger and Procaccia (1983). The technique was validated, in that paper, by computing values of ν from sequences derived from known maps and systems of equations. The resulting values were successfully compared to values computed with other methods.

Their analyses considered the Henon map, the Kaplan-Yorke map, the logistic equation, the Lorenz equations, the Rabinovich-Fabrikant equations and the Zalaviskii map. The value of ν for the Hénon map was 1.21 and they used 15000 values in their sequence to deduce it. The largest value of ν was 2.18 for the Rabinovich-Fabrikant equations. To determine this value they utilized 25000 data points.

If one were to embed a random number control sequence for each of these computations, Table 1 indicates that these values of M would be more than adequate to accommodate the values of D necessary for determining these values of ν. That is, $D<D_c$ for the control sequence, for all necessary D. Indeed, if the critical behaviour of the control sequence is used as a criterion, Table 1 indicates that far fewer values would be sufficient for these calculations. This agrees with the observations of Grassberger and Procaccia, who also commented that only a few thousand points appeared to be necessary to reveal the value of ν in their calculations. The control sequence criterion also supports a computation involving experimental data (Brandstater et al. (1983)), where a time series of velocities, with approximately 33000 values, arising from Couette-Taylor flow was used to determine a value of ν of approximately 3.

The computation of ν from geophysical or climatological data sets differs from the problem of computations in these cases, primarily because the amount of data in geophysical data sets are not easily increased; whereas, in experiments or direct calculations from known equations, insufficient data set size can normally be remedied by returning to the lab or the computer. Nevertheless, the question of the true value of ν in various climatological contexts remains.

Unfortunately, no computations from climatological data sets seem to pass the control criterion presented here. The values of N in Table 1 represent the values in actual published accounts of computations of ν involving climatological data sets. In each case, the resulting value of ν exceeds the critical embedding dimension of the control sequence.

In one of the papers (Fraedrich (1986)), where the results of a computation of random number control sequence is displayed together with the actual results, the random number control sequence is indicated as not filling space for values of D=4, agreeing remarkably well with the corresponding value for D_c from Table 1! In Table 1, the value N=184 is reported to be the actual number of observations used to compile the data set used by Nicolis and Nicolis (Grassberger (1986)). The value of N=12084 was reported to be too small for determining ν for a different data set, and it was proposed (Essex et al. (1987)) that D_c 5 for that value of N, solely on the basis of the data from that set. However, the computations for that paper were performed in a slightly different way.

5. LIMITATIONS OF THE ESTIMATE OF D_c

The values of D_c determined by eqs. (7) and (8), while reasonable, are probably somewhat conservative. The lower bound where depopulation occurs was assumed to be the mean nearest neighbour distance. However, the interval between the least nearest neighbour distance, r_{ln}, and the mean nearest neighbour distance, r_n is not entirely depopulated, even though it is depleted. Scaling could be detected in the interval (r_{ln}, r_n) under favourable circumstances. In that case, D_c would be larger than the values presented here.

The value of D_c would also be increased if one were to account for correlations. Correlations would cause saturation to occur more gradually; thus values of $r>r_s$ would display approximate scaling in C(r) more closely than in the uncorrelated case. However, true scaling would not be as accurately represented for $r<r_s$.

The estimate does not apply to boundary corrected calculations (Essex et al. (1987)) or to similar average pointwise calculations (Holzfuss and Mayer-Kress (1986)). Both of these techniques can effectively increase the value of D_c attainable for a given N, by reducing the effects of saturation. The estimate also assumes that the time series has, in some sense, fully explored the attractor. Otherwise, the effective value of D_c could be reduced. A problem of this kind was encountered by Grassberger and Procaccia (1983b) in connection with the Zalavaskii map. The estimate also does not strictly consider the case where $\nu<D$, although one can take the position that the random control could be representative. Although, the value of D_c in this regime must stricly speaking depend upon the nature of the attractor itself.

6. CONCLUSION

Many refinements could be undertaken to improve the estimate of D_c presented here. However, it would be more complex and more specific to particular systems. It is unlikely that the essential features would be greatly altered; that is, strong, super-exponential dependence on D_c by N, and the primacy of N over other system properties in determining D_c.

The results in Table 1 do not refute the calculations of ν from climatological data sets, even if they do not support those calculations because of the conservative nature of the estimate of D_c. However, Table 1 clearly puts the climatological calculations into a more equivocal category than that of the original calculations by Grassberger and Procaccia (1983), which eqs. (8) and (9) fully support.

The determination of dimensionality of the physical system determining the climatological state remains physically important, but the dimensionalities in question tend to be somewhat higher than the test examples used for validating the original calculations of ν. The resulting data requirements are much greater for a corresponding quality in the result. Unfortunately, the data is not often available, and because of the super-exponential growth in data requirements with dimension deduced here, there never may be enough data for higher dimensional physical systems. New methods will need to be developed in order to confidently analyze the dimentionality of geophysical data sets.

7. ACKNOWLEDGEMENTS

I wish to thank Dr. M.A.H. Nerenberg for his assistance and encouragement. This work was supported by the Alexander von Humboldt-Stiftung, the Natural Sciences and Engineering Research Council of Canada, and the Deutsche Forschungsgemeinschaft.

8 REFERENCES

Brandstater, A., J. Swift, H. L. Swinney, A. Wolf, J. D. Farmer, J. Jen, and P. J. Crutchfield, 1983: Low-dimensional chaos in a hydrodynamic system. Phys. Rev. Let., 51, 1442.

Denker, M., and G. Keller, 1986: Rigourous statistical procedures for data from dynamical systems. J. Stat. Phys., 44, 67.

Essex, C., T. Lookman, and M. A. H. Nerenberg, 1987: The climate attractor over short timescales. Nature, 326, 64.

Fraedrich, K., 1986: Estimating the dimensions of weather and climate attractors. J. Atmos. Sci. 43, 419.

Grassberger, P., 1986: Are there really climate attractors? Nature, 322, 609.

Grassberger, P., and I. Procaccia, 1983a: Characterization of strange attractors, Phys. Rev. Let. 50, 346.

Grassberger, P., and I. Procaccia, 1983b: Measuring the strangeness of strange attractors. Physica, 9D, 189.

Hense, A., 1986: On the possible existence of a strange attractor for the southern oscillations. Contr. to Atmos. Phys., 60, 1987.

Holzfuss, J., and G. Mayer-Kress, 1986: An approach to error-estimation in the application to dimension algorithms, in Dimensions and Entropies in Chaotic Systems. p. 114, ed. G. Mayer-Kress, Springer- Verlag, Berlin.

Nerenberg, M. A. H., T. Lookman, and C. Essex, 1988: On the existence of low dimensional climate attractors, in Scaling, Fractals and Non-linear Variability in Geophysics, (this volume), ed. S. Lovejoy and D. Schertzer, D. Reidel, Dordrecht, Netherlands.

Nicolis, C., and G. Nicolis, 1984: Is there a climate attractor? Nature, 311, 529.

Takens, F., 1981: Detecting strange attractors in turbulence, in Lecture Notes in Mathematics, 898, 366, Springer-Verlag, Berlin.

Takens, F., 1985: On the numerical determination of the dimension of an attractor, in Lecture Notes in Mathematics, 1125, 99, Springer-Verlag, Berlin.

ON THE DETERMINATION OF THE CODIMENSION FUNCTION

D. Lavallée, D. Schertzer* and S. Lovejoy
Department of Physics,
McGill University,
3600 University st., Montréal, Québec
Canada, H3A 2T8

ABSTRACT. Motivated by the necessity of developing new multifractal analysis techniques to characterize empirical fields by scale invariant (sensor resolution independent) codimension functions, we introduce a new method, PDMS, to directly estimate the codimension of the singularity spectrum c(γ) and we also indicate the theoretical (or practical) limits of this method as well as its consequences for the determination of highest values of c(γ). These properties also have implications for the behaviour of K(h) – related to c(γ) by a Legendre transformation – in particular for large h. The characteristic behaviour of c(γ) and K(h) are illustrated respectively by the estimation of the scaling properties of the probability distribution and of statistical moments of simulated fields, obtained with multiplicative self similar cascade processes.

1. INTRODUCTION

For several years now, the concepts of fractals and multifractals have increasingly served as new tools in investigating and understanding the behaviour of atmospheric and other geophysical fields. It has been particularly important in characterizing and mathematically linking two fundamental properties of turbulent flows: their intermittency (high variability of the field) and scale invariance (observed over a wide range of time and/or space scales). By scaling (or scale invariance) we mean that quantities associated with the fields at different scales are related by transformations involving only the scale ratios. When all the statistical properties of these quantities can be described by a unique exponent of the scale ratio, we have simple scaling; see Schertzer and Lovejoy (1988), Lovejoy and Schertzer (this volume) for brief reviews. In general however the complex behaviour of the fields could not be reduced to simple scaling (as already pointed out by Kolmogorov (1962) and Obukhov (1962) since they found multiple scaling in their log-normal[1] model for intermittency), which leads recently to the concept of multifractal dimensions, introduced by Hertschel and Procaccia (1983), Grassberger (1983) and Schertzer and Lovejoy (1983). Multiple scaling takes into account that region of varying intensities scale with different exponents and are characterized by different (fractal) dimensions (a dimension function). The latter describe how the various intensities level of the fields are distributed - or projected - on a given space of observation. Subtracting the dimension of the space of observation from the dimension we obtain the codimension, a quantity not only scale invariant but also independent of the dimension of the space in which the field is embedded, and could be then regarded as the fundamental function characterizing the system.

The codimension function is usually determined by examining the statistical properties of the fields. This could be done by estimating directly the probability distribution of the field 's intensities or by looking at the behaviour of their statistical moments. The codimension functions obtained in both case are related to each other by a Legendre transformation as discussed in Frisch and Parisi (1985) and Halsey et al. (1986). Until now, the different attempts to estimate the codimension functions of experimental data

*EERM/CRMD, Météorologie Nationale, 2 Ave. Rapp, Paris 75007, France

1 Corresponding to the case $\alpha = 2$ discussed below.

D. Schertzer and S. Lovejoy (eds.), Non-Linear Variability in Geophysics, 99–109.

belong to one of these approaches. The codimension function of the radar rain reflectivity fields (Schertzer and Lovejoy, 1987 a,b), of the energy dissipation fields (Meneveau and Sreenivasan, 1987) and of lidar rain drops (Lovejoy and Schertzer, this volume) have been obtained by studying the scaling behaviour of their statistical moments. On the other hand Box Counting – used recently by Hubert and Carbonnel (1988) to analyze rain gauge data – or Functional Box Counting – introduced by Lovejoy et al. (1987) and used by Gabriel et al. (1988) for satellite data – are techniques developed to determine the codimension function from the probability distributions of the fields. Unfortunately the two last methods, which are designed to work on sets, can only be applied indirectly to physical fields.

In section 2, we propose a new method to estimate the codimension function, the PDMS – Probability Distribution-Multiple Scaling – essentially based on the scaling properties of the probability distribution. We indicate how the finite number of samples impose upper limits on the measured codimension function . In section 3, its consequences for the behaviour of the (highest) statistical moments of the fields, are also discussed in parallel with the spurious (or pseudo-) scaling.

2. PDMS: PROBABILITY DISTRIBUTION-MULTIPLE SCALING

It is now well established (in other paper in this book but also in Halsey et al. (1986), Meneveau and Sreenivasan (1987), etc.) that the multiple scaling properties of the field intensities ε_λ are correctly describe by the following two equations:

$$\varepsilon_\lambda \propto \lambda^\gamma, \quad \lambda > 1 \tag{2.1}$$

$$\Pr(\varepsilon_\lambda \geq \lambda^\gamma) \propto \lambda^{-c(\gamma)} \tag{2.2}$$

where λ is ratio of the largest scale of interest to the scale of homogeneity, and γ is positive it is the order of the singularity: $\varepsilon_\lambda \rightarrow \infty$ when $\lambda \rightarrow \infty$. The second equation indicates that the probability distribution of singularities of order higher than a value γ, is related to the fraction of the space occupied by them, as determined by their codimension function $c(\gamma)$. This equation holds to within slowly varying functions of λ (e.g. logs). The function $c(\gamma)$ and $K_D(h)$, used in this paper, are related to the $f(\alpha)$ singularity spectrum and the $\tau(q)$ introduced by Frisch and Parisi (1985) and used by Halsey et al. (1986), by these simple relations $c(\gamma) = d - f(\alpha)$ and $K_D(h) = -\tau(q)$; when $\alpha = (d - \gamma)$, $h=q$ and d is the dimension of the space in which the set is embedded. It could be show that the hth statistical moments of the field intensities will generally have a scaling exponent $K(h) = (h-1)C(h)$ and then using the definition for the statistical moments we obtain the following relation:

$$<\varepsilon_\lambda{}^h> \propto \int \lambda^{h\gamma} \lambda^{-c(\gamma)} \, d\Pr(\gamma) \propto \lambda^{(h-1)C(h)} \tag{2.3}$$

In the limit $\lambda \rightarrow \infty$, the scaling exponents are related by the the following Legendre transformation:

$$K(h) = (h-1)\, C(h) = \max_\gamma (h\gamma - c(\gamma)) \tag{2.4}$$

$$c(\gamma) = \max_h (h\gamma - K(h))$$

where C(h) (a monotically increasing function) is the codimension function associated to the statistical moments of the field intensities, as $c(\gamma)$ is the one associated to the probability distribution.

Let us now consider how these properties arise and how they could be understood in terms of cascade processes, firs postulated by Richardson. Without losing generality, we will restrict ourselves studying the dynamical cascade (i.e. energy flux). Although we discuss only this case, the cascade process could also be used to describe passive scalars (e.g. scalar variance) or pseudo scalars (the helicity flux, discussed in

Levich and Shtilman (this volume)) and have been also used to study showers of cosmic rays (Bialas and Peschanski, 1986). Here, the scale invariant quantity is the energy flux density ε whose ensemble spatial average is constant (i.e. scale independent) but nevertheless highly intermittent. This condition on the ensemble spatial average of the energy flux density corresponds to the canonical case. In this process the flux of the field at large scale multiplicatively modulates the various fluxes at smaller scales, the mechanism of flux redistribution is repeated at each cascade step (self similarity). Figure 1 gives a schematic illustration of the discrete cascade process in two dimensions: a large eddy of characteristic length l_0 with an initial energy flux density ε_0, is broken up (via non-linear interaction with other eddies or internal instability) into smaller sub eddies of characteristic length $l_1 = l_0/\lambda$ ($\lambda = 2$, is the scale ratio between to consecutive cascade step), transferring in the process to each sub-eddy a fraction of its energy flux density $\mu\varepsilon_0$.

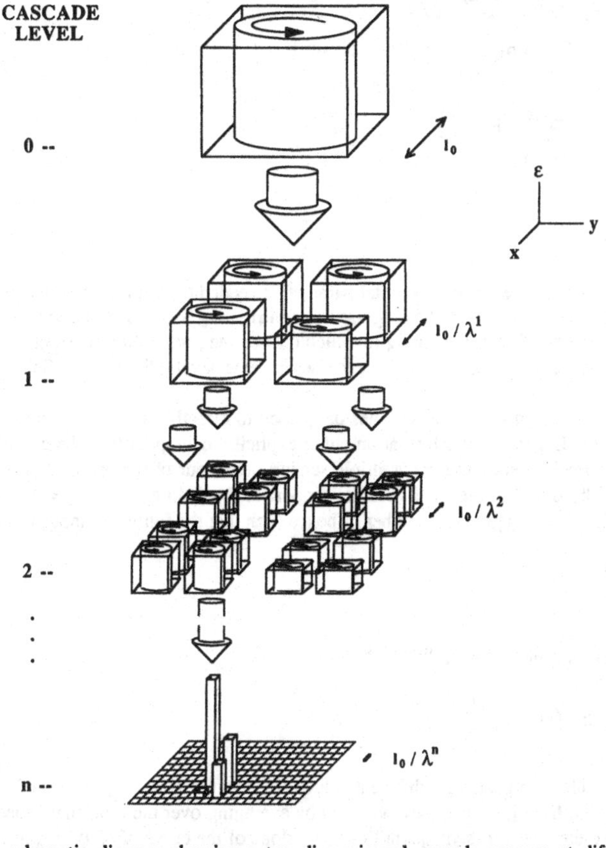

Fig. 1 : A schematic diagram showing a two-dimensionnal cascade process at different levels of its construction to smaller scales. Each eddy breaks up into four sub-eddies, transferring a part or all its energy flux.

The process could be repeated, and after n iterations (or n cascade steps) we obtain sub-eddies with length of homogeneity $l_n = l_0/\lambda^n$ and energy flux density ε_n:

$$\varepsilon_n = \mu\varepsilon_1\, \mu\varepsilon_2\, \mu\varepsilon_3 \ldots \mu\varepsilon_n\, \varepsilon_0$$

when $\mu\varepsilon_i$ is the fraction of energy flux density to sub-eddy of ith cascade step. The last quantity is a typical "bare" quantity as long as n is finite, conversely "dressed" quantity are obtained by integrating a completely developed cascade processes, experimental data are then best approximated by the second one; see Schertzer and Lovejoy (1987a, b). As the cascade proceeds to smaller scale, high values of ε_n appears, concentrated on smallest volume l_n^d, which is habitually recognized as a basic characteristic of intermittency. In the limit of the cascade scale going to zero ($l_n \rightarrow 0$), the energy flux densities ε_n take singular values of all orders. Such behaviour is characteristic of multiple scaling, and could be expressed by relation (2.1).

For a generalization of discrete cascade processes to the continuous cascade process and its applications see Wilson et al. (this volume). In particular Schertzer and Lovejoy (1987a, b) shows that in this case the codimension functions belongs to universal classes:

$$c(\gamma) = \begin{cases} C_1\left(\frac{\gamma}{C_1\alpha'} + \frac{1}{\alpha}\right)^{\alpha'} & \alpha \neq 1 \\ C_1 \exp\left(\frac{\gamma}{C_1} - 1\right) & \alpha = 1 \end{cases} \tag{2.5}$$

$$K(h) = \begin{cases} \frac{C_1\alpha'}{\alpha}(h^{\alpha} - h) & \alpha \neq 1 \\ C_1\, h\, \mathrm{Log}(h) & \alpha = 1 \end{cases} \tag{2.6}$$

$$\frac{1}{\alpha} + \frac{1}{\alpha'} = 1$$

where C_1 and α ($0 \leq \alpha \leq 2$) are the fundamental parameters needed to characterize the processes. The first one, C_1 is the fractal codimension of the singularities contributing to the average values of the field, and the second one, α the Lévy index indicating to which classes the probability distribution belongs. These fundamental properties of multifractal cascade processes are extensively discussed in Schertzer and Lovejoy (this volume).

The PDMS is a new multifractal techniques developed to directly estimate the codimension function $c(\gamma)$, without using the Legendre transformation either explicitly or implicitly. The codimension function is essentially determined by studying the multiple scaling behaviour of the probability distribution . The last one is obtained by dividing the total volume, support of the field ε_λ, into $N_\lambda = \lambda^d$ disjoint boxes of volume λ^{-d}, then defining $N_\lambda(\gamma)$ as the number of boxes such that the following inequality is verified:

$$\frac{\log \varepsilon_\lambda}{\log \lambda} \geq \gamma$$

then equation (2.2) is approximated by this relation:

$$\Pr(\varepsilon_\lambda \geq \lambda^{\gamma}) \propto \frac{N_\lambda(\gamma)}{N_\lambda}$$

The operation could be repeated for different values of γ, and for decreasing values of scale ratio λ. The corresponding rescaled field intensities are obtained by averaging over the field intensities ε_λ at the finest resolution; and are therefore "dressed" quantities. The slope of the curve of $N_\lambda(\gamma)/N_\lambda$ as a function of the scale ratio λ on a log-log graph, for a given value of γ, will give the corresponding codimension function $c(\gamma)$. This procedure bypassed the problem of the correct but non trivial normalization of eq. (2.2).

Using this techniques, we estimate the codimension function of unidimensional ($d = 1$) simulated fields generated with the help of discrete log-normal ($\alpha = 2$) multiplicative cascade process. The field has been simulated over 10 cascade steps ($n = 10$), i.e. the scaling ratio between the largest and smallest length is 2^{10}.

The eq. (2.5) with $\alpha = 2$ gives the theoretical expression of the codimension function:

$$c(\gamma) = \frac{C_1}{4}\left(\frac{\gamma}{C_1} + 1\right)^2 \tag{2.7}$$

The free parameter C_1, $0 \leq C_1 \leq d$[1], is chosen to equal to 0.125 and the number of independent samples N, used to estimate to codimension function, is equal to 10 in the first example and to 1000 in the second. The curves of the $\log(N_\lambda(\gamma)/N_\lambda)$ as function of $\log(\lambda)$ (λ decreasing by a factor of 1/2, from 1024 to 2) are given in fig. 2 for different values of γ. The straightness of the curves indicates that the scaling is accurately respected.

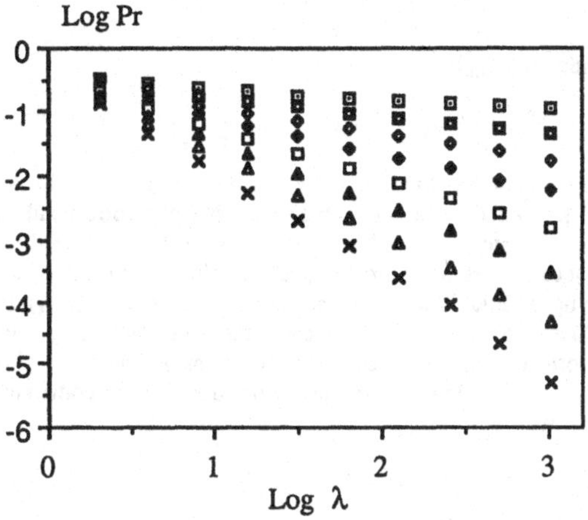

Fig. 2 : Illustration of the scaling behaviour of the probability distribution of 1000 independent samples of density fields induced by log-normal ($\alpha = 2$) multiplicative cascade process over a scale range $2^{10} = 1024$. From bottom to top, each curve corresponds to an order of singularity γ going from 0.1 to 0.8, in increments of 0.1. Note that the scaling is accurately followed.

In fig. 3, the curves of the estimated codimension functions, with $N = 10$ (black symbols) and $N = 1000$ (white symbols), are compared to the theoretical functions (continuous curves) given by eq. (2.7). They are in good agreement up to a given value of γ, which obviously increases with the samples size.

This behaviour could be understood if we recall that in the canonical case, a finite number of samples, each of finite scale ratio, will impose an upper bound on the highest values of the order of singularities γ that could be generated by the multiplicative cascade processes, and conversely the accessible values of γ will be limited by the finite sample size used to evaluated it. This implies that the estimate of the frequency of occurrence of the largest value of γ, from which we estimate the codimension function, could not be performed on only one sample, as they will only appear in some samples events. Obviously, increasing the number of samples will increase the probability of their observation. The codimension function $c(\gamma)$ is then a measure of the fraction of the space, formed by the total number of samples of dimension d (corresponding.roughly to the hyper space of the probability distribution)), occupied by the singularities of order equal or superior to γ.

---

1 The case $C_1 > d$ yields degenerate cascades discussed by Schertzer and Lovejoy (1987 a, b).

Following this and noticing that $c(\gamma)$ is an increasing function for positive values of γ (indicating that the largest singularities are the rarest) one will find that the maximum value of γ, denoted γ_{max}, observed at least once in the N independent samples of volume d (with λ^d sub-boxes on each sample) is approximated, within a normalization constant, by:

$$N \lambda^d \lambda^{-c(\gamma_{max})} \sim 1 \qquad (2.8)$$

Introducing the definition of the dimension of sampling D_S for N samples:

$$\lambda^{D_S} = N$$

$$D_S = \frac{\log N}{\log \lambda}$$

Using eq. (2.8) we obtain the following relation for γ_{max}:

$$c(\gamma_{max}) \approx d + D_S \qquad (2.9)$$

The last equation shows that the larger the sampling dimension D_S – or the dimension d – the larger will be the spectrum of accessible values of γ. Sometimes it is possible to evaluate $c(\gamma)$ corresponding to γ greater than γ_{max}, but they are then poor estimates. The values of $c(\gamma_{max})$ can be considered as an indication of the maximum values of $c(\gamma)$ that can be estimated from a given sample size and resolution λ.

Going back to fig. 3, the lower and the upper dotted lines corresponds to the values of $c(\gamma_{max})$, for a number of samples N = 10 and N = 1000 respectively. In both case, the value of $c(\gamma_{max})$ is an appropriate upper limits for the maximum correct values of the estimated codimension functions. Notice also in fig. 3, that increasing the number of samples improve the precision to which the codimension function is determined even at small values of γ.

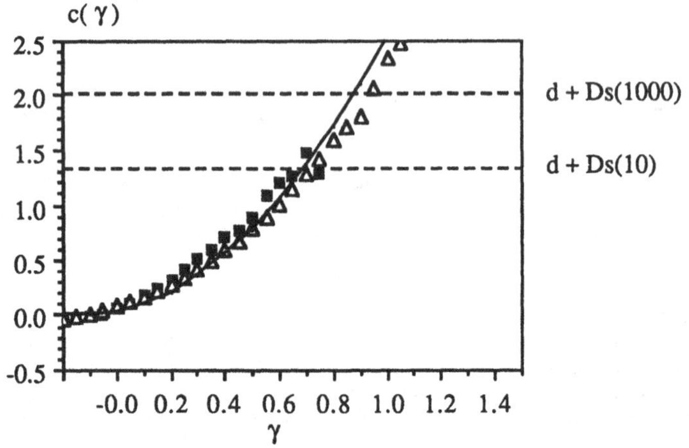

Fig. 3 : The curves of the estimated $c(\gamma)$, using the PDMS technique, with 1000 independent samples (white symbols) and 10 independent samples (black symbols) compared to their theoretical values given by the continuous curve. The horizontal dashed lines, indicate the estimated upper limits of $c(\gamma)$ imposed by the finite values of the dimension of sampling D_s. In both cases, the estimated $c(\gamma)$ is in good agreement with the theoretical curve when $c(\gamma) < d+D_s$.

3. UPPER LIMITS OF THE ESTIMATED CHARACTERISTIC FUNCTION K(h)

We have seen at the preceding section that a finite number of samples restricts the breath spectrum of the accessible γ, and hence of $c(\gamma)$, to those smaller than the maximum values γ_{max} and $c(\gamma_{max})$ respectively. Now we will see its consequences on the determination of statistical moments, in particular for the trace-moments, and how it will affect the behaviour of the codimension function C(h) associated to them. Following Schertzer and Lovejoy (1989) we define the hth trace moment of the energy flux densities over a set A of dimension d by these relations:

$$Tr_{A_\lambda}\, \varepsilon_\lambda{}^h = < \int_{A_\lambda} \varepsilon_\lambda{}^h\, d^{hd}x > \propto \lambda^{K_D(h)} \tag{3.1}$$

with:

$$K_D(h) = (h-1)\,[C(h) - d] \tag{3.2}$$

and approximated for discrete fields by this summation:

$$Tr_{A_\lambda}\, \varepsilon_\lambda{}^h \approx < \sum_{A_\lambda} \varepsilon_\lambda{}^h\, \lambda^{-hd} > \tag{3.3}$$

Using this we estimate the trace moments of fields, generated the by log-normal ($\alpha = 2$) multiplicative cascade process already discuss in section 2. The statistical ensemble averages are replaced by sums over N independent samples of the fields.

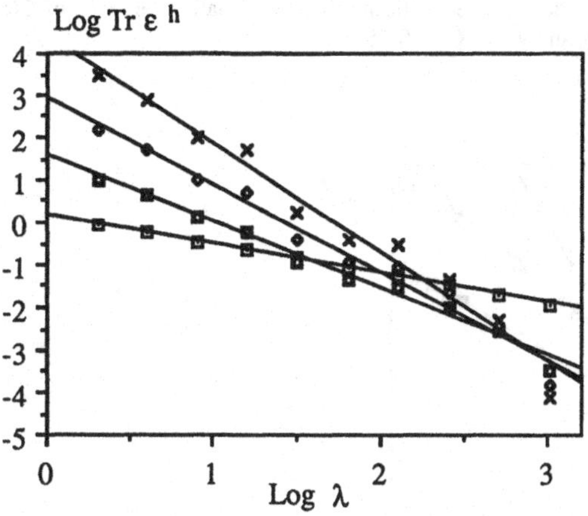

Fig. 4 : The estimated h^{th} trace moments of 1000 independent samples of density fields induced by log-normal ($\alpha = 2$) multiplicative cascade process with scale range equal to 2^{10}. The curve correspond at h = 2, 5, 7 and 9 going from the bottom to the top. Here also the scaling is accurately followed.

In fig. 4, the curves of the log ($Tr_{A\lambda}\, \varepsilon_\lambda{}^h$) against $\log(\lambda)$ are illustrated for several values of h (for a number of sample N = 1000); here also the scaling is well respected for all the range of λ (from 1 to 1024). The slopes of these curves gives the estimated codimension functions, with N = 10 (black symbols) and N = 1000 (white symbols), are compared in fig. 5 to the theoretical curve (continuous line) of K(h), obtained with eq. (2.6) and $\alpha = 2$. We see that for small h the three curves are in good agreement, but for large h the estimated values of K(h) no longer correspond to the theoretical parabola, but are asymptotically straight. This behaviour could be understood as a failure in the Legendre transformation, given by eq. (2.4), essentially caused by the finite sample size. The Legendre transformation implies a dual correspondence between the orders of moments h and the orders of singularities γ (given by the following relation $h = c'(\gamma)$ and $\gamma = K'(h)$). So as long as γ is smaller then γ_{max}, the corresponding h and K(h) are in agreement with the theoretical curve given by the Legendre transformation. The finite number of sample implies a γ_{max}, but we have no restriction of this kind on the order of the moments h, that could be increase indefinitely (in practice its value stay finite), corresponding to γ greater then γ_{max}, and as for $c(\gamma)$ the estimated K(h) is no longer in agreement with the codimension function of the system. It is clearly not possible to estimate correctly the codimension function of singularities never encountered.

The asymptotic linear behaviour of K(h) could then be approximated by the following relation:

$$K(h) \approx h\,\gamma_{max} - c(\gamma_{max}) \tag{3.4}$$

(This relation could be deduced from eq. (2.3), using the asymptotic expansion for general Laplace integrals when the maximum values of the exponent $(h\,\gamma - c(\gamma))$ lies at one of the endpoints of the integral and its derivative is nò longer equal to zero; see, e.g. Bender and Orszag (1978).) The asymptotic slopes and the y axis intercept, estimated for the 5 last points K(h) in fig. 5, are equal to 0.54 and 1.01 with sample size N = 10, and to 0.74 and of 1.24 when N = 1000. Using PDMS to determine the $c(\gamma)$ of the same fields, one found that $c(0.55) \approx 1.1$ when N = 10 and $c(0.75) = 1.3$ when N = 1000. So the values given by the estimation of the asymptotic slopes of K(h) are in good agreement with that one obtained with PDMS. However the values of γ_{max} observed for both examples are respectively 0.75 and 0.95 for N equal to 10 and 1000, and are slightly greater than the values given by the asymptotic slopes of K(h). But – as already pointed out – these highest values of the order of singularity are extremely rare events and their probability of occurrence too low to contribute significantly to the statistical moments, which is why the values of γ_{max} in eq. (3.4) will be smaller than the one observed with the PDMS. Finally, the slope of C(h) is evaluated for values of h between 0.1 and 2, and yields $C_1 \approx 0.15$.

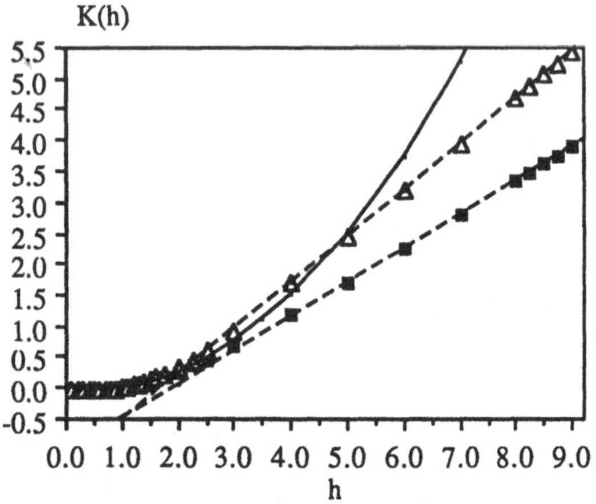

Fig. 5 : The curves of the estimated moment scaling function K(h) for 1000 independent samples (white symbols) and 10 independent samples (black symbols) compared to theirs theoretical values given by the continuous curve. The dashed lines are the asymptotic slopes of K(h) evaluated for the five last values of h.

This underestimation of the codimension function should not be confused with that resulting from the divergence of the trace moments, which leads to spurious scaling. Since C(h) is a monotonically increasing function, the scaling exponents $K_D(h)$ of the trace moments given by eq. (3.2) have two zero: h = 1 implies the scale independence of the averaged fields, and $h = h_d > 1$, defined by the following equality:

$$C(h_d) = d \tag{3.5}$$

So when h is greater than this critical value, the scaling exponent $K_D(h)$ is positive, and in the limit of $\lambda \to \infty$ the trace moment, given by eq. (3.1), will diverge. The divergence of trace moments and its interpretation as a Hausdorff measure was first pointed out in Schertzer and Lovejoy (1987 a,b). However statistical moments of the "bare" quantities always remains finite. But for the dressed fields, resulting from the integration of a completely developed multiplicative cascade processes (i.e. in the limit $\lambda \to \infty$), the divergence of moments implies a break down in the law of large numbers, and then the usual procedure of estimating their statistical properties fails for $h > h_d$. This divergence occurs because the dimension d of the volume over which the field is integrated is not large enough to smooth out rare high order singularities, and rescaling the field by averaging it at larger scales will not "kill" these singularities, or attenuated them enough, since the dimension d is not affected by this procedure. (The behaviour of the statistical moments of dressed fields for some special cases of discrete multiplicative cascade processes and h integer is derived in Lavallée (1990).)

As soon as the number of samples is large enough so that h_{max} (i.e. the order of moments corresponding, by the Legendre transformation to γ_{max}) is greater than h_d, it implies that rare high singularities are encountered, and spurious scaling will be observed. However it is important to notice how to distinguish the spurious scaling that underestimation of K(h) due to the under sampling. In the latter case, the estimated $K_D(h)$ becomes more and more negative, as can be seen on the graph of the $\log(Tr_{A_\lambda}\, \varepsilon_\lambda{}^h)$ versus $\log\lambda$ in fig. 4 where their slopes decreases to lower and lower negative values, as h increase. On the contrary spurious scaling will be observed when the slope of $\log(Tr_{A_\lambda}\, \varepsilon_\lambda{}^h)$ versus $\log\lambda$ goes to zero for h > 1, that implies $K_D(h)$ became less negative as h goes to h_d.

The following relation for h_{max}:

$$h_{max} = \left(\frac{d + D_s}{C_1}\right)^{1/2} \tag{3.6}$$

is obtained for $\alpha = 2$, using eq. (2.9) and the Legendre transformation.

Using discrete log-normal ($\alpha = 2$) multiplicative cascade process with 10 cascade steps and $C_1 = 0.72$, we found $h_d = 1.388$ using eq. (2.7) to solve eq. (3.5) with d = 1, and the number of samples N = 2000, is chosen so that $h_{max} = 1.7$ is greater than h_d. In the preceding example, with $C_1 = 0.125$, and for N = 1000, $h_{max} = 4$ which is smaller than $h_d = 8$. We then use eq. (3.3) to estimate the trace moments of the fields at different scale ratio, and as usual to estimate the scaling exponents by their slopes. The estimated codimension function C(h) is compared to its theoretical values (continuous curve) in fig. 6. For large values of h, the estimates stay bounded and are therefore poor estimator of the theoretical C(h) which grows linearly with h. This behaviour is recognized as the spurious scaling. For small values of h, C(h) is in agreement with the theoretical curve, the slope of C(h) evaluated for values of h between 0.1 and 0.9, and gives C_1 equal 0.75.

So to increase the range of values of h for which one can expect to correctly determine the scaling exponents, we must first increase N, until $h_{max} > h_d$, and then measure (or project) the fields on spaces of increasing dimension d. In practice this means that experimental data will have to be taken with instruments of the largest dimension possible (e.g. d = 4 for isotropic processes varying in space and time). Finally we would like to point out that while the spurious scaling can only be observed on "dressed" fields, the finite number of samples will modified the behaviour of both the "bare" and of the "dressed" fields and can be observed on either. One can find other examples of the application of PDMS as well as trace moments to geophysical fields in Lovejoy and Schertzer (this volume).

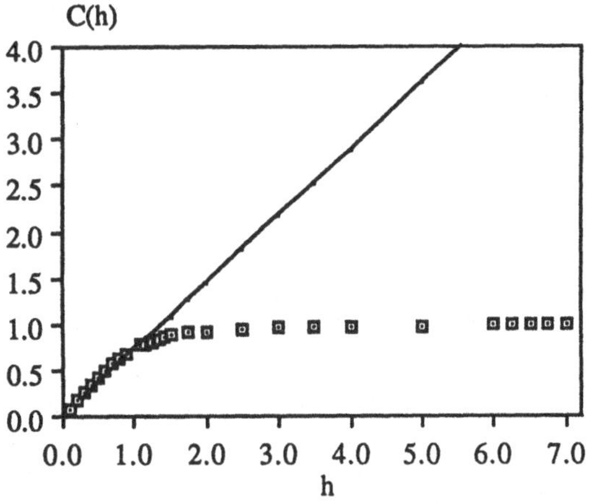

Fig. 6 : Illustration of the consequences of spurious scaling of the "dressed" quantities for the estimated C(h), that stay bounded for large values of h, in disagreement with the linear behaviour of the theoretical C(h) given by the continuous curve. The estimated C(h) is obtained by trace moment analysis of 2000 independent samples of density fields induced by log-normal ($\alpha = 2$) multiplicative cascade process, the scale ratio $l = 2^{10}$.

4. CONCLUSIONS

The results obtained in this paper underline the importance of phenomenological stochastic models - in particular multiplicative cascade processes - not only in understanding turbulence, but also in verifying the relevance and limits of conceptual tools developed for analyzing their scaling properties. We introduce a new method to directly estimate the codimension function ($c(\gamma)$) which is directly related to the probability distribution. The PDMS method avoids using Legendre transformations and their implicit assumption of the converge of all statistical moments. Since we generally expect the observed "dressed" moments to diverge for sufficiently high orders, this method is a significant improvement: it is insensitive to spurious scaling. We also discuss how a finite number of samples reduces the range of accessible γ, and how to characterize it with the dimension of sampling D_S. Its effects on K(h), the Legendre transform of $c(\gamma)$, are also studied in parallel with spurious scaling. This clarifies their differences and shows how the asymptotic behaviour of the real scaling exponents are modified.

We obtained specific estimates of the maximum order of singularities that can be evaluated with a given sample size and resolution. These limits give precise information about the reliability of exponent estimates and suggests that many $K_D(h)$ (or equivalently $\tau(q)$) values cited in the literature should be critically re-examined, especially in the (common) case where straight-line behaviour is observed at large h. This asymptotic linearity could also introduce a discontinuity, or a sudden change in the derivative of $K_D(h)$. Since there is a formal analogy between thermodynamics and "flux dynamics", it has often been claimed on the basis of observed discontinuities that phase transitions occur. We discuss several mechanisms by which discontinuities will arise even when no "phase transition" is present. This suggests that the analogy with thermodynamics may be largely formal in nature.

5. ACKNOWLEDGMENTS

We acknowledge fruitful discussions with A. Davis, P. Gabriel, J.P. Kahane, P. Ladoy, E. Levich, A. Saucier, G. Sarma, G. Sèze, Y. Tessier, R. Viswanathan, and J. Wilson.

6. REFERENCES

Bialas, A., R. Peschanski, 1986: Moments of rapidity distributions as a measure of short-range fluctuations in high-energy collisions. Nucl. Phys. B, B273, 703-718.

Bender, C.M., S.A. Orszag, 1978: Advanced mathematical methods for scientists and engineers. McGraw-Hill, New York, NY.

Frisch,U., G. Parisi, 1985: A multifractal model of intermittency. in Turbulence and predictability in geophysical fluid dynamics and climate dynamics, Eds. Ghil, Benzi, Parisi, North-Holland, 84-88.

Gabriel, P., S. Lovejoy, D. Schertzer, and G.L. Austin, 1988: Multifractal analysis of resolution dependence in satellite in satellite imagery. Geophys. Res. Lett., 15, 1373-1376.

Grassberger, P., 1983: Generalized dimensions of strange attractors. Phys. Lett., A 97, 227.

Halsey, T.C., M.H. Jensen, L.P. Kadanoff, I. Procaccia, B. Shraiman, 1986: Fractal measures and their singularities: the characterization of strange sets. Phys. Rev., A 33, 1141-1151.

Hentschel, H.G.E., I. Proccacia, 1983: The infinite number of generalized dimensions of fractals and strange attractors. Physica, 8D, 435-444.

Hubert, P., J. P. Carbonnel, 1988: Caractérisation fractale de la variabilité et de l'anisotropie des précitations tropicales. C. R. Acad. Sci. Paris, 2-307, 909-914.

Kolmogorov, A. N., 1962: A refinement of previous hypothesis concerning the local structure of turbulence in viscous incompressible fluid at high Reynolds number. J. Fluid Mech., 13, 82-85

Lavallée, D., 1990: Ph. D. Thesis, University McGill, Montréal, Canada.

Lovejoy, S., D. Schertzer, A.A. Tsonis, 1987a: Functional box-counting and multiple elliptical dimensions in rain. Science, 235, 1036-1038.

Lovejoy, S., D. Schertzer, 1990: Multifractal analysis techniques and the rain and cloud fields from 10^{-3} to 10^{-6}m. (this volume).

Levitch, E., I. Shtilman, 1990: Helicity fluctuations and coherence in developed turbulence. (this volume).

Mandelbrot, B., 1974: Intermittent turbulence in self-similar cascades: Divergence of high moments and dimension of the carrier. J. Fluid Mech., 62, 331-350.

Obukhov, A., 1962: Some specific features of atmospheric turbulence. J. Geophys. Res, 67, 3011-3014

Meneveau, C., K. R. Sreenivasan, 1987: Simple multifractal cascade model for fully developed turbulence. Phys. Rev. Lett., 59(13), 1424-1427.

Schertzer, D., S. Lovejoy, 1983: On the dimension of atmospheric motions. Preprint Vol., IUTAM Symp. on Turbulence and Chaotic phenomena in Fluids, Kyoto, Japan, IUTAM, 141-144.

Schertzer, D., S. Lovejoy, 1984: On the dimension of atmospheric motions. Turbulence and chaotic phenomena in fluids, Ed. Tatsumi, Elsevier North-Holland, New York, .505-508.

Schertzer, D., S. Lovejoy, 1987a: Physical modelling and analysis of rain and clouds by anisotropic, scaling multiplicative processes. J. Geophys. Res., 92(D8), 9693-9714.

Schertzer, D., S. Lovejoy, 1987b: Singularités anisotropes, divergences des moments en turbulence: invariance d'échelle généralisé et processus multiplicatifs. Annales Math. du Qué., 11, 139-181

Schertzer and Lovejoy 1988: Multifractal simulation and analysis of clouds by multiplicative process. Atmospheric Research, 21, 337-361

Schertzer, D., S. Lovejoy, 1990a: Scaling nonlinear variability in geodynamics: Multiple singularities, observables and universality classes. (this volume).

Schertzer, D., S. Lovejoy, 1990b: Nonlinear variability in geophysics: multifractal simulations and analysis. in Fractals: physical origins and properties, edited by L. Pietronero, Plenum, New York, NY, pp.49-79.

Wilson, J., D. Schertzer, S. Lovejoy, 1990: Physically based cloud modelling by multiplicative cascade processes. (this volume).

MULTIFRACTAL ANALYSIS TECHNIQUES AND THE RAIN AND CLOUD FIELDS FROM 10^{-3} TO 10^6m

Shaun Lovejoy, Daniel Schertzer*
Department of Physics, McGill University,
3600 University st., Montreal, Qué., H3A 2T8,
CANADA

ABSTRACT. We discuss the scaling properties of the rain and cloud fields over the range of ≈1mm to ≈1000km. We find that these fields are multifractal; i.e. the weak and intense regions scale differently, involving multiple fractal dimensions. We re-evaluate several early (mono-dimensional) analyses and argue that failure to account for the multifractal nature of the fields has in several instances lead to spurious breaks in the scaling symmetry. A related result is that area-perimeter exponents no longer yield the dimension of the perimeters, and that the interpretation of the distribution of isolated cloud or rain areas must be modified.

Empirically, we use two recent multifractal analysis techniques: Probability Distribution/Multiple Scaling (PDMS) and Trace Moments to analyse blotting paper traces of rain drop impacts (≈1mm to 128cm), lidar reflectivities from raindrops (≈3m to 540m), and satellite data from both 100m to 100km and 8km to 256km. The conclusions support the idea that the atmosphere has a multifractal structure over this large fraction of the meteorologically significant length scales.

Applications of the results discussed here include resolution independent methods of remote sensing, anisotropic space/time transformations (generalizations of Taylor's hypothesis of frozen turbulence) which we show holds in rain, as well as a method of correcting radar for (mono)fractal effects not included in the standard theory of radar measurements of rain.

1. INTRODUCTION

1.1. The need for systematic study of the scaling properties of atmospheric fields

In spite of its obvious importance for understanding and predicting the atmosphere, no systematic study of scale dependence of atmospheric fields has yet been undertaken. The development of new in situ and remote measurement techniques coupled with rapid advances in computing power now provide the impressive sources of data needed to attempt such a study. What is perhaps most important of all is the equally rapid series of advances in our understanding of non-linear dynamical systems, scaling and (multi)fractals. In particular, multifractal measures are much more relevant in geophysical applications than fractal sets since geophysical quantities are best described as measures, with empirical data being functional approximations[1] to the latter which depend on the resolution of the sensor. Multifractal measures are characterized by their scale-invariant (co)-dimension function which is an exponent function describing the variation of the probability distributions with scale[2]. The geometry of sets and their associated fractal dimensions are secondary; the scale invariant dynamics (characterized by the generator of the measure) play the primary role.

*EERM/CRMD, Météorologie Nationale, 2 Ave. Rapp, Paris 75007, FRANCE.

[1] Since the underlying multifractal measures are singular, the measurements (which are typically spatial and/or temporal integrals) do not "approximate" the measurements in a simple way.

[2] Since the codimension is related to the lof of the probability distribution, it is formally analogous to the entropy - see Schertzer and Lovejoy (1990a) for more on this "flux dynamics".

D. Schertzer and S. Lovejoy (eds.), Non-Linear Variability in Geophysics, 111–144.

In this paper we describe two fairly new multifractal analysis methods (the "Probability Distribution/Multiple Scaling" technique and the "Trace Moment" technique), applying these methods to satellite radiance fields, as well as lidar and blotting paper measurements of rain drops overall, spanning the range of scales of ≈1mm to ≈1000km. Finally, in appendices, we re-examine some conventional (mono-dimensional) analysis techniques (such as area-perimeter relations), and show how they can be fit into a multifractal framework.

The need for new data analysis techniques can be appreciated by considering that virtually the only commonly used geophysical data analysis technique that enables one to directly compare the small and large scale statistical properties of fields is Fourier analysis[1]. In spite of its obvious importance, it should be recalled that the resulting energy spectra are only second order statistics[2], and are not particularly robust (i.e. when applied to highly intermittent data, large samples may be needed to obtain good estimates of the ensemble averaged spectra). If the fields were mono-dimensional fractals, then the scaling of the second order moments (characterized by the spectral exponents) would provide nearly complete information about the scaling properties of the field. However, this is generally not the case; multifractal techniques are required to allow for systematic study of the scaling of moments of all orders (or equivalently of weak and strong fluctuations separately), over wide ranges of scale.

In a series of papers - Lovejoy (1981, 1982), Lovejoy and Schertzer (1983, 1985, 1986, 1990a), Lovejoy and Mandelbrot (1985), Schertzer and Lovejoy (1983, 1984, 1985, 1986, 1987a,b, 1988, 1989, 1990, Lovejoy et al. (1987), and Schertzer and Lovejoy (this volume) -, we have argued that we may expect the atmosphere to exhibit scaling, fractal structures. Although we do not wish to repeat these arguments in detail, the basic idea may be simply expressed. If we consider scaling as a symmetry principle (i.e. the system is unchanged under certain scale changing operations), then in the spirit of modern physics, we may tentatively assume (a first approximation), that the symmetry is respected except for symmetry breaking mechanisms.

In the atmosphere, the scaling symmetry is obviously broken at one extreme by the finite size of the earth, and at the other extreme, at scales of ≈1mm, by damping due to viscosity. Atmospheric boundary conditions such as topography (see Lovejoy and Schertzer (1990a) for an analysis) are also multiply scaling are are not expected to introduce a characteristic length to break the symmetry. This leaves a wide range of factor 10^9-10^{10} in scale where scaling symmetries might hold. In this context, it is worth recalling that it has only been in the last few years that the full generality of the scale invariant symmetry principle has been realized. This has enabled us to go far beyond the qualitative (and very restrictive) ideas of fractals as self-similar geometric objects with a single fractal dimension. It can now be quantitatively understood as a system composed of two totally distinct elements unified by the formalism of Generalized Scale Invariance (see Schertzer and Lovejoy (1983, 1985, 1986), and especially in (1987a, b)). The first element is a scale changing operator T_λ which reduces scales by the factor $\lambda \geq 1$; scaling implies $T_{\lambda_1\lambda_2} = T_{\lambda_1}T_{\lambda_2}$ and hence that T_λ has the form $T_\lambda = \lambda^{-G}$ where G the generator of the (semi-group) of scale changing operators (=1 for self-similar, isotropic systems). A power law form of this type ensures that the small and large scales are related only by (dimensionless) scale ratios; hence that over this range, the system has no characteristic size. An immediate consequence is that the operation required to go from one scale to another can be far more complex than simple geometrical magnifications (e.g. self-similarity), hence the rejection of self-similarity does not imply a rejection of scaling. The standard meteorological argument that since the large scale is "apparently" two dimensional due to the stratification caused by gravity, and the small scale is "apparently" three dimensional, is therefore not relevant to the issue of scaling. The second element of a scale invariant system is a multifractal measure that is invariant under the application of T_λ. In turbulence, the basic scale invariant multifractal measure is presumably the energy flux density, since this quantity is exactly conserved by the non-linear terms in the Navier-Stokes equation. In the atmosphere, other scale invariant multifractal measures will be necessary to account for other conserved quantities.

---

[1] Although formally autocorrelation functions contain the same information, they are usually only used to obtain the 1/e point, which when combined with the ad hoc assumption of exponential decorrelations (and hence the absence of scaling), is used to obtain the decorrelation length. When there is scaling the autocorrelations decay algebraically rather than exponentially and the decorrelation length is no longer the 1/e point, but is rather the outer scale of the scaling regime which in geophysics can be many orders of magnitude larger.

[2] They are simply related to the Fourier transform of the autocorrelation function.

GSI allows for so much generality, that not only can the stratification of the atmosphere be easily accounted for in a single scaling regime (intermediate between two and three dimensional turbulence), but also, differential rotation (associated for example with cloud "texture") and even variable Coriolis parameters can be dealt with. It also provides a natural framework for analysing and modelling the space/time strucutre of various fields and hence for investigating the issues of prediction and predictability[1]. We are now faced with the task of restricting our attention to the sub-classes of scaling appropriate to the atmosphere (for example by restricting the class of generators G in some way). Ultimately, such restrictions will have to be derived theoretically from dynamical principles. Unfortunately, the relation between the scale invariant symmetry and the non-linear atmospheric dynamics is far more difficult to discern than in other areas of physics (particularly quantum mechanics) where due to the linearity of the equations, the dynamics and symmetries are synonymous. We therefore give special attention to the empirical characterization of both scaling and symmetry breaking.

1.2. Developing Resolution-independent measurement techniques

Below, we restrict our attention to studies of the rain and associated cloud fields. From our perspective - aside from their intrinsic interest - the rain reflectivity, cloud radiance and lidar[2] reflectivity fields have the advantage of being among the best measured meteorological fields: radar, satellites and lidar all provide excellent remotely sensed data spanning wide ranges of time and space scales. High quality in situ and aircraft data are also available. Naturally, a full statistical description of either field requires knowledge of properties of the measures defined by the drop volumes (V_i) and their distribution in space ($\underline{r}_i$), whereas the above techniques do not measure these quantities directly. Some reasonably direct information is available from both the lidar and blotting paper methods discussed in sections 3,4; however, for investigating the large scales ($\approx$1-10^3km) we rely on the indirect information supplied by radar and satellite data. For example, the radar measures the ("effective") reflectivity[3] (Z_e) of the (n) drops in the (microwave) pulse volume:

$$Z_e \propto \left| \sum_i^n V_i e^{i\underline{k}\cdot\underline{r}_i} \right|^2 \qquad (1.1)$$

(where $\underline{k}$ is the radar wave vector), whereas, the (volume averaged) rainrate (R) is a different measure:

$$R \propto \sum_i^n w_i V_i \qquad (1.2)$$

where w_i is vertical velocity of the ith drop. At visible and infra red wavelengths, the relationship between the rain/cloud measures and the reflected/emitted radiation fields is even more indirect[4], however such measurements still give us valuable information about the scaling symmetries of the rain field. For example, symmetry breaking in the latter will be evident in the former. More generally, the multiple scaling of these fields implies that the values of sensor averages depend critically on the sensor resolution, hence for example, if remotely sensed data are to be used for estimating rain or cloud amounts, then the resolution dependencies of both the in situ (ground truth) data, and remotely sensed data must be systematically removed (e.g. by using the associated scale invariant codimension functions). Such a

---

[1] In geophysical flows, each distance scale has a corresponding time scale, hence scaling is expected in space-time. Section 4 gives empirical evidence in rain showing that time is "stratified" with respect to the horizontal space coordinate i.e. space-time transformation should be anisotropic, and appendix E gives a theoretical discussion.

[2] A lidar is the laser analogue of a radar.

[3] Normally, the cumbersome adjective "effective" is dropped and and the symbol Z is used rather than the more correct Z_e. See section 3. for more details.

[4] See Davis et al. (this volume) for more discussion of this radiative transfer problem.

resolution-independent approach to remote sensing will require both careful measurements of the statistical properties of the fields over wide ranges in scale, combined with stochastic modelling of the fields themselves (see e.g. Wilson et al this volume).

2. MULTIFRACTAL MEASURES, CODIMENSION FUNCTIONS AND THE PROBABILITY DISTRIBUTION / MULTIPLE SCALING (PDMS) TECHNIQUE

2.1. Discussion

Based on studies of certain fractal sets obtained either as purely geometric constructs, or associated with certain stochastic processes, Mandelbrot [1982] used these sets as models of the geometry of various natural systems. However, few natural systems are sets (they are usually best treated as fields or measures), and it soon became clear - Hentschel and Proccacia (1983); Grassberger (1983); Schertzer and Lovejoy (1983); Benzi et al. (1984), Frisch and Parisi (1985) - that such measures are fundamentally characterized not by a single dimension, but by a dimension function (sometimes called the "spectrum of singularities"). Furthermore, this dimension function is simply related to the probability distribution. In fractal sets, the concept of fractal dimension is important because it is invariant under transformations of scale. In fractal measures, the notions of scaling (or scale invariance) and the generator of the measure are more basic.

Geophysical systems typically have variability extending from a large "external" scale L down to very small scales η (often 1 mm or less) and are therefore usually observed (literally "measured") at scales (l) with scale ratio $l/\eta >> 1$. It is therefore natural to consider the underlying phenomenon as a fractal measure, and the empirically accessible measurements (e.g., satellite photos) as a series of associated functions (denoted $f_\lambda(r)$), whose properties will depend greatly on the averaging scale ratio $\lambda = L/l$ (e.g. l is the size of a pixel). Using the external scale to define scale ratios in this way, we find that qualitatively, the relationship of a series of higher and higher resolution images (i.e. f_λ as $\lambda \rightarrow \infty$) to the underlying multifractal measure is that as the resolution increases, the structures are increasingly sharply defined, are found to occupy a decreasing fraction of the image, while simultaneously increasing in value (e.g., brightening) to compensate. Since over our range of interest, there is no characteristic scale, this behaviour is algebraic and can be expressed as follows (algebraic relations are for the moment valid to within proportionality constants and log corrections; see below)

$$\Pr(f_\lambda > \lambda^\gamma) \approx \lambda^{-c(\gamma)} \tag{2.1}$$

where Pr means probability, γ is the order of singularity associated with the (nondimensionalized, see below) pixel value f_λ, and $c(\gamma)$ is the associated codimension (the dimension of the underlying space (d) minus the corresponding dimension $d(\gamma)$). Equation (2.1) is the general characterization of multifractal fields and is the generic result of multiplicative cascade processes (Schertzer and Lovejoy (1987a,b)). This equation shows that $c(\gamma)$ is directly related to the probability distribution. This fact will be used below as the basis for empirically estimating $c(\gamma)$. Qualitatively, γ is the resolution-independent characterization of the intensity of the feature with brightness f_λ, whereas, $c(\gamma)$ is the resolution independent characterization of the image fraction occupied by features with brightness f_λ.

For those who are familiar with multifractals, it is worth noting here that we have denoted the orders of singularities by the symbol γ because the atmospheric quantities of interest are modelled by densities of multifractal measures (such as f) and γ gives the orders of these singularities directly. In other systems such as phase space portraits of strange attractors (e.g., Halsey et al. (1986)), it is more usual to treat the singularities of the measures (rather than their densities) usually denoted by the symbol α; the relation between α and γ being $\gamma = d - \alpha$ where d is the dimension of space in which the process occurs. Furthermore, we use the codimension function $c(\gamma)$ rather than a dimension function since we are really interested in a family of measures each identical except for the dimension of the space in which it is embedded (in some applications it is even useful to take the latter as a fractal set, e.g. the global meterological measuring network), and the codimensions specify the probabilities independently of the latter. In contrast, in studying strange attractors, d is usually kept fixed and the dimension is denoted usually $f(\alpha)$. We therefore have[1] $f(\alpha)=d-c(d-\alpha)$.

We can now appreciate some of the difficulties encountered in many of the early studies, where multifractal phenomena were analyzed with methods originally designed for studying sets (e.g., area-perimeter relations, distribution of areas, dimensions of graphs, box counting: see appendices A and B). Even before the analysis begins, experimental measuring devices integrate the underlying measure scale ratio λ, converting it into a (spatially or temporally discretized) function. This function is then converted into a set with the same resolution, typically with the help of thresholds. Finally the geometric properties of the resulting set are characterized by (at most) a few exponents (e.g., dimensions, area-perimeter exponents) essentially by degrading the resolution of these sets. Although careful and systematic study of the properties of the sets as functions of scale and threshold (such as with "functional box counting" - Lovejoy et al., 1987; Gabriel et al., 1988) can be used to estimate $c(\gamma)$, such methods are indirect and are less satisfactory than other methods such as trace moments (Schertzer and Lovejoy, 1987a) or the PDMS method (Lavallée et al., this volume). For comparison, functional box-counting exploits (2.1) by transforming the function f_λ into an exceedance set (see Appendices A, B) and covering the latter with larger and larger boxes. The fraction of the scene covered by boxes of scale λ is the probability in (2.1). The method works by degrading the resolution of the exceedance sets, rather than of the measures themselves. The approach described below is more straightforward and statistically robust, since it is defined directly by the measures f_λ rather than via associated sets. In contrast, the use of what might be termed "monofractal" analysis techniques (i.e., techniques designed primarily for analyzing sets) can easily lead to seemingly contradictory results, and even to spurious breaks in the scaling (see appendices).

2.2. The Probability Distribution Multiple Scaling (PDMS) technique

We seek to directly apply (2.1) to determining the scale invariant codimension function $c(\gamma)$. We have already introduced the dimensionless scale ratio λ to nondimensionalize the scales, we must now discuss how to nondimensionalize f. This is conveniently done by using the large-scale average f_1, as a reference value for the measure since $\lambda=1$ implies $l=L$, the external scale of the image[2]. Using an overbar to denote the values of the function normalized in this way, we write: $\bar{f}_\lambda = f_\lambda/f_1$ Theoretically, the latter should really be the ensemble (i.e., climatological) average of the random process at scale ratio $\lambda=1$; the sample average being an approximation to the latter[3]. We therefore obtain

$$\Pr\left(\bar{f}_\lambda > \lambda^\gamma\right) = F\lambda^{-c(\gamma)} \quad (\lambda\geq 1) \tag{2.2}$$

where F is a prefactor which is only a function of γ and $\log\lambda$ (for example, if F contains a $(\log\lambda)^\Delta$ dependency, Δ is called a "sub-codimenison"). Taking logs and rearranging, we obtain

$$c(\gamma) = -\frac{\log \Pr\left((\log \bar{f}_\lambda)/(\log \lambda) > \gamma\right)}{\log \lambda} - \frac{\log F}{\log\lambda} \tag{2.3}$$

Hence, plotting the normalized log probability distribution ($-\log\Pr/\log\lambda$) against the normalized log intensity ($\log \bar{f}_\lambda/\log\lambda$) we obtain the resolution (λ) independent function $c(\gamma)$. To empirically test this multiple scaling behavior, we therefore take our empirical field and successively degrade it by averaging, obtaining a series of functions $f_{\lambda_1}(r)$, $f_{\lambda_2}(r)$, $f_{\lambda_n}(r)$, which with decreasing λ simulates the results of sensors with successively lower resolutions with $L/\eta>\lambda_s\geq\lambda_n>1$, with λ_s the ratio of external scale to the sensor scale (if we define L as the external scale of the image, then this is the largest ratio accessible from

---

[1] Do not confuse this function f or the value α (which we cite here purely for reference) with the quite different f, α used in this paper.

[2] In Lovejoy and Schertzer (1990c), a slightly different normalization was used.

[3] In two dimensions, $\bar{f}_1 = \frac{1}{L^2}\left\langle \int_L\int_L f\,dx \right\rangle$ where "< >" indicates ensemble averaging, L is the entire image size.

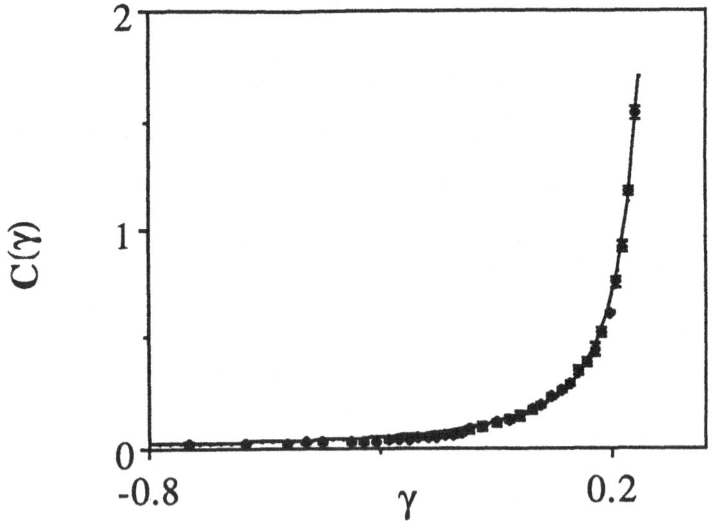

Fig. 1a: PDMS estimates of the function c(γ) from the five visible GOES images over 1024X1024 km at 8km resolution discussed in the text. The points indicate the mean c(γ) curve obtained by averaging the 6 individual c(γ) functions obtained at 8, 16, 32, 64, 128, 256 km scales (the histograms associated with 512, 1024 km did not have enough points, and were not used). The error bars indicate one standard deviation (and were of average magnitude ±0.011) showing that the c(γ) function was nearly scale invariant over this range. The solid line is the least mean squares fit to the universal form (eq. 2.4) (α'= -1.70, α=0.63). The standard error of the fit was ±0.011.

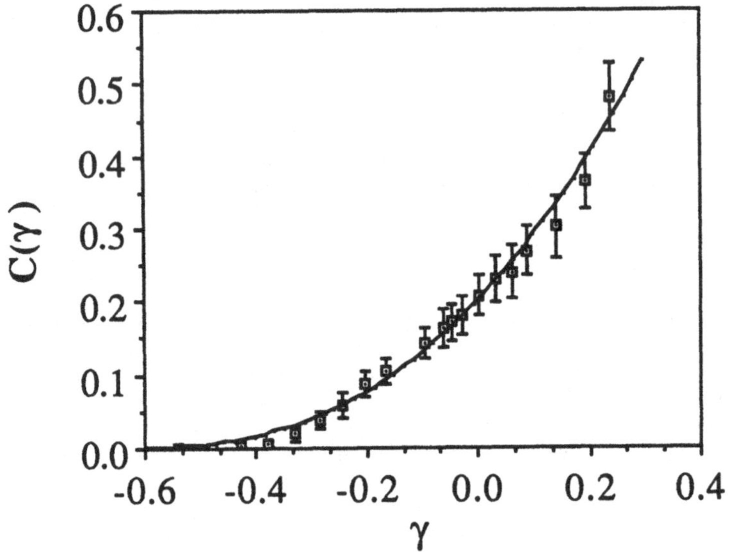

Fig. 1b: PDMS estimates of the function c(γ) from the five infra red GOES images discussed in the text with the same range of scales and format as fig. 2.1a. The mean standard error was ±0.023, and the best fit regression to eq. 2.4 yielded α'=-2.52, α=1.66. The standard error of the fit was ±0.015.

the data set). Successive factors of 2 can be easily implemented recursively[1]. Note that we must not nonlinearly transform our radiance field (e.g., by transforming from radiances to equivalent blackbody temperatures), since this does not simulate the result of a lower resolution sensor. Furthermore, the normalization based on f at scale ratio $\lambda=1$ implicitly assumes that the probability distribution in (2.2) is either from a single scene or from several independent scenes. If single scenes are used, then we cannot obtain information on the codimension for values of c>d(A), since the corresponding structures would have negative dimensions. However, when many realizations are available, the effective dimension of the sample can be larger, and higher values of $c(\gamma)$ can be determined; see Lavallée et al. (this volume) for discussion of this "sampling dimension." Finally, if many dependent samples are used as in the use of time series of images in A. Seed et al. (manuscript in preparation, 1990), then an "effective" external scale L can be determined from regression (as can f_1).

The Probability Distribution/Multiple Scaling (PDMS) technique refers to the direct exploitation of (2.2, 2.3) to obtain $c(\gamma)$. This direct method has a number of advantages when compared to the conventional route (via the moments K(h) followed by Legendre transformation; see for example Halsey et al. (1986)) not the least of which is that it avoids the problem of estimating high order moments which may in fact diverge. The PDMS method can be implemented in various ways. In Lavallée et al. (this volume), histograms of all the values of $f_\lambda/\bar{f}_1$ at the various resolutions λ were produced, taking for the value of $\bar{f}_1$ the mean of all the sample spatial averages at scale $\lambda=1$ (the number of "scenes" / satellite pictures, etc.). From the histogram, the largest to smallest values were summed to yield the probability distribution. Finally, $c(\gamma)$ was determined as the absolute slope of plots of log Pr against log λ for given values of γ. This method has the advantage of readily taking into account the slowly varying prefactor F, since Log F is simply the intercept at log $\lambda=0$. See Lavallée et al. (this volume) for a much more complete discussion of this method including theoretical considerations and numerical simulations.

In this paper, we used a slightly different method inspired by "functional box-counting" (Lovejoy et al. (1987)), in which the probability distributions at various scales λ was determined differently. The data at highest resolution (λ_s) was covered with a series of lower resolution grids (the "boxes") as explained above. However, rather than using the average value over each box (and create histograms of these averages), we used the maximum value in each box scale λ (denoted $\max_\lambda(f_{\lambda_s})$). Since the function was not averaged, the singularity corresponding to each maximum value was simply estimated as $\gamma=\log(\max_\lambda(f_{\lambda_s}))/\log\lambda_s$. The corresponding $c(\gamma)$ for each resolution λ was then estimated as $-\log\Pr/\log\lambda$ (we assumed $\log\lambda >> \log F$, ignoring the prefactor in (2.2)). Finally, we took the average $c(\gamma)$ function over a series of resolutions λ, indicating the scatter with one standard deviation error bars.

Figures 1*a*, 1*b* show the results when this technique is applied to five visible and five infrared GOES[2] pictures over Montreal, respectively. The original (raw) satellite pictures were first resampled on a regular 8 x 8 km grid over a region of 1024 x 1024 km. As can be seen, all the distributions are nearly coincident, in accord with the multifractal nature of the fields. To judge the closeness of the fits, we calculated the mean $c(\gamma)$ curves as well as the standard deviations for 8, 16, 32, 64, 128, and 256 km, finding that the variation is small, being typically about ±0.02 in (γ), which is more accurate than estimates obtained using functional box counting on similar data (Gabriel et al. (1988) found accuracies of about ±0.05).

2.3. Universality classes of $c(\gamma)$

We have already argued that the resolution independent codimension function $c(\gamma)$ is of considerably more interest than particular values of the function. Continuous cascade models, allow us to go even further (Schertzer and Lovejoy 1987a,b, this volume) since it can be shown that $c(\gamma)$ falls into the the following universality classes (e.g. functional forms):

---

[1] Note that we must not non-linearly transform our radiance field (e.g. by transforming from radiances to equivalent black body temperatures), since this does not simulate the result of a lower resolution sensor. Typically $\lambda_s >> 1$ since the sensor resolution is of the order of meters or more whereas the fields typically vary over distances of the order of millimeters.

[2] Geostationary Operational Environment Satellite.

$$c(\gamma) = c_0(1 - \frac{\gamma}{\gamma_0})^{\alpha'} \tag{2.4}$$

where α' and c_0 and γ_0 are the fundamental parameters describing the process characterizing respectively the generator of the cascade, the intermittency, and smoothness of the process. The generator is characterized by the Lévy index α with $1/\alpha + 1/\alpha' = 1$ with the value $\alpha' = \alpha = 2$ corresponding to the Gaussian case, $1<\alpha<2$ with $\alpha'>2$, and $0<\alpha<1$ with $\alpha'<0$. Equation 2.4 can be regarded as a kind of "central limit" theorem for multiplicative processes. When the quantity of interest is conserved by the cascade[1] the cascade generators are normalized and only two parameters are required to characterize the universality class:

$$c(\gamma) = C_1(\frac{\gamma}{C_1\alpha'} + \frac{1}{\alpha})^{\alpha'} \tag{2.5}$$

The value of the parameter α is of particular interest since it is associated with qualitatively different types of cascades[2]. For example, when $\alpha<2$, the generator of the process takes on values near 0 so frequently that all the negative moments diverge (a consequence of "extremal " Lévy generators - see Schertzer et al., 1988; Schertzer and Lovejoy, 1990b, this volume). This yields processes which often have large holes (regions with extremely low values), and may be good candidates for generators in cloud and rain models.

The difficulty in testing these ideas empirically is that the key parameter α' characterizes the concavity of $c(\gamma)$ which is only pronounced when γ and $c(\gamma)$ vary over a substantial range. From the point of view of non-linear regression, to fit c_0, γ_0, α' to the data we find that γ_0 and α' are highly correlated and hence parameter estimates are not very sharp. In Gabriel et al. (1988), functional box-counting was used yielding less accurate estimates of $c(\gamma)$ than those obtained here. The issue was side-stepped by assuming $\alpha'=2$ and testing the consistency of the data with that hypothesis.

Here we improve on these results by determining α' by a least squares regression on the mean of the 8 to 256km curves in fig. 2.1a,b. Maximum likelyhood estimates for the parameter α were found to be: $\alpha=0.63\pm0.035$ and $\alpha=1.66\pm0.37$ for the visible and infra red data respectively[3]. Fig. 2.2a shows the best fit and mean visible and infra red curves. The standard errors in the fit are ±0.011 and ±0.015 respectively. In Lovejoy and Schertzer 1990c, an efficient graphical method of estimating the parameters is outlined which exploits the special properties of the universal $c(\gamma)$ function eq. 2.5: $c(C_1)=C1$, $c'(C_1)=1$ which can be easily verified by inspection. This means that for these (conserved) codimension functions have the property that straight lines with slope 1 passing through the origin will be exactly tangent to $c(\gamma)$ at the value $\gamma=C_1$ independently of the value α. Once C_1 has been determined in this way, α can be determined by specifying one other value of the curve ($c(0)$ usually works well) and solving an algebraic equation for α. For the (non-conserved) quantities analysed below an additional parameter specifies a left/right shift, but the idea is the same. See Lovejoy and Schertzer (1990c) for more details.

The α parameters estimated at these two wavelengths are significantly different since they correspond to α', $\gamma_0<0$ and $\alpha'>2$, $\gamma_0>0$ respectively; it is therefore of some interest to corroborate these findings. The simplest (qualitative) method of distinguishing these two cases, is to consider the graph of $\ln c(\gamma)$. Taking derivates, we find:

$$\frac{d^2 \ln c(\gamma)}{d\gamma^2} = \frac{-\alpha'}{\gamma_0{}^2} \frac{1}{(\gamma/\gamma_0+1)^2} \tag{2.6}$$

---

[1] For example the energy flux to smaller scales in the dynamical turbulent cascade, or the concentration variance flux in passive scale cascades.

[2] C_1 is the co-dimension of the "support" of the measure corresponding to the unique dimension introduced by Mandelbrot (1974).

[3] The large difference in the maximum likelyhood errors cited here is due at least in part to the fact that we directly estimate α' and $\Delta\alpha = (1+\alpha)^2\Delta\alpha'$ hence this effect alone accounts for a factor 2.7 in difference.

hence for $\alpha'\geq 2$ ($1<\alpha\leq 2$), ln c(γ) is everwhere convex, whereas, for $\alpha'<0$ ($0<\alpha<1$), it is everywhere concave. This sharp contrast is readily confirmed in fig. 2.2b. These analyses are important not only because they confirm cascade theories, but also because they will help calibrate stochastic rain and cloud models.

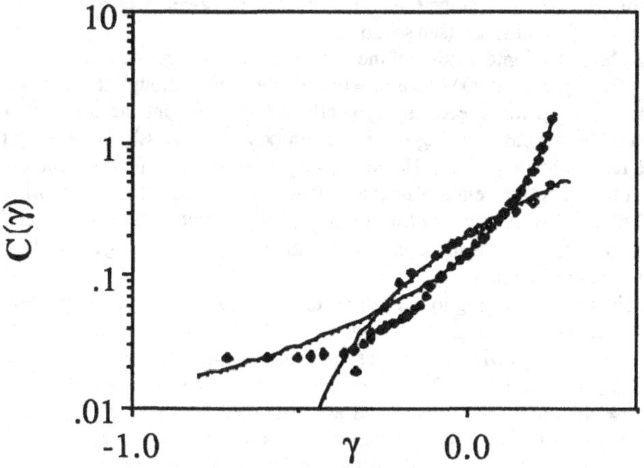

Fig. 2. The same as fig. 1a,b except that the c(γ) axis is in log coordinates. Note that in accord with the best fit values of α (the lines), the visible curve is concave, while the infra red curve is convex.

3. MULTIFRACTAL RAIN MEASURES AND RADAR REFLECTIVITIES

3.1. Multifractal Marshall-Palmer experiment

In the previous sections, we ignored the particulate nature of rain, investigating the properties of the radiance fields without attempting to interpret the latter in terms of the rain (or cloud) measures themselves (i.e. (V_i, $\underline{r}_i$). In this section, we attempt to obtain (V_i, $\underline{r}_i$) directly, enabling the statistical properties of Z, R and (in principle) all the radiative transfer characteristics to be determined. The method used is a modern day version of the famous Marshall-Palmer (1948) experiment which was the first to measure the size distribution of rain drops. This experiment – which in the form of a semi-empirical Z-R relation, still provides the quantitative basis of most radar meteorology – consists in using chemically treated blotting paper (that changes from pink to blue when wetted) to record the impacts of rain drops. The drop volume (V) can be fairly accurately estimated from the radius ρ of the coloured stains on the blotting paper. By dropping carefully calibrated drops down the four floors of the stairwell of the Macdonald physics building at McGill University, Marshall and Palmer showed that $V\propto\rho^2$ (with little statistical scatter) which is the relation expected if the penetration depth of the water into the blotting paper is constant. We then show how these results can be exploited to obtain corrections to radar measurements of rain.

Our modern rerun was performed with the help of B. Miville and T, Pham; two third year honours physics students, as part of their honours physics laboratory project. They reran the calibration procedure using the same stairwell as Marshall and Palmer, (finding the same exponent in the V-ρ relation). The improvements were a) the use of much larger (128X128 cm) squares of blotting paper (rather than the 16X24cm size used in the original experiment), b) the digitisation of the results, c) unlike Marshall and Palmer who were interested only in drop sizes (and assumed that the latter were distributed uniformly in three dimensional space, and hence uniformly over the blotting paper), we also recorded the position ($\underline{r}_i$) of the drops (for some more information on the experimental set-up, see Lovejoy and Schertzer (1990b)). By exposing the blotting paper for very short times (≈1s), we attempted to obtain a horizontal intersection of the $(V_i, \underline{r}_i)$ measure. If the latter were isotropic, then knowledge of the statistics of intersections would yield statistical information on the full multifractal $(V_i, \underline{r}_i)$ measure for all subsets with dimension ≥1 (this codimension is the "dimensional resolution" of the blotting paper – see Lovejoy et al. (1986a,b) for a discussion). However, the anisotropy requires us to be more prudent; for example, we must use the elliptical dimension of rain (=2.22±0.07 rather than 3) estimated in Lovejoy et al. 1987 in order to extrapolate to the properties of $(V_i, \underline{r}_i)$ in space (see section 3.2).

In order to sample the horizontal intersection of the rain field, the blotting paper must be exposed for as short a time as possible. In this case, an exposure of ≈1s was obtained, although this is not as short as might be hoped given that rain drop fall speeds are typically 2-5m/s. To put the problem in context, consider very long exposures. In this case, (taking the rain as an (x,y,z,t) process), the blotting paper will record the projection of the rain on the x-y plane. However, the properties of projections and intersections are quite different. Here, the projection relation indicates that any component of the multifractal rain measure with dimension $D \geq 2$ will lead to planar projections (i.e. the projection has dimension 2, and the blotting paper gets wet everywhere), whereas the intersection will have $D_\cap < 2$ as long as $D < d_{el}$ (where d_{el} is the elliptical dimension of the (x,y,z,t) process).

Fig. 3.1 shows the points corresponding to the centres of the circular blobs on the blotting paper; in this case there are 452 of them. These were digitised along with their radius (to an accuracy of 0.5mm). 452 is a relatively small number of drops with which to estimate dimensions (see the paper by Essex in this volume), however, since $D \leq 2$, it is sufficient. We also analysed two other rain events (with 1293, 339 drops respectively), but these we analysed by hand and are discussed later on. The statistically most sensitive analysis method is to estimate the correlation dimension (as was done in Lovejoy et al. (1986a) for the meteorological measuring network). This is done by considering the function $\langle n(l) \rangle$ which is the average number of other drops in a radius l around each drop. Since there are 452X451/2=101,926 drop pairs, this function contains a great deal of information about the drop clustering. We then define the correlation dimension D_2 as $\langle n(l) \rangle \propto l^{D_2}$. Fig. 3.2 shows that over the range 2mm $\leq l \leq$ 40cm, that $D_2 \approx 1.83$ (the subscript refers to the fact that the drops are embedded in a two dimensional space). The large l behaviour deviates below the line because many of these large circles go outside the blotting paper and are therefore biased downwards. At the small scale end, we also obtain a bias due to the finite number of points; for example, clearly $\langle n(l) \rangle \geq 452^{-1}$. We therefore take this as evidence that rainfall is scaling over this range.

In order to extrapolate the $\langle n(l) \rangle$ result from the measured (horizontal) intersection to the full x,y,z space, the strong horizontal stratification of the rain process due to gravity must taken into account. Introducing the codimensions with respect to two and three dimensional embedding spaces $C_3=3-D_3$, $C_2=2-D_2$ (≈0.17 here) and using the formalism of "generalized scale invariance" (Schertzer and Lovejoy, 1985a,b; 1987a,b), in x,y,z space, we expect $\langle n(l) \rangle \propto l^{D_3} = l^{3-C_3}$ with:

$$C_3 = C_2 \frac{3}{d_{el}} \tag{3.1}$$

where d_{el} is the "elliptical" dimension of the rain process characterizing the stratification, estimated (Lovejoy et al., 1987) to have the value $d_{el}=2.22\pm0.07$ in rain (d_{el} would be three if the space was isotropic, and two if it was completely stratified into flat layers). Using eq. 3.1, and expressing n in terms of the volume $\upsilon = l^3$ we obtain:

$$\langle n(\upsilon) \rangle \propto \upsilon^{(1-C_2/d_{el})} \tag{3.2}$$

Hence, using the above values of C_2, d_{el} in eq. 3.2 the drop density $\langle n(\upsilon) \rangle / \upsilon$ is no longer constant as demanded by Poisson statistics, but decreases as $\upsilon^{-0.08}$.

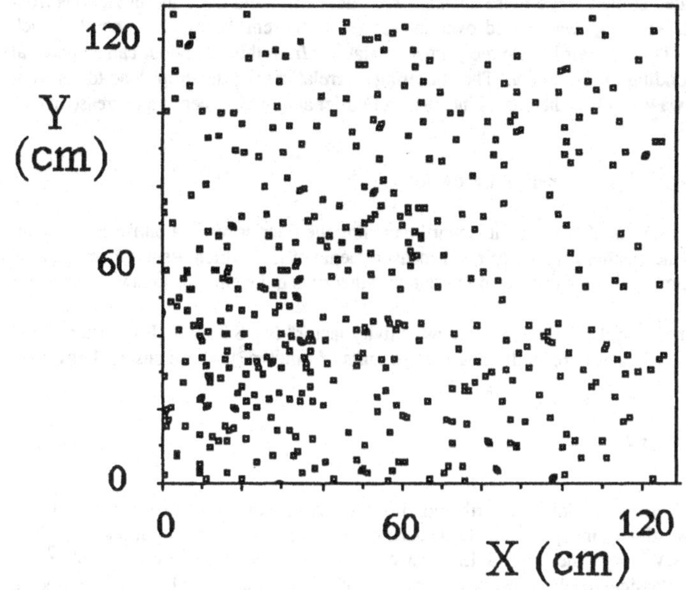

Fig. 3.1: Each point represented the centre of a raindrop for the 128X128 cm piece of chemically treated blotting paper discussed in the text. There are 452 points, the exposure was about 1s.

The effect of this drop clustering on radar signals will be to introduce some degree of coherent scattering from the drops. To quantify this, consider a radar at the origin that emits a pulse of electromagnetic waves with wavevector $\underline{k}$, that fills a volume $\upsilon = l \times r\theta \times r\theta$ where r is the range, θ the angular width of the radar beam, l is the pulse length. The power recieved at the radar depends on various instrumental characteristics including the transmitter, antenna geometry etc. Putting these factors into a multiplicative constant (ignored below) and statistically averaging (indicated by angle brackets), the radar measures the "effective radar reflectivity factor" (Z_e) whose statistical average (indicated by the brackets "< >") is:

$$<Z_e> \propto \frac{<|A|^2>}{\upsilon} \tag{3.3}$$

Where

$$A = \sum_{j}^{n(\upsilon)} V_j \, e^{i2\underline{k}\cdot\underline{r}_j} \tag{3.4}$$

and $\underline{r}_j$ is the position vector of the jth drop and the factor 2 arises because the beam makes a round trip.

Therefore, accoring to eq. 3.3, 3.4, the radar measures the square of the modulus of the Fourier component of the drop distribution (weighted by the drop volumes) at wavevector $\underline{k}$. We therefore calculate the Fourier transform ($Z_e(\underline{k})$), and plot (fig. 3.3) the angularly and radially integrated quantity

$$|\underline{k}|^{-2} \int_{S|\underline{k}|} Z_e(\underline{k}')\, d^2\underline{k}' \propto |\underline{k}|^{\zeta} \tag{3.5}$$

where $S_{|\underline{k}|}$ is the circle of radius $|\underline{k}|$ and empirically, we find $\zeta \approx -0.12$ (we integrate over circles in Fourier space in eq. 3.5 to take advantage of the presumed statistical horizontal isotropy of the drop distribution). The factor $|\underline{k}|^{-2}$ was included so that white noise (Poisson statistics) will yield $\zeta=0$, and deviations from white noise will be easy to see. $|\underline{k}|$ was varied over the range $\frac{2\pi}{128} 50\ \mathrm{cm}^{-1} \geq |\underline{k}| \geq \frac{2\pi}{128}\mathrm{cm}^{-1}$ (which corresponds to the range 2 - 128 cm). Only extremely small deviations from this behaviour can be noted at high wavenumbers corresponding to about 4cm. The clustering (correlations) of the drops lead to coherent scattering, thus for a given wavenumber, there is an increase in Fourier amplitudes, hence a decrease in ζ as compared to white noise.

3.2. Estimating range dependent biases in radar reflectivities

Before continuing with our analysis of this data, it is worth pursuing the Fourier result a bit further, since it introduces corrections into the standard theory of radar measurements of rain which assume homogeneity (white noise). The calculation of the corrections to the standard theory is quite simple, so that we sketch it below.

We have already introduced the "effective radar reflectivity factor" (eq. 3.3, 3.4). It is customary to introduce the "radar reflectivity factor" (usually measured in units of mm^6/m^3) whose ensemble average $\langle Z\rangle$ is defined by:

$$\langle Z\rangle = \frac{\langle n(\upsilon)\rangle}{\upsilon} \langle V^2\rangle \tag{3.6}$$

If the drops are uniformly randomly distributed (i.e. they have Poisson statistics), the phases $2\underline{k}\cdot\underline{r}_j = \phi_j$ in eq. 3.4 are statistically independent. Considering the complex sum A as a random walk in phase space, as long as $\langle V^2\rangle<\infty$ the central limit theorem applied to (3.4) implies: $\langle |A|^2\rangle = \langle n(\upsilon)\rangle\langle V^2\rangle$ and hence the classical result $\langle Z_e\rangle=\langle Z\rangle$. However, if the drops are distributed over a fractal, we have partially coherent scattering and we expect drop correlations to yield an anomalous exponent:

$$\langle |A|^2\rangle \propto \langle n(\upsilon)\rangle^{2H} \langle V^2\rangle \tag{3.7}$$

where $H=1/2$ for completely incoherent scattering, and $H\neq 1/2$ when some degree of coherent scattering is present. Hence:

$$\langle Z_e\rangle \propto \langle Z\rangle\langle n(\upsilon)\rangle^{2H-1} \tag{3.8}$$

In order to evaluate H from the blotting paper we used the following procedure. First, in order to reduce statistical scatter, we take $|\underline{k}|$ fixed and exploit the statistical isotropy of the drops in the horizontal by averaging over wavevectors in 19 equally spaced directions, adding more and more terms in the sum (A) by choosing drops at random from the 452 available. Fig. 3.4 shows that convergence to a power law is obtained for $n\geq 16$. Varying $|\underline{k}|$ in 10 equal logarithmic increments through the scaling region, from $\frac{2\pi}{128}\mathrm{cm}^{-1}$ to $\frac{2\pi}{1.28}\mathrm{cm}^{-1}$ (corresponding to distances of 1.28 to 128cm), we obtained $2H=1.24\pm0.09$ where the error is the standard deviation of the 2H values estimated from each of the values of $|\underline{k}|$.

We can now combine this result with our previous formula (3.2) for $n(\upsilon)$ to obtain the volume (and hence range) dependence of $\langle Z\rangle$, $\langle Z_e\rangle$. Recalling that $\upsilon = l\theta^2r^2$ and keeping only the r dependence, combining eq. 3.2, 3.6, 3.8, and using the notation $\langle Z\rangle \propto r^{\xi}$, $\langle Z_e\rangle \propto r^{\xi_e}$ we obtain:

$$\xi = -\frac{2C_2}{d_{el}} \tag{3.9}$$

$$\xi_e = 4H\left(1 - \frac{C_2}{d_{el}}\right) - 2$$

Taking $C_2 \approx 0.17$, $H=0.62$, $d_{el}=2.22$ yields $\xi= -0.15$, $\xi_e = 0.28$ (recall that the standard values are $C_2=0$, $H=1/2$, $d_{el}=3$, hence $\xi=\xi_e=0$). To judge the overall magnitude of these effects, consider a weather radar such as the 10cm wavelength one at McGill, with minimum range (limited by ground echoes) of $\approx$10km, and maximum range $\approx$240km. Comparing near and far range, we obtain a variation in $<Z_e>$ of $\approx 24^{0.28} \approx 2.4$, and a corresponding variation in $<Z>$ of $24^{-0.15} = 0.6$. These effects are somewhat larger in magnitude than those due to absorption (by humidity, O_2, and by the drops themselves) and should be taken into account during radar calibration from rain gages. However, more study is needed since the above corrections do not fully take into account the scaling proerties of rain. They could be termed "monofractal" corrections since they involve a small number of paramters (H,C), and yield corrections for the mean reflectivity factor. In Lovejoy and Schertzer (1990c), we discuss other (multifractal) corrections which correct the higher order moments (or equivalently, the probability distributions via corrections to $c(\gamma)$).

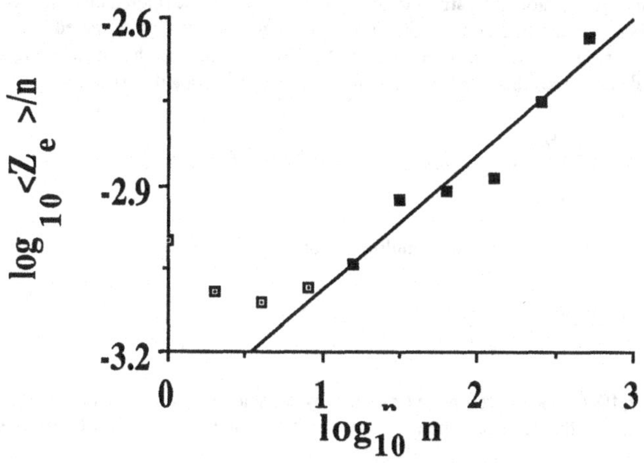

Fig. 3.4: The variation of the mean (non-dimensionalised) reflective radar reflectivity of the distribution in fig. 3.1 normalized by the number of drops (n) used in calculating the sum (eq. 3.3, 3.4), as a function of the number of drops (n). The curve is calculated as indicated in the text and involves averaging over 19 angles in Fourier space, and 10 logarithmically space wavelengths from 1-128cm. The straightline shows the asymptotic power law behaviour which is obtained for $n \geq 16$, with slope = 2H-1 = 0.24 (H = 0.62). Note that white noise (Poisson statistics) would yield a flat curve (slope zero, H = 1/2). The increase is due to some degree of coherent scattering.

3.3. Trace moment analysis

Physical applications of scaling typically involve scaling functions (or measures) rather than sets (see Schertzer and Lovejoy (1987a,b) for more detailed discussion). In spite of this, the traditional emphasis has been on fractal sets and their dimensions, even though it requires us to first transform the measure into a set (e.g. by thresholding) and then to construct Hausdorff measures from the set by covering it with "balls". As argued earlier, it is more natural to to treat the fractal measure directly; we do this by introducing a series of measures (introduced in Schertzer and Lovejoy, 1987a,b), called "trace moments". This approach is similar to that of Hentschel and Proccacia (1983), Grassberger (1983), Halsey et al. (1986). For an early application of the same technique to rain (including an empirical evaluation the codimension function in rain), see Schertzer and Lovejoy (1985b) where it is refered to there as an "integral structure function".

We again cover the set S (dimension D) with disjoint boxes B_λ, scale λ (diameter $l = L/\lambda$, L again being the largest scale of interest). We then "homogenize" f over the various B_λ's writing:

$$f_\lambda = \frac{\int_{B_\lambda} f(\underline{r}) d^D \underline{r}}{\lambda^{-D}} \quad (3.10)$$

f_λ is the "homogenized" ("smoothed") f over B_λ used in section 2 (the denominator λ^{-D} is proportional to the area (D=2), volume (D=3) etc. of B_λ). Rather than study the effect of smoothing on the probability distributions of f_λ (characterised by $c(\gamma)$), we can study its effect on the moments of different orders:

$$\langle f_\lambda^h \rangle \approx \lambda^{(h-1)C(h)} \quad (3.11)$$

Where the "< >" indicates ensemble averaging. The equality holds when f_λ has been normalized so that $\langle f_\lambda \rangle = 1$, otherwise the symbol "≈" in eq. 3.11 indicates proportionality. The exponent $(h-1)C(h)$ quantifies the rate at which the various moments are smoothed with decreasing λ (increasing l). As long as the statistical moments converge, a box-counting approximation allows $C(h)$ to be interpreted as a codimension function for the moments (not to be confused with $c(\gamma)$). This can be seen by introducing a Hausdorff measure called an "h^{th} order trace moment[1], resolution λ, of f over S" denoted: $\langle Tr_\lambda f(S)^h \rangle$:

$$\langle Tr_\lambda f(S)^h \rangle \approx \sum_i^{N(\lambda)} \langle f_\lambda^h \rangle (\mathrm{diam} B_\lambda)^h \approx \lambda^D \lambda^{(h-1)C(h)-hD} \approx \lambda^{(h-1)d(h)} \approx \lambda^{-K_D(h)} \quad (3.12)$$

since in the above the sum is approximated by multiplication by $N(\lambda) \approx \lambda^D$ and:

$$\langle f_\lambda^h \rangle (\mathrm{diam} B_\lambda)^h = \langle \left(\int_{B_\lambda} f(\underline{r}) d^D \underline{r} \right)^h \rangle \quad (3.13)$$

the trace moments $\langle Tr_\lambda f(S)^h \rangle$ can therefore be estimated by the statistical moments of f spatially integrated over boxes of size L/λ. We then estimate the exponent $K_D(h)$ (= $-(h-1)d(h)$, $d(h)=D-C(h)$) by studying the scaling of the latter.

Furthermore, $C(h)$ is a dual codimension function related to $c(\gamma)$ via a Legendre transformation:

$$c(\gamma) = \min_h \left(h\gamma - (h-1)C(h) \right) \quad (3.14)$$

3.4. Trace moments for drop volume and probability measures

We may now use eq. 3.12 to estimate $K_D(h)$ for the blotting paper (V_i, $\underline{r}_i$) rain measure. Two measures with obvious physical significance are: $f_V=(V_i, \underline{r}_i)$ and $f_n=(1, \underline{r}_i)$ where

$$\int_S f_V(\underline{r}) d^D \underline{r} = \text{the total amount of water in S} \quad (3.15)$$

and

$$\int_S f_n(\underline{r}) d^D \underline{r} = \text{the total number of drops in S} \quad (3.16)$$

---

[1] The "≈" sign on the left of eq. 3.12 is introduced because the covering yielding the "infimum" in the Hausdorff measure defining the trace moment has been approximated by a covering of $N(\lambda)$ disjoint boxes - see the discussion of the box-counting approximation in appendix B.

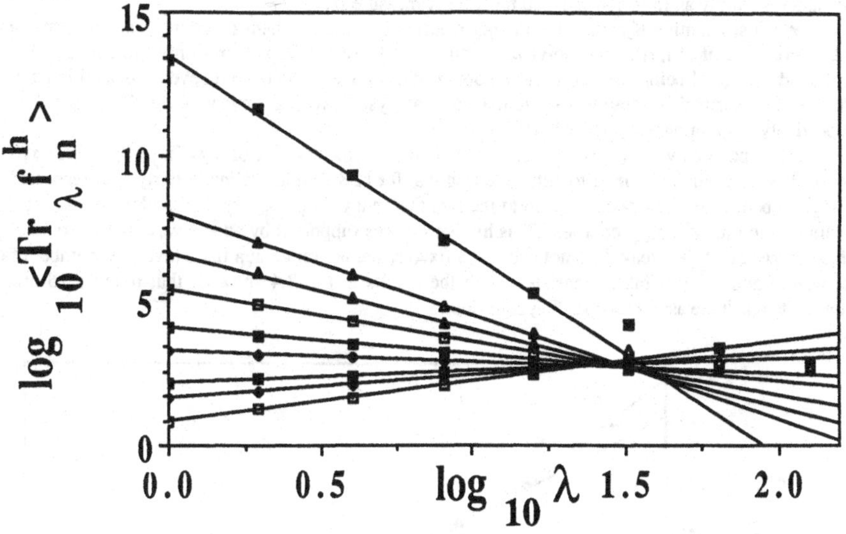

Fig. 3.5a: A log-log plot of the (unnormalized) $<Tr_\lambda\ f_n^h>$ which is proportional to $\lambda^{-K_{nD}(h)}$ v.s. $\lambda=L/l$ where λ is the scale ratio, L the external size (=128cm), l the size of homogeneity and f_n is the number measure discussed in the text. Note that l=128 cm and that convergence to power laws occurs only for lengths $l\geq$4cm (λ<32). The curves, top to bottom, are for h = 5, 3, 2.5, 2, 1.5, 1.2, 0.8, 0.6, 0.3, and the (negative) slopoes are the corresponding values of the function $K_{nD}(h)$. Straight lines were fit for $l\geq$8cm.

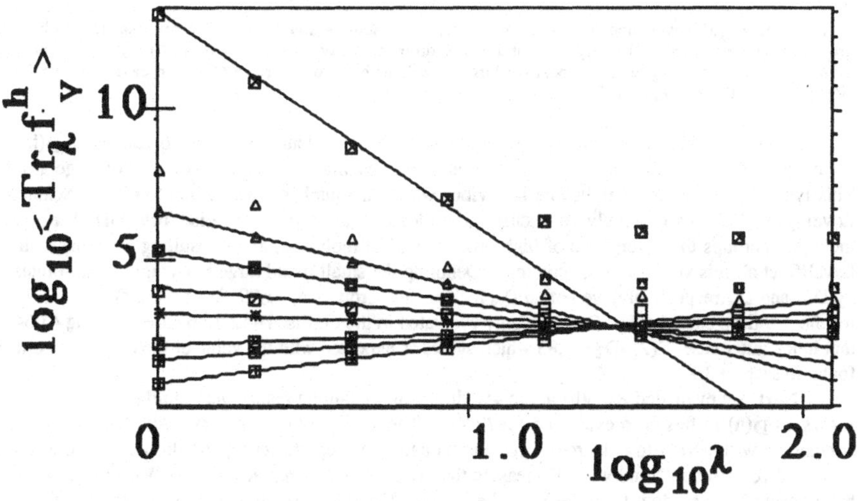

Fig. 3.5b: Same as 3.5a but for the volume measure discussed in the text. Straight lines fitted for $l\geq$8cm.

The latter is simply an (non-)normalized probability measure f_p.

We first examine $K_{nD}(h)$ (the number measure scaling exponent function), since it is the simplest characterization of $(V_i, \underline{r}_i)$; it is equivalent to studying the set of drop centres only. In our case, it has the further advantage of being the only measure obtained from more than one rain event. In addition to the 452 drop event described in section 3.1, we manually analysed two other events (with 1293, and 339 drops respectively), obtaining $K_{nD}(h)$ for $h>0$.

All exposures were for ≈ 1s, fig. 3.5a, b shows a typically plot of $\log\langle Tr_{\lambda} f(S)^h\rangle / \langle Tr_{\lambda_s} f(S)^h\rangle$ vs. $\log \lambda$ with λ ranging from 1 to 128. We find that for $l \geq 8$cm that scaling is fairly well respected. The change in behaviour for $l \leq 4$cm is due to the fact that many drops per resolution element are required to obtain asymptotic scaling properties. This hypothesis was supported by an analysis of the two other cases which showed that the break does not occur at a fixed scale, but rather at a fixed (average) number of drops per box of about 16. This is consistent with the results in fig. 3.4 showing that roughly 16 drops are required to reach the asymptotic scaling behaviour.

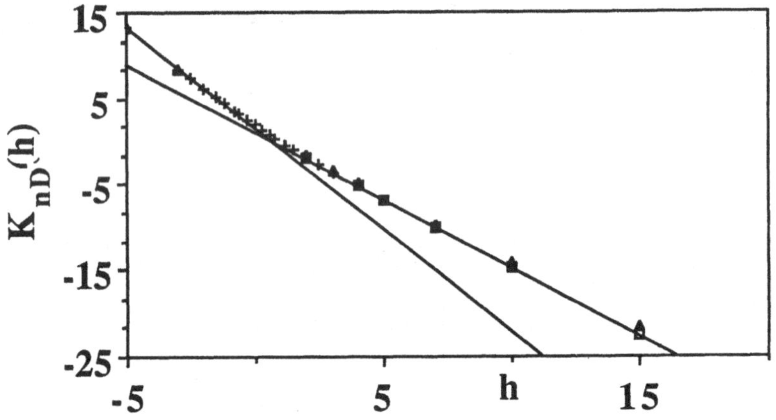

Fig. 3.6: $K_{nD}(h)$ estimated from (bottom to top) a manually analysed 1293 drop case (h>0 only), the 452 drop (digital) case analysed in fig. 3.5, and a 339 drop manually analysed case (h>0). The straight lines are asymptotic fits to the negative and positive large (absolute) h regions for the 452 drop case, yielding (absolute) slopes $D_{-\infty} = 4.06$ and $D_{\infty} = 1.56$.

Fitting the trace moments for $l \geq 4$cm, we obtained excellent fits. Fig. 3.6 shows $K_{nD}(h)$ for the three cases indicating that $K_{nD}(h)$ is curved at small absolute h, but fairly straight for large absolute h. This type of asymptotic straight-line behaviour was also found for radar reflectivities in Schertzer and Lovejoy (1987a). Theoretically, this behaviour (at least for positive h) can arise from a number of sources, including perhaps the divergence of high order statistical moments, and a resulting spurious scaling (see Lavallée et al. this volume for discussion). Denoting the small h and large h asymptotic (absolute) slopes by $D_{-\infty}$ and D_{∞}.respectively, we obtain D_{∞} for the 1293 drop case, =1.79, for the 452 drop case =1.56, and for the 339 drop case, =1.51. $D_{-\infty}$ was only evaluated in the digitised 452 drop case, yielding 4.064. Note that in the latter case, $K(2)=D_2 \approx 1.83$ which agrees as expected with the value of the correlation dimension found in section 3.1.

Next, we evaluated $K_{VD}(h)$ in the 452 drop case, obtaining results very similar to $K_{nD}(h)$. Fig. 3.7 shows $K_{VD}(h)$ in this case evaluated for $l \geq 8$cm. The similarity of this curve with the $K_{nD}(h)$ curve is compatible with a very low degree of correlation between drop clustering and drop size. To test this, we evaluated yet another, "randomized" measure that we obtained by replacing V_i in $(V_i, \underline{r}_i)$ by a V_i' obtained by choosing V_i at random from the V_i in the sample. Fig. 3.7 shows that except for very large absolute h, the functions $K_{VD}(h)$ and $K_{V'D}(h)$ are nearly the same. The large h behaviour does appear to be somewhat different however, since the (randomized) asymptotic absolute slope of $K_{V'D}(h)$ yielded $D_{\infty} = 1.47 \pm 0.03$

while that of the (unrandomized) $K_{VD}(h)$ yielded $D_{\infty} = 1.56$. We interpret this as indicating that correlations between drop position and size are most important for the larger drops (since they contribute most to the higher moments).

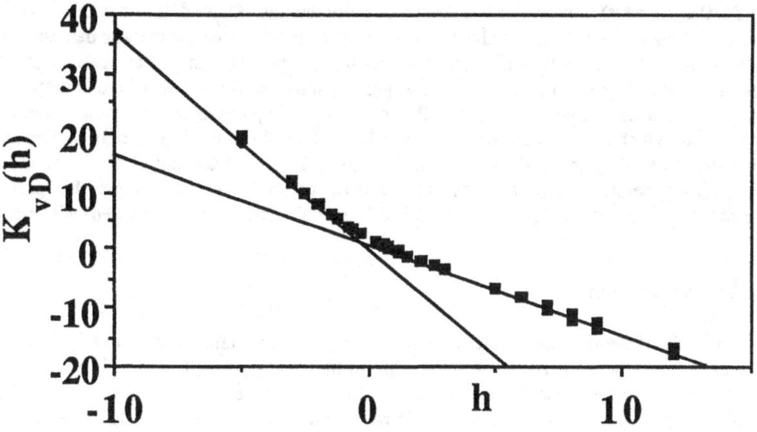

Fig. 3.7: $K_{VD}(h)$ for the 452 drop case (lower curves) and $K_{V'D}(h)$ for two randomized cases discussed in the text (upper curves). The straight lines are asymptotic fits to the negative and positive large (absolute) h regions for the 452 drop case, yielding for $K_{VD}(h)$, $D_{-\infty} = 3.65$, $D_{\infty} = 1.56$.

4. LIDAR MEASUREMENTS OF RAIN DROPS

4.1. Extending the blotting paper results in time and space

In an attempt to extend the blotting paper results of section 3 to much greater time and space scales, we undertook to examine the reflexions of high-powered laser pulses (lidar) from rain drops. Lidar has generally been used to detect very small particles (such as aerosols) whose tiny size (typically <1μm) precludes their measurement by microwaves. Even with high power and telescopic receivers, the return signal is typically very weak and relatively long pulse lengths (e.g. 50m) and averages over many repeated pulses are used to improve the signal to noise ratio. Lidars have also been used to study clouds, where the drops are typically 1-20μm in size. Here the problem has not been the weakness of the return signal, but rather the high density and efficiency of the scatterers which attenuates the signal so strongly that typically only the cloud edge is detected.

Our approach was somewhat different. By studying large drops (rain-typically 0.5-3mm in diameter), with high power and short range, we can obtain excellent signal to noise ratios, even with very small pulse lengths (here ≈ 3m) and averaging over pulses is not required (of course, the measurements had to be performed after dark). Also, unlike the typical lidar configuration in which the beam is directed upwards, here, we pointed it in the horizontal. Furthermore, the lower density of rain drops as compared to cloud drops means that attenuation is not a serious problem. Although an ultimate objective is to reduce the pulse length to centimetres or smaller (lasers with these characteristics are readily available), and to measure the position and reflection of individual drops, our electronics were insufficiently fast. The set-up involved a YaG laser with 10ns pulse duration (0.1J per pulse, at 10Hz) with a frequency doubler so that it emitted in the green region of the spectrum. Although the pulse volume is too large to isolate individual drops completely unambiguously (due to beam spreading, the pulse volume at a 10m distance is $20cm^3$, while at 1km, it is $0.2m^3$) the return signal still yields valuable information on the time/space structure of rain. The experiment (including the assembly of the laser, telescope, photomultiplier and triggering circuits) was performed by Alex Powell and John Weisnagel, as part of their 3rd year honours physics lab project.

The data were digitised and archived on floppy disks in two basic formats, the overall limitation being the floppy disk capacity. In the first format, the object was to produce roughly square (x,t) plots so that 180 consecutive pulses were stored (0.1s between pulses) with a downrange resolution of 180 pulse lengths (=180X3=540m). In the second format, much longer series of pulses were available (5000 over a period of 5000X0.1=500s), but over only a couple of pulse lengths. Other data sets obtained included (z,t) experiments where z is the vertical coordinate. We expect systematic analysis of these data sets to yield not only information on the multiple scaling of rain over these small time and space scales, but also on the statistical anisotropy of (x,y,z,t) space. This is important since the hypothesis of the isotropy of (x,y,z,t) space constitutes a statistical version of the well-known Taylor hypothesis of "frozen turbulence" which is often invoked but which has not been directly tested[1]. In the following, we give only preliminary results which over the range of scales studied, confirm the scaling in both space and time. These are however, consistent with a generalization of Taylor's hypothesis of frozen turbulence in which the space and time scales are statistically the same if one of the axes is differentially stratified (compressed) with respect to the other.

4.2. Data analysis and results

An example of the raw data photomultiplier voltage (in light rain) is shown in fig. 4.1. The basic features to note are a) the steep rise in the first 50m, b) the steady power law fall-off of the mean signal. The initial rise is due to the telescope focusing effect (the close range was out of focus), while the fall-off, roughly following r^{-4} law, is expected for single particle scattering (if there is no attenuation, then the outgoing signal falls off as r^{-2}, similarly for the reflected signal).

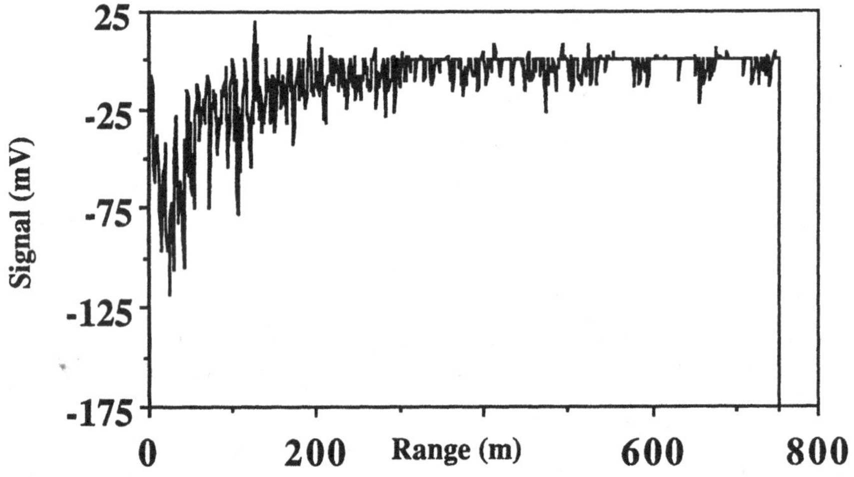

Fig. 4.1: An example of an uncorrected lidar pulse return. The reflected intensity is measured by the magnitude of the (negative) photomultiplier voltage (in millivolts). Note the initial dip in the signal due to poor telescope focus at near range, as well as the fairly rapid (r^{-4}) fall-off further downrange. Positive values are due to electronic noise and must be reset to zero. The far range rerurn is from a distant apartment building.

To remove both of these obvious range dependencies, we eliminated all data closer than 50m, and applied a simple r^4 correction. A perspective plot of a typical (range corrected) sequence of pulse returns, as is shown in fig 4.2, indicating that there is no longer any obvious range dependence. In fact, a subtle range effect must still exist because as the pulse spreads away from the lidar, there is a tendency for the pulse

[1] See appendix E for discussion.

volume to include more and more drops, and hence increase the signal. Apparently, this effect has been somewhat offset by a decrease due to some attenuation. In any case, we do not expect these effects to yield a serious bias in our analyses since the same mix of near and far data was used in trace moments estimated at each scale. This is the same argument as that used in Lovejoy et al. (1987) with regard to the effect of the spreading of the radar beam.

4.3. Trace moment estimates of correlation dimensions

Although the pulse volume is generally too large to lead to completely unambiguous conclusions about number and size of the drops in a pulse volume, it was still small enough that at light rain rates, many of the pulse volumes gave very low return signals (signals compatible with either aerosol scattering or other noise sources this is true of most of the return signals with voltage ≥0 in 4.1). There was therefore some indication that by putting a fairly low threshold on the return signal, that a crude separation could be made into rain and no-rain pulse volumes. In the following, we therefore analyse only the set of pulse volumes exceeding the sample average signal. The properties of this set were then investigated by the trace moment analysis described in section 4.3.

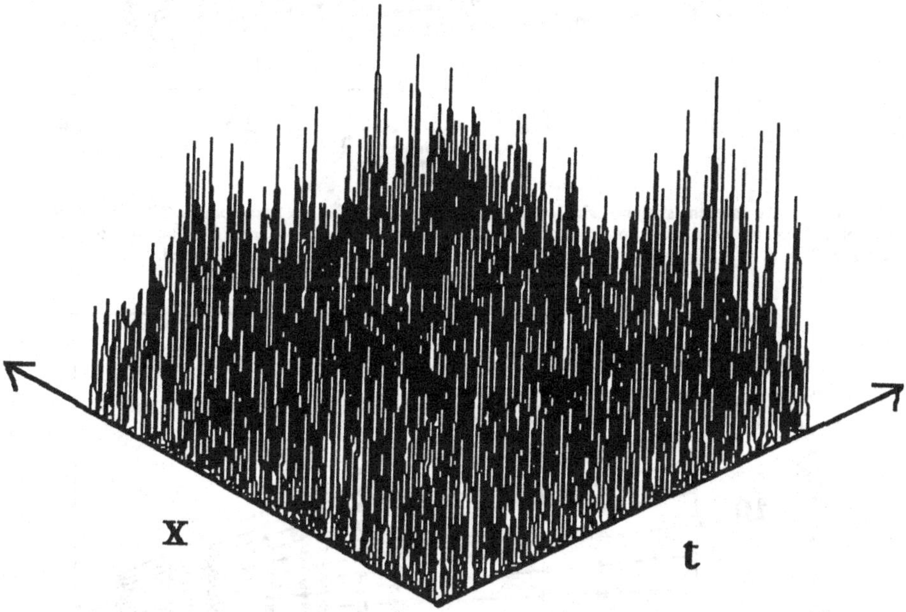

Fig. 4.2: A (x,t) perspective plot of the log of the absolute (range corrected) lidar reflectivity (vertical coordinate) indicating the extreme variation, in both space and time, as well as the approximate absence of remaining range biases (180X180 points are shown).

Fig. 4.3 shows such an analysis on a time sequence involving 5000 pulses (=500s). The telescope was focused at near range allowing pulse volumes only 10m from the lidar to be studied. At this range, these were 0.3mm in diameter and therefore sufficiently small so that often no scattering took place. In this case, due to the proximity - the scattering occured in a McGill parking lot - the above statement could be readily verified by eye: at a given range reflections occured only at fairly irregular intervals. In fig. 4.3, note how accurately the scaling is followed. Using the slopes to estimate $K_D(h)$, and the formula $d(h) = -K_D(h)/(h-1)$ we find that the dimensions of these 1-D sequences were typically in the range

0.9-0.85. The resulting codimensions (0.1 - 0.15) are comparable to those found in section 4.1 for the (x,y) distribution (0.17) but a proper analysis of the difference in space/time codimensions has not yet been performed, however, the results here are sufficient to conclude that space and time do not scale in the same way - see appendix E for a theoretical discussion.

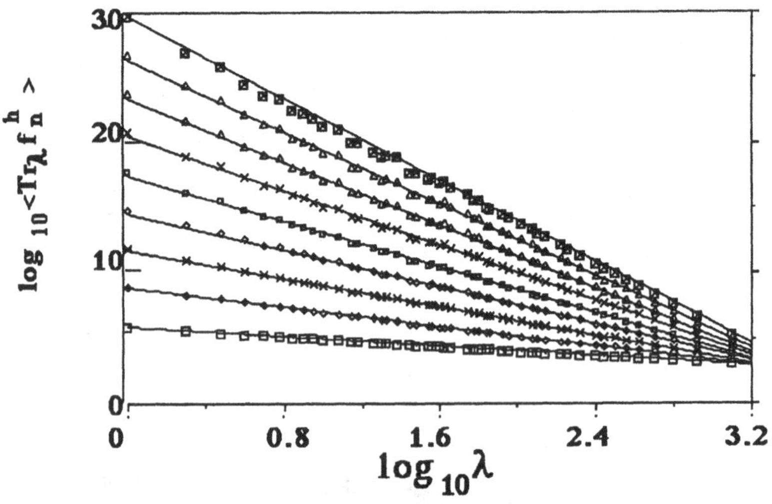

Fig. 4.3: Trace moment analysis (similar to that in figs. 3.5a,b) for the time domain (5000 pulses over a period of t_0=500s) for those range corrected returns that exceeded the average. Curves from top to bottom are for h=10, 9, 8, 7, 6, 5, 4, 3, 2 respectively. Note that the scaling is extremely accurately followed.

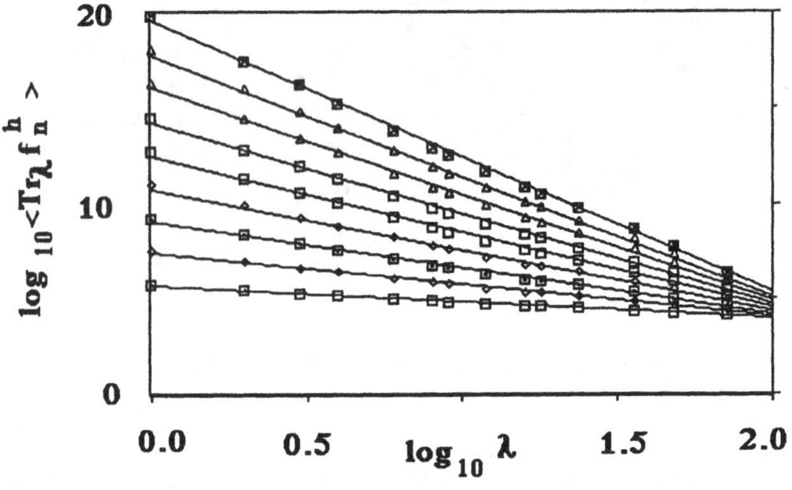

Fig. 4.4: Trace moment analysis for downrange domain (each pulse return is divided into 180 pulselength sections, 3m apart (L=540m) for those range corrected returns that exceeded the average. Curves from top to bottom same as for fig. 4.3. Note that the scaling is extremely accurately followed.

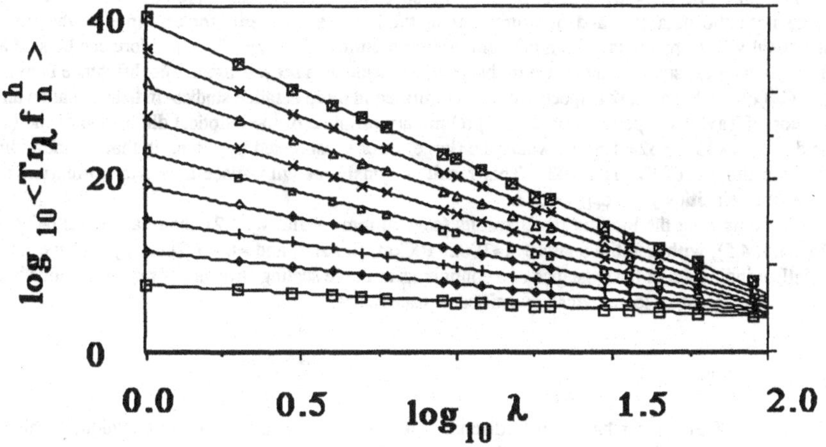

Fig. 4.5: Trace moment analysis for the (x,t) domain (180 pulses, 0.1s apart in time, space resolution 3m) for those range corrected returns that exceeded the average. Curves from top to bottom same as for fig. 4.3. Note that the scaling is extremely accurately followed. This data set is the same as that shown in fig. 4.4, except that analysis was performed on "squares" in (x,t) space rather than by intervals (downrange) only.

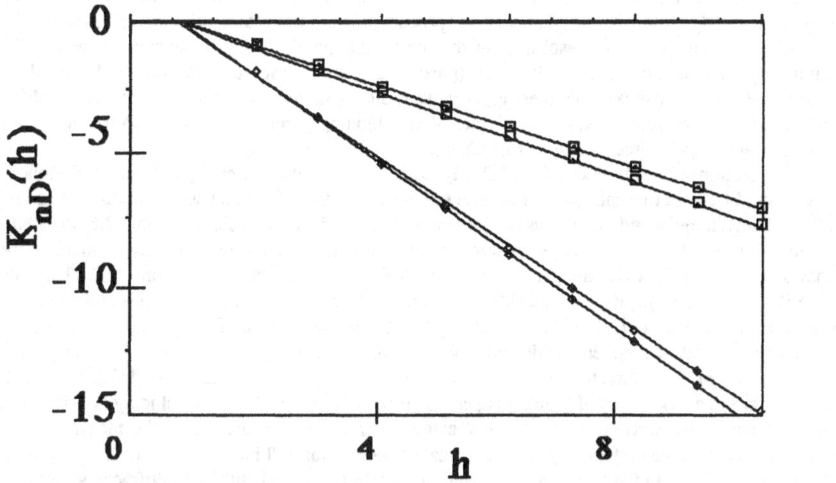

Fig. 4.6: Comparisons of the functions $K_{nD}(h)$ obtained (bottom to top) from lidar (x,t space), blotting paper (x,y space) lidar (t space), lidar (x space). The two upper lines are steeper because they are embedded in a two dimensional space, while the lower lines in a one dimensional space.

In fig. 4.4 we analyse down-range returns for another storm, using the r^4 correction and again thresholding at the sample averaged threshold. Again, the scaling is accurately followed over the entire range of scales from 3 to 540m, with similar dimensions. Fig. 4.5 shows the results for the 180X180 (x,t) field described above, again showing the scaling. Fig.4.6 compares the functions $K_{nD}(h)$ obtained from 3.5a, 4.3, 4.4, 4.5 yielding estimates of D_∞ = 1.56 (x,y, blotting paper), 1.68 (x,t, lidar), 0.84 (t, lidar),

0.76 (x, lidar), or, C_∞ = 0.44, 0.32, 0.16, 0.24 respectively. These results are not immediately comparable since both the data sets and measurement methods were different (for example, if the x,y space is statistically isotropic, then from a theorem on intersections, $C_\infty(x,y)= C_\infty(x)$). More can be said about the lidar (x) and (x,t) space results since in this case, the same data set was used. The difference between $C_\infty(x)$ and $C_\infty(x,t)$ (0.24 and 0.32 respectively) - if confirmed in more detailed studies- indicates that an anisotropic version of Taylor's hypothesis holds. A preliminary estimate of the elliptical dimension of the (x,t) space is: $d_{elx,t} = 2\times0.24/0.32= 1.5\pm0.3$ where the (large) error is estimated by assuming that the individual errors in the estimates of C_∞ is ±0.03. This result is consistent with space-time transformation involving turbulent velocities yield $d_{elx,t} \approx 5/3$.

Combining the blotting paper results (several mm- ≈1m fig. 4.2), the lidar results (≈3m - 0.5km, figs. 4.4, 4.5), with satellite analyses (≈100m-100km, fig. A.2, and ≈8 - 512km, fig. 2.1) we we have an excellent indication of the existence of a continuous multiple scaling structure of rain and clouds over nearly the entire range of meteorologically significant space scales.

5. CONCLUSIONS

The object of this paper has been to discuss and illustrate new data analysis techniques which directly analyse multifractal measures, determining the codimension function characterizing the singularities of various orders (the PDMS technique), as well as a related (dual) codimension function characterizing the scaling of the various statistical moments (the trace moment technique). Unlike ("mono-fractal") methods adapted to studying the geometric properties of sets (which involve single fractal dimensions), both methods involve creating a series of lower resolution fields and allow direct determination of the entire (co)-dimension function. Since geophysical phenomena are typically measures (or fields when averaged over the resolution of a sensor), mono-fractal techniques operate by first transforming the fields into sets, typically by applying thresholds. This indirect procedure is unsatisfactory since the values of the thresholds depend fundamentally on the resolution of the sensor; in appendix A we re-examine several of the most commonly used mono-fractal techniques (particularly area-perimeter relations and area distribution exponents[1]) in this context. We then apply these multifractal techniques to infra red and visible satellite surface and cloud radiances, radar reflectivities of rain, blotting paper data on rain-drop size and position, and lidar data on the space-time structure of raindrops.

In section 2, we applied the Probability Distribution/Multiple Scaling (PDMS) technique showing how it could be used to establish the multiple scaling of satellite cloud and surface radiances in visible and infra-red wavelengths and determine the resolution independent (normalized) probability distribution (the codimension function). We further showed how this function fits into theoretically predicted universality classes, estimating the important parameters characterizing the generators of the process. This is important since it helps both to justify and to calibrate, the modelling of clouds by continuous cascade process (c.f. Wilson et al. this volume). The PDMS technique is appealing because of its simplicity, and the fact that it directly corresponds to a systematic degradation of the data resolution.

In section 3, we directly studied the $(V_i, \underline{r}_i)$ (= volume, position) measure defined by raindrops as they intersected a large piece of blotting paper. Using a variety of techniques including Fourier analysis and trace moments, we were able to not only establish the multiple scaling of the measure, but also to quantitatively evaluate systematic corrections that the behaviour will introduce into radar reflectivities. The corrections arise due to the long-range correlations in rain associated with its multifractal structure. On the one hand, these correlations lead to a degree of coherent scattering that enhances the radar signal above its conventionally assumed level. On the other hand, the sparseness of the raindrops leads to a decrease in their density with scale which tends to act in the opposite sense. Quantitative evaluation of the relative

---

[1] The re-evaluation of the mono-dimensional techniques involved a number of new findings. First, that the area-perimeter exponent will not in general be the dimension of the perimeter, but rather the ratio of the perimeter dimension to half the dimension of the area (in appendix A- we empirically show that this correction is often of the order of 25% or more). Second, the distribution of areas has an exponent which is generally expected to be <1, since unlike many simple geometric fractal sets, in fractal measures, we do not expect fragmentation of structures, to play a very important role in determining the fractal dimensions (appendix D).

importance of the two effects shows that it is the former which dominates, yielding a systematic bias that may easily reach 50% in typical weather radar systems.

The blotting paper analysis was unfortunately too limited to permit more systematic investigation of the rain structure- particularly due to the lack of vertical resolution and small sample sizes. This limitation prompted the final study reported here, which involved very short pulselength lidar (laser) measurements of rain drops. Although we could not completely unambiguously distinguish pulse volumes with one or more drop from those without any, an approximate separation was achieved showing that over the range 0.1s to 500s, and 3 - 540m, in space, that the distribution of raindrops is scaling. We also evaluated the multiple fractal dimensions of the set of rain-filled pulse volumes. Although these results are preliminary, they clearly established the scaling in this important range of time/space scales. If we combine the analyses of section 2,3,4 and appendix A (not to mention quite a few other analyses in the literature), we find a convincing case for the multiple scaling of rain and clouds over the entire range of millimeters to nearly a thousand kilometers.

In light of these findings, we can affirm that the only convincingly documented evidence of scale breaking to date is at the viscous scale (≤1 mm) and at planetary scales (≥1000km). However, as with any theoretical idea, no amount of empirical evidence could ever <u>prove</u> the validity of the scaling hypothesis. However, the lack of strong evidence for breaks in the scaling means that the latter- if they exist- are likely to be relatively weak. This evidence, coupled with theoretical arguments in favour of such scaling -not to mention its unifying and simplifying power- certainly make the hypothesis more attractive than ever.

6. ACKNOWLEDGMENTS

We acknowledge discussions with, G.L. Austin, A. Davis, P. Gabriel, J.P. Kahane, P. Ladoy, D. Lavallée, E. Levich, A. Seed, A.A. Tsonis, R. Viswanathan, J. Wilson.

APPENDIX A: THRESHOLDING AND MULTIFRACTAL MEASURES: A RE-EVALUATION OF SEVERAL MONO-DIMENSIONAL ANALYSIS TECHNIQUES

A.1. Discussion

We have argued (especially in section 2), that emphasis on the geometric properties of scale invariant processes has lead to excessive emphasis on the study of sets, and their fractal dimensions. For example, using a threshold on satellite cloud and radar rain data, Lovejoy (1981, 1982) determined area-perimeter and area-distribution exponents. Although limited to the range 1-1000km, and to a single cloud and rain intensity level and meteorological situation, Lovejoy (1982) showed that scaling could hold over a wide range of meteorologically significant length scales. Since then, these and related methods have been used by a number of other investigators (e.g. Carter et al., 1986; Ludwig and Nitz, 1986; Rhys and Waldvogel, 1986; Welch et al., 1988; and several authors in this volume), occasionally yielding apparently contradictory results. For example, the value of the area-perimeter exponent of 1.35 found in Lovejoy (1982) was at first considered a fundamental constant, which was subsequently found to be not always reproducible; Rhys and Waldvogel (1986) and Cahalan (this volume) found generally higher values that depended on the realization ("meteorological situation"), and Carter et al. (1986) estimated the dimension of the "graph" of infra red radiance intensity from clouds, obtaining yet another value of the dimension. However, once the multifractal nature of the fields is appreciated, these results can be easily explained. In particular, the interpretation and significance of the area-perimeter and area distribution exponents, graph dimensions, as well as their relationship to the fractal dimensions of the rain or cloud regions themselves must be re-examined. This is done below; the most important results are a) the area-perimeter exponent will in general not be equal to the fractal dimension of perimeter b) the range of possible values and significance of the area distribution exponent is different than that obtained by simple arguments on geometric sets. We then re-examine these early studies in this multifractal context.

A.2. Multifractal exceedance sets, perimeter sets and graphs

Consider the function $f_\lambda(\underline{r})$ obtained by averaging a multifractal measure over scale l, scale ratio λ, in a region of the plane[1] $\Re$ size R×R (i.e. with large scale L=R). As mentioned in section 2, the underlying measure is most directly studied by considering how the properties of f vary as we change λ (e.g. by successively degrading our sensor resolution). However most applications of remotely sensed data involve studying the properties of f_λ at fixed λ. When, as is often the case, these exponents are obtained by using thresholds (T_λ) on f_λ to define sets, we find that the results will depend directly on λ via the multifractal relation $T_\lambda=T_1\lambda^\gamma$ where T_1 is the large (e.g. image scale resolution) value of the field (T_1 is at the same resolution as f_1). Our exponents will therefore depend (via T_λ) on both the sensor resolution (λ), and the "meteorlogical situation" (i.e. stochastic realization) of the process, and hence be of limited utility. Below, we fix the function resolution λ, and write simply $f(\underline{r})$, T.

Define the (closed) exceedance set $S_{T\geq}$ as the set of points satisfying $f(\underline{r}) \geq T$. If $f(\underline{r})$ is a scaling function, the (Hausdorff) dimension $D(S_{T\geq})$ of $S_{T\geq}$ will be a nonincreasing function of T, since $S_{T\geq} \supset S_{T'\geq}$ for T'>T and the dimension of a subset must ≤ to the dimension of the entire set (this property is so basic that it holds for all definitions of dimension of which we are aware, including topological dimensions)[2].

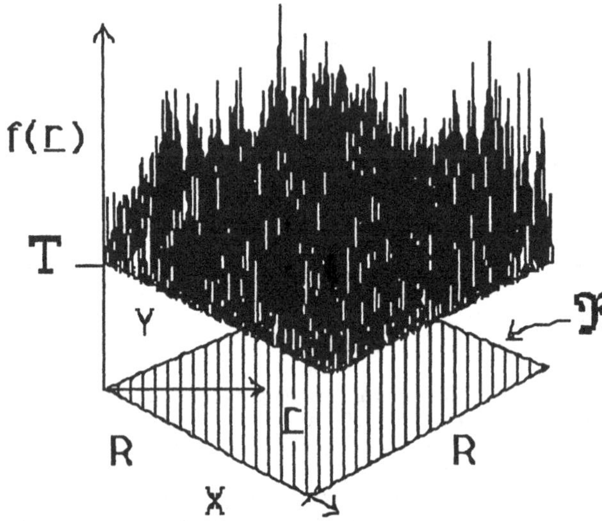

Fig. A.1a: Schematic illustration showing some of the definitions used in the text. G_T is the set of points in the black (spiky) region above the plane $f(\underline{r})$=T.

Consider next the "graph" (G) of $f(\underline{r})$ defined as those points in three dimensional ($\underline{r}$, $f(\underline{r})$) space. As before, we may define G_T as the subset of G such that $f(\underline{r})\geq T$ (see fig. A.1 for an illustration, and appendix D for more discussion). Consider now the perimeter set of $S_{T\geq}$, denoted p_T. p_T is the "border set" of $S_{T\geq}$, more properly defined as the "T-crossing set" of G with the plane $f(\underline{r})$=T (in analogy with the expression "zero-crossing" used in the theory of stochastic processes). This is the set of points $\underline{r}$ such that arbitrarily small neighbourhoods of $\underline{r}$ contain some points such that $f(\underline{r})<T$ and some such that $f(\underline{r})\geq T$. Using basic notions about sets, we can now give a definition of p_T. Define the (open) complement of $S_{T\geq}$ as: $\bar{S}_{T\geq} = \Re - S_{T\geq}$. p_T is thus the set of "contact" points of $\bar{S}_{T\geq}$ which are required to close it, yielding $[\bar{S}_{T\geq}]$. We thus obtain:

[1] Generalizations to higher dimensional spaces are straightforward and will not be explicitly considered.

[2] Complications arising from non-self-similar, anisotropic scaling (Generalized Scale Invariance, see e.g. Schertzer and Lovejoy, 1987), will not be considered here.

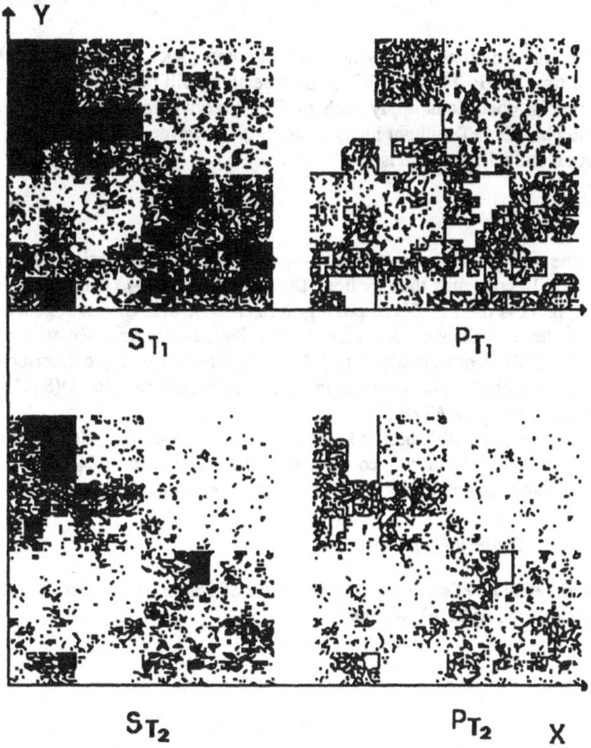

Fig. A.1b: Schematic illustration showing the definition of $S_{T\geq}$, p_T used in the text, for two different thresholds $T_1<T_2$. The illustration was produced using a multiplicative cascade process (an "alpha model") with a cascade discretised into 8 steps of a factor two in scale each step.

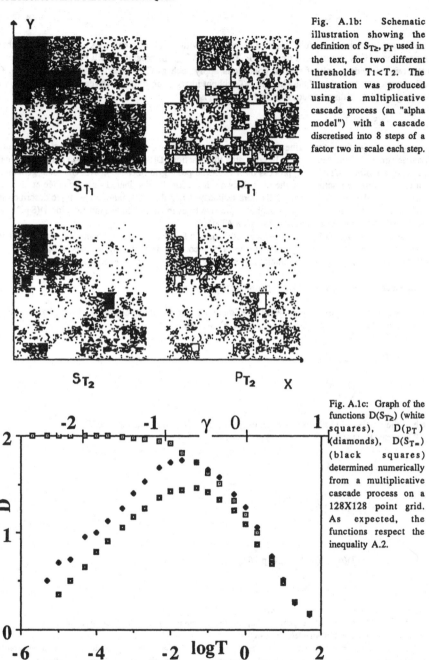

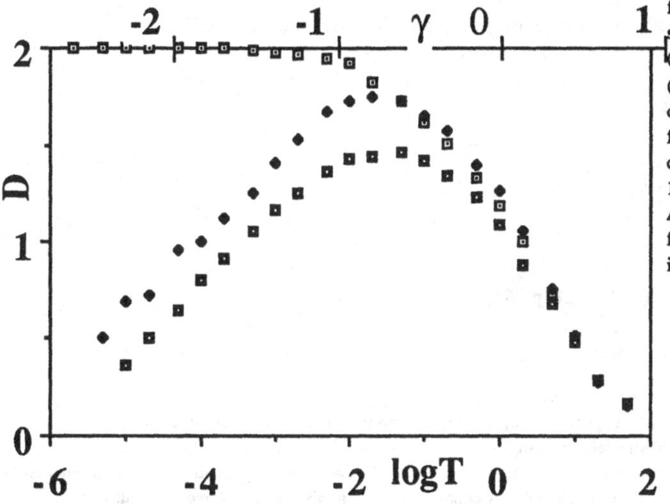

Fig. A.1c: Graph of the functions $D(S_{T\geq})$ (white squares), $D(p_T)$ (diamonds), $D(S_{T=})$ (black squares) determined numerically from a multiplicative cascade process on a 128X128 point grid. As expected, the functions respect the inequality A.2.

$$p_T = [\bar{S}_{T\geq}] \cap S_{T\geq} \tag{A.1}$$

It is important not to confuse this zero-crossing set with the set of points $\underline{r}$ such that $f(\underline{r})=T$ which is the intersection of G with the plane $f(\underline{r})=T$, $(G\cap T)$ denoted $S_{T=}$. In general, p_T will be different from $S_{T=}$ with the two coinciding only if special conditions apply such as G is everywhere continuous (i.e. G typically jumps from one side of the plane $f(\underline{r})=T$ without intersecting it[1]). However, in what follows, we see that in general, G is discontinuous, and all we obtain is:

$$D(S_{T\geq}) \geq D(p_T) \tag{A.2}$$

(since $S_{T\geq} \supset p_T$) with the actual value of $D(p_T)$ however depending critically on the topological structure (i.e. connectedness) of the set. Fig. A.1c compares the functions $D(S_{T\geq})$, $D(p_T)$, $D(S_{T=})$ for a numerical simulation of a multifractal cascade process on a 128X128 point grid (mean=1), with Gaussian generator (except for the extreme fluctuations, the intensities are log-normally distributed - see Lavallée et al. (this volume) for more details). The dimensions were estimated using the box-counting technique described in appendix B[2]. In order to get a large enough sample size to estimate the dimension function $D(S_{T=})$, the latter was estimated from the sets $S_{T+\Delta T\geq}-S_{T>}$ with $\Delta T=0.2T$.

Note that as $T\rightarrow\infty$, $D(S_{T\geq})$ can $\rightarrow D_\infty >0$ (this occurs for example in the "α model", Schertzer and Lovejoy (1983,1985)), although in general[3] (especially in continuous cascades), we expect $D_\infty =0$. In the latter (more general) case $D(p_T)$ may initially increase with T, although it must eventually decrease.

A.3. The relation between $D(P_T)$ and area-perimeter exponents

We can now relate the areas and perimeters by eliminating R in the above equations. Using box-counting to estimate the Hausdorff dimensions (see appendix A), we obtain:

$$N_s(r) \approx \left(\frac{R}{r}\right)^{D(S_{T\geq})} \tag{A.3}$$

$$N_p(r) \approx \left(\frac{R}{r}\right)^{D(p_T)}$$

where N is the number of rXr boxes required to cover the set (which is of linear size R).

The areas $A_T(r)$ and perimeters $P_T(r)$ are given by:

$$A_T(r) = N_s(r)r^2 = \left(\frac{R}{r}\right)^{D(S_{T\geq})} r^2 \tag{A.4}$$

$$P_T(r) = N_p(r)r = \left(\frac{R}{r}\right)^{D(p_T)} r$$

eliminating R, we obtain:

$$P_T = A_T^{D(p_T)/D(S_{T\geq})}\, r^{1-2D(p_T)/D(S_{T\geq})} \tag{A.5}$$

or:

---

[1] $p_T = G\cap T$ does however generally apply to the mono-dimensional processes (such as fractional Brownian motion) discussed in Mandelbrot (1982).

[2] The procedure of transforming f_L into sets with threshold L, and measuring the dimension as a function of T using box-counting, constitutes the functional box-counting technique discussed in Lovejoy et al. (1987).

[3] See Schertzer and Lovejoy (1987a, this volume), for a discussion of this point, including universality classes of $D(S_{T\geq})$.

$$P \propto (\sqrt{A})^{\xi_T} r^{1-\xi_T} \tag{A.6}$$

where $\xi_T = 2D(p_T)/D(S_{T\geq})$. When $D(S_T)=2$ (i.e. the exceedance sets are not fractal), we obtain $\xi_T = D(p_T)$ which is the relationship discussed in Mandelbrot (1982), and applied to cloud areas in Lovejoy (1982). Alternatively, using the value ξ_T rather than $D(p_T)$ we are in error by the ratio $2/D(S_{T\geq}) \geq 1$.

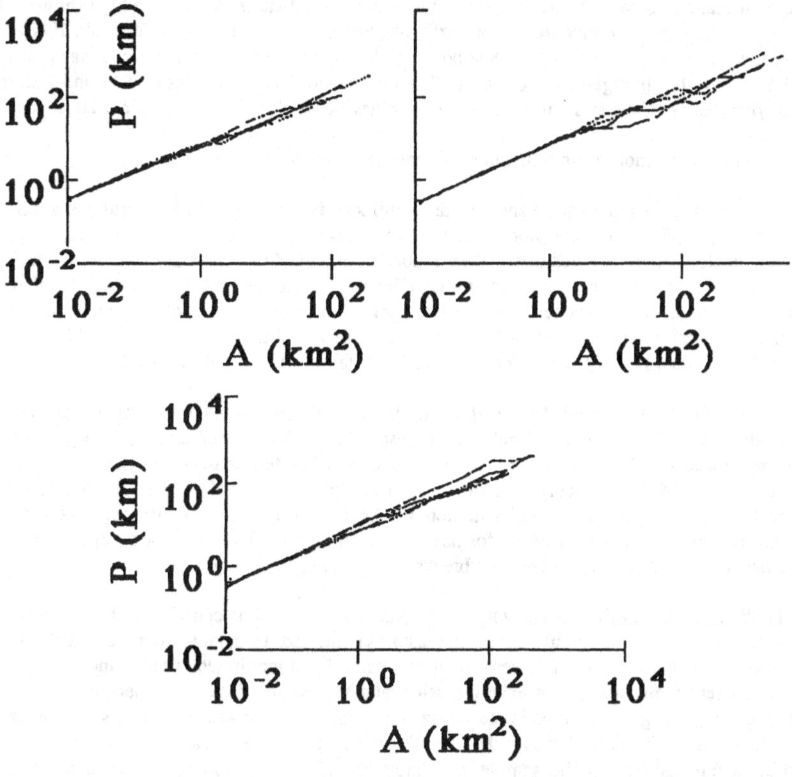

Fig. A.2: Area-perimeter curves for 9 different LANDSAT scenes (28m resolution) replotted from Welch et al. (1988). Although, as expected, different scenes yield different slopes, there is no evidence for a break in the scaling over the entire range of 10^{-2} to (nearly) 10^4km^2.

Recently, Welch et al. (1988) have used the same method to analyse LANDSAT data with 28m resolution to study visible cloud radiance fields. Although the authors claim that two straight lines with slightly different slopes (what they call "bifractal" behaviour) can be fit to their graphs (with a break at about 1km^2), single straight lines (one for each scene) do excellent jobs, as readers may verify for themselves (see fig. A.2 reproduced from their fig. 8). In any case, since the latter applied their analyses to single scenes, there will be statistical fluctuations due to the finite sample size. Furthermore, basic theoretical ideas about multifracals (as well as cascade simulations performed in Lavallée et al. (this volume)) indicate that for multifractal fields fluctuations are generally likely to be very large (strong intermittency) for individual realizations. In any case, it would be interesting (following Gabriel et al. (1988)) for the authors to reanalyse their data to try (for each scene seperately) to statistically reject the hypothesis that individual straight provide good fits. For the moment, we interpret these data as giving strong support for the scaling of cloud radiance perimeters between 10^{-2} and 10^4 km^2 for a variety of cloud types (i.e. precisely through the range of $1km^2$ where Cahalan (this volume) claims evidence for a break).

The above analysis shows that we should not be surprised by the empirical finding that area-perimeter relations give fairly constant exponents typically in the range $1.3 \leq \xi_T \leq 1.6$, since both $D(p_T)$ and $D(S_T)$ are likely to decrease slowly with increasing T, and hence the ratio $\frac{D(p_T)}{D(S_T)}$ may be expected to remain relatively fixed (see Yano and Takeushi (this volume) and Cahalan (this volume); the latter finds a slight increase of ξ_T with T, yielding typical low (dim) cloud values of 1.5, and high (bright) cloud values of 1.6). Note that this correction is not negligible; empirically, we find (Gabriel et al., 1988) that even for very low values of T, that $D(S_T) \approx 1.8$ at both visible and infra red wavelengths, and may easily decrease to ≈ 1.5 for very bright regions or cold tops. These values lead to corrections of 1.11 and 1.33 respectively. Applying these to Cahalan's typical range of ξ_T values we obtain $D(p_T) \approx 1.35$ and 1.20 respectively.

A.4. Other implications of multifractals for analysis methods

The preceding sub-sections have shown that multifractal fields are generally considerably more difficult to analyse than their monodimensional counterparts. Two other techniques that have been used to study scaling in cloud fields are area distribution exponents and the dimensions of graphs. Both techniques must be applied with considerable care to multifractal fields. In the former case (discussed in detail in appendix C) the interpretation of the exponents is quite different for multi and monofractals. In the latter (appendix D), shortcuts in performing box-counting that assume continuity of the process (which holds in many common monofractal processes but not generally in multifractal ones) can (and do) lead to completely erroneous results

Another aspect of multifractals that must be carefully considered in analysing data, is that unlike the monodimensional case, the thresholds that correspond to a given fractal dimension depend directly on the resolution with which the basic fractal measure is averaged by the measuring device. Even if the resolution is constant, the dimension corresponding to a given threshold will also depend on the realization of the process (e.g. the "meteorological situation"). As argued earlier, this difference between mono and multifractal processes is very basic; for now, we briefly discuss how these new dependencies can lead to practical difficulties, including apparent breaks in the scaling symmetry.

i) Pooling data from different realizations: Pooling data (e.g. in box counting, in area-perimeter graphs, or in area histograms) is a useful way of increasing sample size to obtain better statistical estimates of the parameters. However, it must be performed very carefully in multifractal fields, since it can "mix" fractals with different dimensions. For example, Rhys and Waldvogel (1986) consider area-perimeter relations obtained by pooling areas and perimeters of radar rain areas over consecutive images in time, using the same threshold. Since the dimensions for the fixed threshold will in general vary in time, this mixes fractals with different dimensions. If the sample was large enough, this would not be serious, since the largest area-perimeter exponent would eventually dominate the rest. However, if the fractals have nearly the same dimensions (as they do in their study), the convergence is extremely slow, and finite samples, will yield either large spreads or non-linear log-log plots, that can easily (and erroneously) be interpreted as breaks in the scaling. These results must therefore be re-analysed before any conclusions about symmetry breaking can be drawn.

ii) Combining spatial averaging with thresholding: The variation of fractal dimension with threshold can lead to artificial symmetry breaking in yet another way. Consider taking a short-cut in estimating areas and dimensions by fixing a threshold and degrading successively the resolution of the set which exceeds the threshold (e.g. by box-counting), calculating p_T from the set at different resolutions. This method will work whether or not the field is multifractal, since it is first converted into a set having a well defined dimension. However, if rather than degrading the set resolution, we degrade the multifractal field itself by simply averaging the field (rather than the set) over larger and larger scales, and then defining the exceedance set with respect to the previous threshold (as in Yano and Takeuchi, this volume), the method will no longer work. To recuperate a set with the same dimension, the threshold must be appropriately decreased to compensate for the fact that averaging over larger scales decreases (smooths) the intense regions (the precise amount of decrease can be quantitatively estimated by associating each threshold with singularities as in the following section). As long as sufficiently low thresholds are used so that the dimension varies relatively little with threshold (i.e. in this range of thresholds, the field is approximately mono-dimensional), this effect will not be too important, but at extreme threshold levels, where the dimension changes more rapidly with threshold (e.g. for the cirrus clouds in Yano and Takeuchi, this volume) this will yield systematic (but totally artificial) breaks in the scaling (the downward curvature observed in their curves).

APPENDIX B: ESTIMATING HAUSDORFF DIMENSIONS BY BOX-COUNTING:

Box-counting techniques have been developed for some time to estimate the dimension of strange attractors and other fractal sets. Lovejoy et al. (1987) showed how it could be applied to multifractal fields by transforming them into the series of sets $S_{T\geq}$ as indicated above.

Box counting arises as a natural method for estimating Hausdorff dimensions. This may be seen by recalling the definition of Hausdorff measures and dimensions. Consider a set (fractal or otherwise) S. Define the "Hausdorff measure of S resolution l relative to the (convex) function g" as follows:

$$\mu_{l,g}(S) = \inf_{\text{all coverings } B_i \text{ such that diam } B_i \leq l} \sum_{i}^{N(l)} g(\mathrm{diam}(B_i)) \qquad \text{(B.1)}$$

where we cover the set with $N(l)$ "balls" B_i of diameter = diam(B_i) such that diam(B_i)$\leq l$. The "inf" is over all possible coverings and is responsible for the practical difficulties in evaluating Hausdorff measures. Next, we define the "Hausdorff measure of S relative to g" as:

$$\mu_g(S) = \lim_{l \to 0} \mu_{l,g}(S) \qquad \text{(B.2)}$$

Next, taking $g(t)=t^h$, we can define the "h-dimensional Hausdorff measure of S", $\mu_h(S)$. Finally, we can define the "Hausdorff dimension" (D(S)) as:

$$D(S) = \inf_h (\mu_h(S)=0) = \sup_h (\mu_h(S)=\infty) \qquad \text{(B.3)}$$

The (non-trivial and extremely general) equality between the inf and the sup (which implies a brutal transition from 0 to ∞ as h is decreased below D(S)), is the property that makes Hausdorff dimensions so useful, since they can be applied to virtually any set. The interpretation of the above property is straightforward. For example, when applied to a planar set S (D(S))=2), it says that the "volume (μ_3(S))=0, and that the "length" (μ_1(S))=∞. The size of the set is then given by $\mu_{D(S)}(S)$ (which for standard sets, yields the Lebesgue measure of S as expected). Note that some sets (such as the trail of a Brownian particle, which, when embedded in a space with d$\geq$2, has D(S)=2), yields $\mu_{D(S)}(S)=0$ even though the set has an infinite number of points. These are "marginal" sets and require using $g(t)=t^{D(S)}(\log t)^{\delta 1}(\log\log t)^{\delta 2}\cdots$ with at least one of the δ's>0. These are the source of the "log corrections" that are sometimes discussed in this context. Note that Mauldin and William (1986) have shown that in a certain class of random fractal sets, that log-correction arise quite naturally. It is therefore perhaps not too surprising that in some cases where the trace moments indicate that power law scaling is well respected (e.g. the lidar data discussed in section 5), that simple box-counting yields curved log $N(l)$ v.s. log l plots compatible with log corrections.

The box-counting approximation is now relatively simple to describe. Ignoring the possibility of logarithmic corrections, (i.e. taking $g(t)=t^h$), and covering the set with disjoint "cubical" balls, size l, so that diam(B_i)=l (i.e a regular grid, discarding the inf), we obtain:

$$\mu_{l,h}(S) \approx \sum_{i=1}^{N_S(l)} l^h = l^h N_S(l) \tag{B.4}$$

taking the limit $l \to 0$, and recalling that $\mu_h(S)$ is finite and non-zero only if $h=D(S)$, we obtain:

$$N_S(l) \approx l^{-D(S)} \qquad \text{as } l \to 0 \tag{B.5}$$

which yields the estimate:

$$D(S) = \lim_{l \to 0} -\frac{\log N_S(l)}{\log l} \tag{B.6}$$

APPENDIX C: AREA DISTRIBUTION EXPONENTS

Consider the problem of relating the distribution of contiguous areas equal to or exceeding T to the set $S_{T\geq}$. If $D(S_{T\geq}) > 1$ then $S_{T\geq}$ will generally be made up of many contiguous subsets (denoted $s_{T\geq}^{(i)}$) each with external scale Λ_i for the ith region. Because of the assumed scaling, the Λ will generally be distributed (to within logarithmic corrections) as:

$$Nr(\Lambda > l) \propto l^{-B_T} \tag{C.1}$$

where "Nr" indicates the number of subsets $s_{T\geq}^{(i)}$ whose size exceeds l. Note that since $S_{T\geq} \supset s_{T\geq}$, $D(s_{T\geq}) \leq D(S_{T\geq})$.

At resolution r, a subset $s_{T\geq}$ with scale l will have area:

$$a_T(r) \approx \left(\frac{l}{r}\right)^{D(s_{T\geq})} r^2 \tag{C.2}$$

Hence, eliminating l in terms of a_T, we obtain:

$$Nr(A_T > a_T) \propto a_T^{-B'_T} \tag{C.3}$$

where $B'_T = B_T/D(s_{T\geq})$. Empirically, B'_T is the most readily accessible area exponent. Lovejoy (1981, and Lovejoy and Mandelbrot (1985) show empirically that for light rain rates, low clouds, $B'_T \approx 0.75$. Lovejoy and Mandelbrot (1985) and Lovejoy and Schertzer (1985) also develop mono-dimensional models with B'_T in the range 0.5 - 0.75. Another reference is Cahalan (this volume) who obtains $B'_T \approx 0.8$ for satellite cloud pictures.

We now seek to relate B_T, $D(s_{T\geq})$ and $D(S_{T\geq})$. To do so, note that the total number of boxes required to cover S_T is:

$$N_s(r) \approx \left(\frac{R}{r}\right)^{D(S_{T\geq})} \propto \int_r^R \left(\frac{l}{r}\right)^{D(s_{T\geq})} dNr(\Lambda > l) \tag{C.4}$$

where $dNr(\Lambda > l) \propto l^{-B_T - 1} dl$ is the number density associated with $Nr(\Lambda > l)$. The above yields:

$$N_s(r) \propto r^{-D(s_{T\geq})} \int_r^R l^{D(s_{T\geq})} l^{-B_T - 1} dl \tag{C.5}$$

taking into account the sign of $D(s_T)-B_T$, this yields

$$N_S(r) \propto r^{-D(s_{T\geq})} \left[\left| R^{D(s_{T\geq})-B_T} - r^{D(s_{T\geq})-B_T} \right| \right] \quad (C.6)$$

The relative values of $D(s_{T\geq})$ and B_T depends on the topological (connectedness) properties of the process. We must now distinguish two cases depending on which is greater:

i) $B_T \geq D(s_{T\geq})$: In this case, in the limit $r \rightarrow 0$, the $r^{D(s_{T\geq})-B_T}$ term dominates the $R^{D(s_{T\geq})-B_T}$ term and the number of boxes/subset is small compared to the total number; fragmentation dominates, and:

$$N_S(r) \propto r^{-B_T} \quad \Rightarrow \quad B_T = D(S_{T\geq}) \quad (C.7)$$

This is the case discussed in Mandelbrot (1982) where geometric generators are used to produce fractal sets which yield $D(s_{T\geq}) < D(S_{T\geq})$. We therefore have $D(S_{T\geq}) = B_T > D(s_{T\geq}) \Rightarrow B'_T = \frac{B_T}{D(s_{T\geq})} > 1$.

ii) $B < D(s_{T\geq})$: In this case, in the limit $r \rightarrow 0$, $R^{D(s_{T\geq})-B_T}$ dominates $r^{D(s_{T\geq})-B_T}$ and the number of boxes/fractal subset is a large fraction of the total; the number of boxes needed to cover the fragments is negligible compared to the number needed to cover individual connected regions, this yields:

$$N_S(r) \propto r^{-D(s_{T\geq})} \quad \Rightarrow \quad D(s_{T\geq}) = D(S_{T\geq}) \quad (C.8)$$

and $B_T < D(s_{T\geq})$ and hence $B'_T = \frac{B_T}{D(s_{T\geq})} < 1$. Each contiguous region has the same dimension as the entire set, and the fragmentation is relatively unimportant. This is the case relevant to multifractal fields, and of interest in geophysical applications.

APPENDIX D: GRAPHS AND THEIR DIMENSIONS

Carter et al. (1986) considered the x-z cross-section (intersection) of the graph G of the infra red cloud radiance as a function of telescope scanning angle (for clarity we use the notation $(\underline{r}, f(\underline{r})) = (x,y,z)$ since $\underline{r} = (x,y)$ and f=z=radiance, x,y are angle variables). Denote this intersection set by $G \cap (xz)$. They then estimate the dimension of G using box-counting to cover the graph of $G \cap (xz)$. However, their method actually implicitly assumes continuity- they use additional boxes to cover not only their experimental points, but also those points on straight lines connecting the latter. Their resulting estimate of $D(G \cap (xz))$ is very near 1 (1.16, and 1.11 depending on the wavelength used), and we may therefore suspect that it is an artifact of their assumption of continuity (since the connecting straight lines have dimension 1, while the experimental points have $D<1$, their method will estimate the maximum of the two). We now discuss this possibility in more detail.

The simplest way to relate $D(G_T)$ to $D(S_{T\geq})$ and $D(S_{T=})$ is recall that $S_{T\geq}$ is the projection of G_T with the x,y plane. The projection set (S_p) of S_1 onto S_2 has dimension:

$$D(S_p) = \min(D(S_1), D(S_2)) \quad (D.1)$$

in this case, we obtain:

$$D(S_{T\geq}) = \min(D(G_T), 2) \quad (D.2)$$

There are now two distinct possibilities:

i) $D(S_{T\geq})=2$:

If the process is non-stationary, such as the (mono-dimensional, Gaussian) fractal Brownian motion processes used by Mandelbrot (1982) and Voss (1983) to model mountains, then the graph is continuous (but non-differentiable), but rarely crosses the plane $f(\underline{r})=T$, and $D(S_{T\geq})=2$, and the projection relation gives us no further information about $D(G_T)$.

ii) $D(S_{T\geq})<2$:

The process is stationary (as are cascade processes, and, presumably most remotely sensed fields), in this case, we will generally have $D(S_{T\geq})<2$, hence the projection relation yields:

$$D(G_T)=D(S_{T\geq})<2 \qquad \text{(D.3)}$$

In this case (which applies to the examples shown in fig. A.1a,b,c), G_T will be discontinuous everywhere. We now consider $G\cap(xz)$ studied by Carter et al. 1986. All the preceeding results apply to $G\cap(xz)$, $S_{T\geq}\cap(xz)$ etc., as long as the dimensions of all the above sets are reduced by one (if the corresponding value becomes negative, it must be reset to zero). Since we have empirical evidence that generally, $D(S_{T\geq})<2$ (see the section 2), we expect $D(S_{T\geq})=D(G_T)$, so that $D(G_T\cap(xz)) = D(S_{T\geq}\cap(xz))<1$, implying that $G\cap(xz)$ is discontinuous everywhere. However, Carter et al. (1986) assumed that $G\cap(xz)$ was continuous, and effectively interpolated their experimental points with a continuous line (dimension 1 which is > $D(G\cap(xz))$), hence we expect them to obtain an estimate D=1 (the maximum of the dimension of the set of points on $G\cap(xz)$ and the set on the line connecting them). Indeed, careful inspection of the Carter et al. (1986) box-counting figures ($N(l)$ v.s.l) indicate that the function $N(l) \approx l^{-1}$ (hence D=1) fits well over most of the range of l (which was only over roughly two orders of magnitude anyway).

This example, illustrates the dangers of approaching the data analysis with, unwarranted theoretical preconceptions about the continuity of the process.

APPENDIX E: Generalized scale invariance and space-time transformation in rain:

In both geophysical and laboratory flows, it is generally far easier to obtain high temporal resolution velocity data at one or only a few points than to obtain detailed spatial information at a given instant. It is therefore tempting to relate time and space properties by assuming that the flow pattern is frozen and is simply blown past the sensors at a fixed velocity without appreciable evolution, and to directly use the time series information to deduce the spatial structure. This "Taylor's hypothesis of frozen turbulence" (Taylor, 1938) can often be justified because in many experimental set ups, the flow pattern is caused by external forcing at a well defined velocity typically much larger than the fluctuations under study. However, in geophysical systems (in particular in the atmosphere and ocean) where no external forcing velocity exists, the hypothesis has often been justified by appeal to a meso-scale gap separating large scale motions (e.g. "weather") and small scale "turbulence". If such a separation existed, it might at least justify a statistical version of Taylor's hypothesis in which a the large scale velocity is considered statistically constant (stationary). Various statistical properties such as spatial and temporal energy spectra would be similar even though no detailed transformation of a given time series to a particular spatial pattern would be possible. Only some kind of statistical equivalence could be made.

If, as argued in this paper and in many of the references cited in the introduction, the meso-scale gap is a fiction, then no large scale forcing velocity can be appealed to in order to transform from space to time, and turbulent velocities must be used instead of amplitude $v_l \approx \langle\varepsilon_l^{1/3}\rangle l^{1/3}$ where l is the scale of the eddy, ε_l is the energy flux through the eddy to smaller scales. Although $\langle\varepsilon_l\rangle$ is scale independent, due to intermittency[1] $\langle\varepsilon_l^{1/3}\rangle \approx l^{\delta}$ where δ is a small correction. Rather than being scale independent, the space-time transformation has a scale dependent velocity[2] $v_l \approx l^H$ with $H\approx 1/3+\delta$. The two geophysically relevant Taylor's hypotheses therefore correspond to H=0 or H=1/3+δ depending on the existence (or not) of the "gap".

The theoretical arguments mentioned above make it clear that the turbulent velocity is likely to be relevant one for space-time transformations, and the clear differences in slope in the trace moments analyses

[1] These are due to multifractal effects, the exact value of will depend on the $K_D(h)$ function for the velocity field, as well as on which value of h is relevant for rain.

[2] Each moment of the rain field will require a different δ. For simplicity, we ignore this complication here.

(figs 4.3-4.5) rule out the constant velocity (H=0) hypothesis[1]. The space-time transformation we infer from the turbulent value of H (≈1/3) can be easily expressed in the formalism of Generalized Scale Invariance. Consider (x,y,t) space, the space-time transformation can be simply expressed by statistical invariance with respect to the following transformation: $x \to x/\lambda$, $y \to y/\lambda$, $t \to t/\lambda^{1-H}$ or using the notation $r=(x,y,t)$, $r_\lambda = T_\lambda r_1$ with $T_\lambda = \lambda^{-G}$ and:

$$G = \begin{bmatrix} 1 & 0 & 0 \\ 0 & 1 & 0 \\ 0 & 0 & 1-H \end{bmatrix};$$

we therefore obtain Trace(G)= 3-H, i.e. by measuring d_{el} or H we can determine G (assuming that there are no off-diagonal elements corresponding to rotation between space and time, and ignoring differential rotation in the horizontal). The isotropic statistical Taylor's hypothesis[2] is $H \approx 1/3$, $d_{el} \approx 8/3$. If we now consider the full (x,y,z,t) space, it has already been shown (Lovejoy et al., 1987) that $d_{el} \approx 2.22$ for the corresponding transformation in (x,y,z) space for radar rain data, hence, $H_z \approx 0.22$, and the corresponding $d_{el} \approx 2.22 + 2/3 \approx 2.89$.

REFERENCES

Benzi, R., G. Paladin, G. Parisi, A. Vulpiani, 1984: J. Phys. A, 17, 3521.

Cahalan, R.F., 1990: Landsat Observations of Fractal Cloud Structure. (this volume).

Carter, P.H., R., Cawley, A. L. Licht, M.S. Melnik, 1986: Dimensions and entropies in chaotic systems. Ed. G. Mayer-Kress, p. 215-221, Springer.

Essex,C., 1990: Correlation dimension and data sample size. (this volume).

Frisch,U., G. Parisi, 1985: A multifractal model of intermittency, Turbulence and predictability in geophysical fluid dynamics and climate dynamics, 84-88, Eds. Ghil, Benzi, Parisi, North-Holland.

Davis, A., S. Lovejoy, and D. Schertzer 1990: Radiative transfer in multifractal clouds. (this volume).

Gabriel, P., S. Lovejoy, D. Schertzer, and G.L. Austin, 1988: Multifractal analysis of resolution dependence in satellite in satellite imagery, Geophys. Res. Lett., 15, 1373-1376.

Grassberger, P., 1983: Generalized dimensions of strange attractors. Phys. Lett., A 97, 227.

Halsey, T.C., M.H. Jensen, L.P. Kadanoff, I. Procaccia, B. Shraiman, 1986: Fractal measures and their singularities: the characterization of strange sets. Phys. Rev. A, 33, 1141-1151.

Hentschel, H.G.E., I. Proccacia, 1983: The infinite number of generalized dimensions of fractals and strange attractors, Physica, 8D, 435-444.

Lavallée, D., D. Schertzer, S. Lovejoy, 1990: On the determination of the codimension function. (this volume).

Lovejoy, S. 1981: Analysis of rain areas in terms of fractals, 20th conf. on radar meteorology, 476-484, AMS Boston.

Lovejoy, S., 1982: The area-perimeter relationship for rain and cloud areas. Science, 216, 185-187.

Lovejoy, S., B. Mandelbrot, 1985: Fractal properties of rain and a fractal model. Tellus, 37A, 209-232.

Lovejoy, S., D. Schertzer, 1985: Generalized scale invariance in the atmosphere and fractal models of rain. Wat. Resour. Res., 21, 1233-1250.

Lovejoy, S., D. Schertzer, 1986a: Scale invariance, symmetries fractals and stochastic simulation of atmospheric phenomena. Bull AMS 67, 21-32.

Lovejoy, S., D. Schertzer, A.A. Tsonis, 1987a: Functional box-counting and multiple elliptical dimensions in rain. Science, 235, 1036-1038.

Lovejoy, S., D. Schertzer, 1988: Extreme variability, scaling and fractals in remote sensing: analysis and simulation, Digital image processing in remote sensing, Ed. J. P. Muller, Francis and Taylor, 177-212.

---

[1] Unfortunately, the value of H estimated (H=0.5±0.3) is for the extreme large moments of the lidar returns, and in any case is not accurate enough to usefully estimate δ.

[2] Since δ depends on the various moments of the rain field considered, the generator G will in fact be stochastic, not constant as assumed for simplicity here (see Schertzer and Lovejoy, this volume).

Lovejoy, S., D. Schertzer, 1990a: Our multifractal atmosphere: a unique laboratory for nonlinear dynamics. Physics in Canada (in press).
Lovejoy, S., D. Schertzer, 1990b: Fractals, raindrops and resolution dependence of rain measurements, J. Appl. Meteor. (in press).
Lovejoy, S., D. Schertzer, 1990c: Multifractals, Universality Classes and Satellite and Radar Measurements of Cloud and Rain Fields, J. Geophy. Res., 95, 2021-2034.
Ludwig, F.L. K.C. Nitz, 1986: Analysis of lidar cross sections to determine the spatial structure of material in smoke plumes. Proc. of Smoke Obscurants Symposium, vol. 10, Harry Diamond Labs; Adelphis, Md., 231-241.
Mandelbrot, B., 1974: Intermittent turbulence in self-similar cascades: divergence of high moments and dimension of the carrier. J. Fluid Mech., 62, 331-350.
Mandelbrot, B., 1982: The Fractal Geometry of Nature. Freeman, 465pp.
Marshall, J.S., W.M. Palmer, 1948: The distribution of raindrops with size. J. Meteor., 5, 165-166.
Mauldin, R.D., S.C. Williams, 1986: Random recursive constructions: asymptotic geometric and topological properties. Trans. Am. Math. Soc., 295, 325-346.
Richardson, L.F., 1922: Weather prediction by numerical process, republished by Dover (1965).
Rhys, F.S., A. Waldvogel, 1986: Fractal shape of hail clouds. Phys. Rev. Lett. 56, 784-787.
Schertzer, D., S. Lovejoy, 1983: On the dimension of atmospheric motions. Preprint Vol., IUTAM Symp on Turbulence and Chaotic Phenomena in Fluids, 141-144.
Schertzer, D., S. Lovejoy, 1985a: The dimension and intermittency of atmospheric dynamics. Turbulent Shear Flow, 4, 7-33, Ed. B. Launder, Springer, NY.
Schertzer, D., S. Lovejoy, 1985b: Generalized scale invariance in turbulent phenomena. P.C.H. Journal, 6, 623-635.
Schertzer, D., S. Lovejoy, 1987a: Singularités anisotropes, et divergence de moments en cascades multiplicatifs. Annales Math. du Qué., 11, 139-181.
Schertzer, D., S. Lovejoy, 1987b: Physically based rain and cloud modeling by anisotropic, multiplicative turbulent cascades. J. Geophys. Res., 92, 9693-9714.
Schertzer, D., S. Lovejoy, 1988: Multifractal simulations and analysis of clouds by multiplicative processes. Atmospheric Research, 21, 337-361.
Schertzer, D., S. Lovejoy 1989: Generalized Scale invariance and multiplicative processes in the atmosphere. Pageoph, 130, 57-81.
Schertzer, D., S. Lovejoy, 1990a: Nonlinear geodynamical variability: multiple singularities, universality, observables. (this volume).
Schertzer, D., S. Lovejoy, 1990b: Nonlinear variability in geophysics: multifractal simulations and analysis, in Fractals: physical origins and properties, Ed. L. Pietronero, Plenum, New York, 49-79.
Schertzer, D., S. Lovejoy, R. Visvanathan, D. Lavallée, and J. Wilson, 1988: Universal Multifractals in Turbulence, in Fractal Aspects of Materials: Disordered Systems, Edited by D.A. Weitz, L.M. Sander, B.B. Mandelbrot, 267-269, Materials Research Society, Pittsburg, Pa.
Taylor, G. I., 1938: The spectrum of turbulence. Proc. Roy. Soc., A164, No. 919, 476-490.
Voss, R., 1983: Fourier Synthesis of Gaussian fractals: 1/f noises, landscapes and flakes. Proceedings, Siggraph Conf., Detroit, p1-21.
Welch, R.M., K.S. Kuo, B.A. Wielicki, S.K. Sengupta, L. Parker, 1988: Marine stratocumulus cloud fields off the coast of Southern California observed by LANDSAT imagery, part I: Structural characteristics. J. Appl. Meteor., 27, 341-362.
Wilson, J., D. Schertzer and S. Lovejoy, 1990: Physically based cloud modelling by scaling multiplicative cascade processes. (this volume).
Yano, J.-I., Takeuchi, Y., 1990: Fractal dimension analysis of horizontal cloud pattern in the intertropical convergence zone. (this volume).

EIGEN STRUCTURE OF EDDIES; CONTINUOUS WEIGHTING VERSUS CONDITIONAL SAMPLING

L. Mahrt
Department of Atmospheric Sciences
Oregon State University
Corvallis, Oregon 97331
U.S.A.

ABSTRACT: A weighting function is applied to potential samples of turbulence structures. Use of the weighting function requires less *a priori* commitment compared to conditional sampling and is less vulnerable to noise at the expense of loss of some structural detail. Eigenvectors of the resulting lagged covariance matrix are computed to examine different structural modes. Comparison with artificial data indicates a degree of independence from the choice of the weighting function.

1. INTRODUCTION

The structure of the main eddies in neutrally or stably stratified turbulent flows is poorly known excluding certain cases where the main eddies are dictated by the geometry of the laboratory setup and perhaps encouraged by low Reynolds numbers. Without some knowledge of the structure of the most coherent eddies, one may be reluctant to specify test patterns (Mumford, 1982) or other criteria required for conditional sampling. Furthermore, conditional sampling analyses can be seriously contaminated by less coherent smaller scale fluctuations.

In the following development, we will apply a weighting function to the potential samples. Application to artificial turbulence with known building blocks indicates that using continuous weighting reduces the impact of noise but also eliminates some of the detail of the building blocks. Since turbulence often contains simultaneous occurrence of different eddy modes, the lagged covariance matrix of the weighted samples will be decomposed into eigenvectors (Busch and Petersen, 1971; Panofsky and Dutton, 1984).

2. WEIGHTED EIGENVECTORS

Consider the multivariate observational vector, f(z, x), where z identifies the realization or sample and x indicates the relative position within the realization. In general, z could represent a three-dimensional position vector or could represent time or any other index which systematically identifies the different realizations. For example, z might indicate the time of the realization while x represents spatial position. In the present analysis, z is the relative position of the observational vector of the sample within a time series while x is the relative position within each sample.

The computation of eigenvectors of the covariance matrix is motivated by maximization of the following measure of similarity

D. Schertzer and S. Lovejoy (eds.), Non-Linear Variability in Geophysics, 145–151.

$$\rho^2(z) = [\int_0^l f(z, x)\phi(x)dx\ ]^2 \ / \int_0^l \phi^2(x)dx \tag{1}$$

where ϕ is the desired function most similar to the observed realizations of f(z, x) and l is the domain of the realization. In the usual eigenvector approach, $\rho^2(z)$ is maximized over the observed samples. With random selection of samples from time or space series, the eigenvectors approach Fourier modes (excluding trend and boundary effects) regardless of the structure of the motions leading to the covariance. It is possible to eliminate this degeneracy and restore phase information by applying a weighting function to the potential samples. This approach includes conditional sampling as a special case. The weighting function is chosen to emphasize some feature of the flow. Examples of the weighting function are presented in the next section.

We wish to maximize the expected value of the similarity measure in eq. 1 subject to the specified weighting function g(z); that is, we wish to maximize

$$\lambda = E[g(z)\rho(z)] \tag{2}$$

Choosing the expected value, E[ ], to be a simple integral average, we obtain

$$\lambda = \frac{1}{L}\int_0^L g(z) \frac{[\int_0^l f(z, x)\phi(x)dx\ ]^2}{\int_0^l \phi^2(x)dx} dz \tag{3}$$

where the realization index z ranges from 0 to L.

To determine the function ϕ which maximizes λ, we proceed to applying the usual calculus of variations (e.g., Panofsky and Dutton, 1984; Chapter 12). This is done by perturbing the function ϕ by amount $\varepsilon\delta\phi$. Replacing $\phi(x)$ in eq. 3 with the perturbed value $\phi(x) + \varepsilon\delta\phi(x)$ and taking the limit

$$\left.\frac{d\lambda(\varepsilon)}{d\varepsilon}\right|_{\varepsilon\to 0} = 0$$

We thus obtain an additional constraint on ϕ which will eventually allow quantitative determination of ϕ. After considerable algebra, we obtain

$$\int \tilde{K}(x, y)\ \phi(y)dy = \lambda\phi(x) \tag{4}$$

$$\tilde{K}(x,y) = \int_0^L g(z)\ f(z, x)\ f(z, y)dz \tag{5}$$

where x and y are two different relative positions within the sample. $\tilde{K}(x, y)$ is simply the usual covariance matrix K(x, y) transformed by the weighting function g(z). This result in eqs. 4 and 5 allows use of the usual mathematics for solution of the eigenvalue problem. All of the convenient properties of eigenvectors follow. For application to actual data (Busch and Petersen, 1971), we must apply the discrete version of eq. 5 in which case z, g(z) and f(z, x) become discrete variables determined by the data resolution.

3. WEIGHTING FUNCTION

In practice, the weighting function can represent any feature of the flow. For example, the weighting function can be expressed in terms of the local buoyancy flux to study the nature of the motions most responsible for the flux in stratified flows. As an alternative example, the weighting function can be chosen to emphasize bursts of small scale activity which might be defined in terms of the VITA method (Chen and Blackwelder, 1978 and others). In the special case of constant weighting (equivalent to no weighting), the covariance matrix is approximately persymmetric (elements of a given diagonal are equal) and the eigen modes degenerate to Fourier modes regardless of the structure of the actual motions.

Here we define the weighting function in terms of similarity of the realization f(z,x) with an indicator function

$$g(z) = \int h(x)\, f(z, x)dx \qquad (6)$$

where h(x) represents the flow signature of interest and the integral applies to the width of the sampling window. Of course with actual data, eq. 6 must be in discrete form.

In an attempt to reduce the dependence of the eigenvectors on the choice of the weighting function, an iterative procedure can be adopted where the indicator function for the ith iteration is assigned to be the first eigenvector of the previous iteration

$$h_i(z) = \phi_{i-1}(x) \qquad (7)$$

Then the weighting function is determined from eq. 6 for each iteration.

The weighting function can be converted to conditional sampling by including only samples where eq. 6 exceeds a critical value. With analysis of time series, an observational rake moves through the time series conditionally selecting samples according to sampling criteria. Overlap criteria are specified to insure that the same event is not selected more than once. In practice additional criteria are often specified to eliminate inadvertent selection of events of no interest.

With continuous weighting (eq. 6), no criteria are specified. In the analysis of time series, several sequential samples of the same event are collected. The sample with the event centered will be given the most weight. Inclusion of the less-weighted uncentered events will lead to smoothing of the eigenvectors although in the compositing of the samples the overall phase will not be lost. Of course with actual turbulence data, the main events are not always obvious.

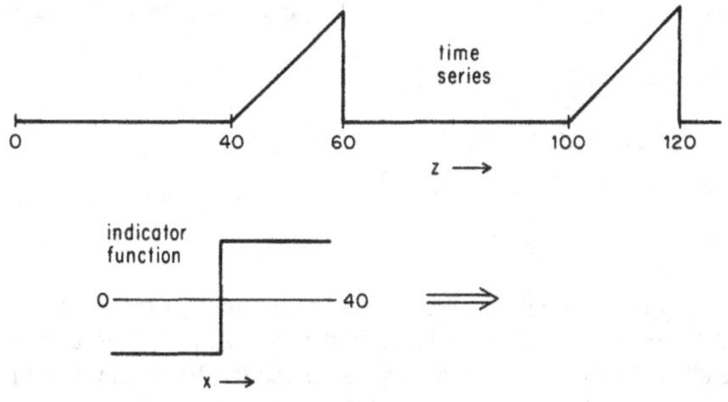

Fig. 1: Ramp structure for the artificial record and indicator function.

Identification of discontinuities by the above methodology is of considerable interest since narrow zones of concentrated gradients seem ubiquitous in turbulence. These zones are sometimes referred to as fronts. Turbulent front-like phenomena have been studied extensively in heated boundary layers where they are often identified in terms of ramp patterns (Antonia and Atkinson, 1976). The studies of Gibson et al.

(1968), Kim et al. (1971), Laufer (1975), Brown and Thomas (1977), Chen and Blackwelder (1978) and Subramanian et al. (1982) have attributed frontal zones of strong shear or temperature gradients to the edges of large eddies or uplifting of near surface fluid by the large eddies (bursting). The formation of sharp eddy boundaries contrasts with the usual concept of eddy generation based on Fourier representation (e.g., Townsend, 1976). To test the ability of the above techniques to determine the structure of fronts, the indicator function will be defined as a simple step function (fig. 1) and applied to artificial turbulence containing discontinuities. This is the case where continuous weighting should perform the worst since the technique's greatest shortcoming is inadvertent smoothing.

4. ARTIFICIAL TURBULENCE

To better understand the capabilities and disadvantages of the above techniques, we will apply the methodology to artificial turbulence consisting of simple building blocks. In some simple cases, the effect of smoothing on the composited structure can be computed analytically. Here we examine the first eigenvector computed from artificial turbulence consisting of 100 regularly spaced ramp functions (fig. 1) and background noise. Using the step indicator function to compute the weighting function (eq. 6), what features of the ramp function can be identified? Note that we have purposely chosen the indicator function to be different than the ramp building block of the artificial turbulence.

Samples are generated by marching a sampling rake, here 40 points wide, through the record. Since the ramp function is asymmetric, only samples with negative values of the weight (eq. 6) will be included in order to study the representation of the ramp discontinuity by the sampling techniques. The sloped part of the ramp could be better captured by using positive values of the weight or increasing the domain of the sampling width.

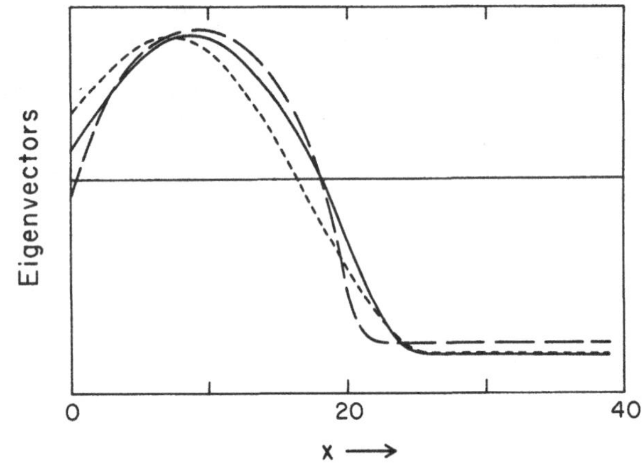

Fig. 2: Leading eigenvectors for the artificial record for the step indicator function (solid line), step function to the 5th power(broken line), and linear indicator function (short dash).

The resulting first eigenvector for continuous weighting for the case of no noise is plotted in fig. 2. The first eigenvector, which explains about 60% of the total variance of the weighted samples, contains a smoothed version of the discontinuity and an indication of the ramp slope on the left-hand side of the plot. The second eigenvector, which explains about 16% of the variance of the weighted samples, is important mainly for uncentered samples.

To sharpen the interrogation of the discontinuity, one can raise the weighting function to an arbitrary exponential power greater than one. With an exponential power of five, improvement of the resolution of the discontinuity occurs with the loss of information on the sloped section of the ramp as is shown in fig. 2.

With iteration (eq. 7), the eigen representation of the discontinuity is shifted to the left. Since the eigenvector of the previous iteration is used as an indicator function, which is smoother than the original step indicator function, the transformation corresponding to the application of this weighting function has the effect of smoothing the perceived ramp and thus smoothing and shifting the representation of the discontinuity by the eigenvectors.

To study the sensitivity of the eigenvectors to the choice of the indicator function for the weighting, additional indicator functions were used. The simplest was a linear dependence. The resulting eigen structure (fig. 2) is similar to, but smoother than, that obtained from use of the step indicator function. Because of the asymmetry of the ramp, the smoothing causes the eigen structure to shift to the left. Use of indicator functions which are shaped more like the ramp function leads to slightly more realistic eigenvectors. However, changes in the eigen structure are always small compared to the specified changes of the indicator function. Thus with artificial time series, the resulting eigenvectors possess a degree of insensitivity to the choice of indicator function. With analysis of actual geophysical turbulence data, the main features of the eigen modes are also found to be only weakly dependent on the choice of indicator function.

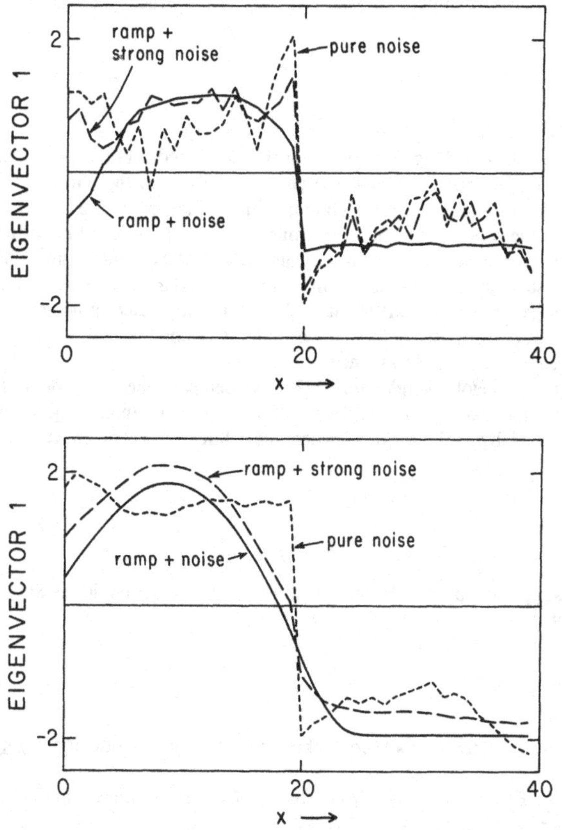

Fig. 3: The first eigenvector for records consisting of ramp structures and white noise for conditional sampling (upper figure) and continuous weighting (lower figure). Shown is the case where noise amplitude equals the ramp amplitude (solid), the noise amplitude is five times the ramp amplitude (long dash), and the case of pure noise (short dash).

While the continuous weighting leads to undesirable smoothing, the resulting eigenvectors are less vulnerable to noise. In this regard, the continuous weighting could be regarded as a low pass filter, although the response function is not specified but is instead determined by the flow structure itself. The partial immunity to noise allows the method to successfully extract structure of the building blocks with a smaller number of samples compared to conditional sampling. This is an important advantage in the analysis of geophysical time series where transport and variance may be dominated by a relatively small number of events.

To study the degree of noise vulnerability of continuous weighting and conditional sampling, these methods were applied to an artificial record consisting of ramp functions and white noise. The results based on continuous weighting are much less affected by the noise (fig. 3). The difference between the two methods is partly due to the general decrease of noise according to the square root of the number of samples; because of the overlap, the continuous weighting generates more samples, in this case by a factor of 20. In addition, conditional sampling is more vulnerable to noise-related phase errors.

As the amplitude of the noise increases beyond the amplitude of the ramp building block, the first eigenvector based on conditional sampling becomes seriously contaminated whereas the first eigenvector based on continuous weighting is still able to extract a smoothed version of the ramp function. For example, when the noise amplitude is five times greater than the ramp amplitude (fig. 3), the first eigenvector based on conditional sampling is similar to the superficial structure extracted from pure noise. In contrast, the five-fold increase of noise exerts only a modest influence on the structure captured with continuous weighting.

5. DISCUSSION

For the cases studied here, the method of weighted eigenvectors identified smoothed versions of building blocks of artificial time series. This identification was not particularly sensitive to the choice of the indicator function required to compute the weighting function. The smoothing corresponding to application of the continuous weighting function reduces the impact of noise while the coherent part of the artificial time series is retained. Application of the method to the actual turbulence data (Mahrt and Frank, 1988) also reduces the influence of smaller scale random motions. In contrast, application of conventional conditional sampling to the artificial or geophysical turbulence data showed much greater vulnerability to contamination by noise. This contamination seriously reduces the usefulness of conditional sampling for relatively short records or records with significant chaotic variance.

The disadvantages of the method of continuous weighting include increased computer time and loss of structural detail. However, the method does not require the degree of a priori commitment required for conditional sampling and is therefore especially useful for flows where the nature of the coherent structures is not well known.

6. ACKNOWLEDGEMENTS

This material is based upon work supported by the Meteorology Program of the National Science Foundation under Grant ATM-8521349.

7. REFERENCES

Antonia, R. A., and J. D. Atkinson, 1976: A ramp model for turbulent temperature fluctuations. Phys. Fluids, 19, 1273.

Antonia, R. A., A. J. Chambers, C. A. Friehe and C. W. Van Atta, 1979: Temperature ramps in the atmospheric surface layer. J. Atmos. Sci., 36, 99-108.

Brown, G. L., and S. W. Thomas, 1977: Large structure in a turbulent boundary layer. Phys. Fluids, 20, S243-S252.

Busch, N., and E. L. Petersen, 1971: Analysis of nonstationary ensembles. Statistical Methods and Instrumentation on Geophysics. Teknologisk Forlag, Oslo, 71-92.

Chen, C.-H. P., and R. F. Blackwelder, 1978: Large-scale motion in turbulent boundary layer: A study using temperature contamination. J. Fluid Mech., 89, 1-31.

Friehe, C. A., J. C. LaRue, F. H. Champagne, C. H. Gibson, and G. F. Dreyer, 1975: Effects of temperature and humidity fluctuations on the optical refractive index in the marine boundary layer. J. Opt. Soc. Amer., 65, 1502-1511.

Gibson, C. H., C. C. Chen, and S. C. Lin, 1968: Measurements of turbulent velocity and temperature fluctuations in the wake of a sphere. AAIA J., 6, 642-649.

Kim, H. T., A. J. Kline, and W. C. Reynolds, 1971: The production of turbulence near a smooth wall in a turbulent boundary layer. J. Fluid Mech., 50, 133-160.

Kovasznay, L. S. G., V. Kibens, and R. F. Blackwelder, 1970: Large-scale motion in the intermittent region of a turbulent boundary layer. J. Fluid Mech., 41, 283-325.

Laufer, J. 1975: New Trends in experimental turbulence reseaerch. Annual Rev. Fluid Mech., 47, 307-326.

Lilly, D. K., 1983: Stratified turbulence and the mesoscale variability of the atmosphere. J. Atmos. Sci., 40, 749-761.

Lumley, J. L., 1981: Coherent structures in turbulence. Transition and Turbulence, Academic Press.

Mahrt, L., 1985: Vertical structure and turbulence in the very stable boundary layer. J. Atmos. Sci., 42, 2333-2349.

Mahrt, L., and H. Frank, 1988: Eigen structure of eddy microfronts. To appear in Tellus., 40A.

Mumford, J. C., 1982: The structure of the large eddies in fully developed turbulent shear flows. Part l. The plane jet. J. Fluid Mech., 118, 241-268.

Panofsky, H. A., and J. A. Dutton, 1984: Atmospheric Turbulence. John Wiley & Sons, 397 pp.

Riley, J. J., R. W. Metcalfe, and M. A. Weissman, 1981: Direct numerical simulations of homogeneous turbulence in density-stratified fluids. Proc. AIP Conf. Nonlinear Properties of Internal Waves, Bruce J. West, Ed., 79-112.

Subramanian, C. S., S. Rajagopalan, R. A. Antonia and A. J. Chambers, 1982 Comparison of conditional sampling and averaging techniques in a turbulent boundary layer. J. Fluid Mech., 123, 335-362.

Townsend, A. A., 1976: The Structure of Turbulence Shear Flow, Chapter 1, Cambridge University Press, 429 pp.

3 - MODELING AND ANALYSIS OF CLOUDS, RAIN, AND OTHER ATMOSPHERIC FIELDS

ATMOSPHERIC BOUNDARY LAYER VARIABILITY AND ITS IMPLICATIONS FOR CO_2 FLUX MEASUREMENTS

L. B. Austin, G.L. Austin, P.H. Schuepp and A. Saucier
McGill Radar Weather Observatory
Box 241, Macdonald College
Ste Anne de Bellevue, Quebec
CANADA H9X 1C0

ABSTRACT. Airborne carbon dioxide flux measurements have been studied. Statistical analysis of the observed great variability suggest that the turbulent eddies producing the flux show extreme values a few of which contribute much of the flux. These extremes are plausibly related to thermal plumes. Averaging of the flux values for at least a few kilometers of flight path was found necessary to allow the unambiguous discrimination of a land-lake interface from the flux data.

1. INTRODUCTION

In recent years considerable interest has been centered on the concentration of carbon dioxide in the atmosphere. This is due to the important consequences of the increasing concentration of carbon dioxide in the atmosphere and the predicted climatological increases in surface temperature. When plants are photosynthesizing they use carbon dioxide from the atmosphere and the flux of carbon dioxide can yield estimates of instantaneous biomass production rate. It is therefore interesting to both agriculture and climatology to be able to measure the pattern of carbon dioxide flux over different terrains, crop types and prevailing meteorological conditions.

If the atmosphere around the plant is at rest then the transfer of carbon dioxide is by molecular diffusion. However, the carbon dioxide is usually constantly replenished by mixing processes due to the turbulent and convective air motion in the boundary layer. The magnitude of this turbulent flux may be estimated by measuring the systematic differences in the carbon dioxide concentrations in the upward and downward parts of the vertical motion of the turbulent eddies. This technique is generally known as the 'eddy correlation method' (Swinbank, 1951). The measurements of vertical wind, concentration and other boundary layer parameters show great variability both in the intensity and the scale of the transferring eddies. Problems associated with this variability encountered in attempting to produce biomass production rate maps from aircraft data were discussed in Austin et al. (1987). It was found that considerable averaging was required in order to produce recognizable images of known features. In this paper we attempt to investigate the origins of these problems in more detail.

2. DATA USED

The carbon dioxide concentration was measured with an open-path analyser. This instrument depends on the differential absorption by carbon dioxide at wavelengths of 4.3 and 4.7μm. A full technical description of the device is given in Brach et al. (1981). The high frequency components of the vertical wind with respect to the aircraft were measured with a nose-mounted gust boom. A Doppler radar measured the low frequency components. The equipment was mounted on a Twin Otter operated by the National Aeronautical

D. Schertzer and S. Lovejoy (eds.), Non-Linear Variability in Geophysics, 157–165.

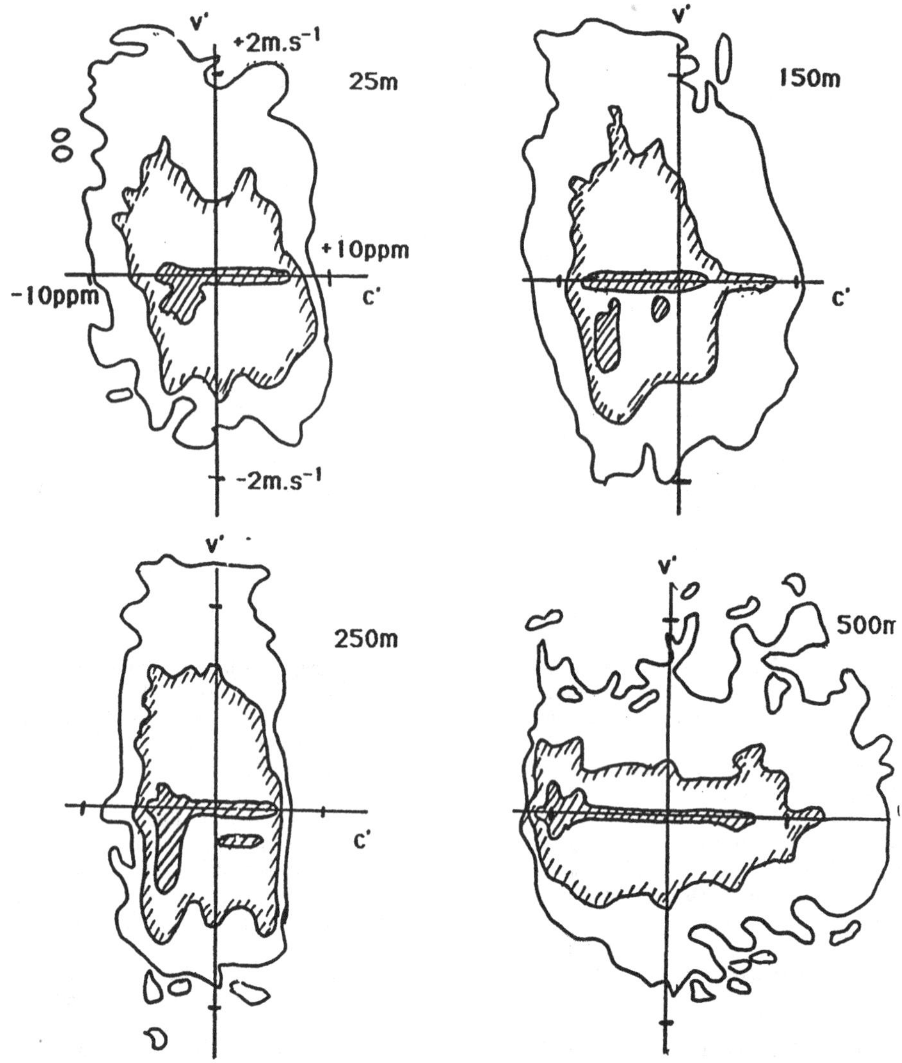

Fig. 1: Scatter plot of vertical wind fluctuation against carbon dioxide concentration fluctuation.

Establishment. Desjardins et al. (1982, 1985) and Alvo et al. (1984) have used this technique to measure carbon dioxide fluxes over various crops and forest land and have found general agreement with ground based measurements. More recently Schuepp et al (1987) have reported simultaneous carbon dioxide and water vapour measurements with the same equipment.

Measurements of the carbon dioxide concentration and vertical wind, together with twenty-eight other parameters, were made every one-sixteenth of a second. During data collection flights the aircraft flew at an altitude of about 50m at a speed of 50 $m.s^{-1}$ in the upwind and downwind directions. Data were obtained from a series of flights made over corn crops in the vicinity of St. Hyacinthe, Quebec and over a prairie land/lake interface. The time of crossing land features can be determined by examining a video tape taken from the front of the aircraft while the data were being collected.

3. BOUNDARY LAYER AIR FLOW

Conceptual, theoretical and experimental models of boundary layers are discussed at great length in the literature of fluid dynamics. Even a brief review of some of this material shows that the shape, size, form and intensity of the eddies depend critically on the conditions producing the boundary layer. Particularly important are the wind shear and stability of the atmosphere. Plumes of hot, moist air are likely to be an important feature of the flow. However the relative amount of moist air compared with dry air is likely to be a strong function of height as well as the relative magnitude of the probable upward and downward wind fluctuations. It is evident that any estimate of the net moisture or heat flux requires averaging over a horizontal distance large compared with the size of the plumes.

A boundary layer regime of this nature would be characterized by small areas of high upward wind fluctuations (and presumably low carbon dioxide concentration) and larger areas of slower downward velocity (with higher carbon dioxide concentration) at high altitudes but a more homogeneously mixed environment at low altitudes. Thus it is to be expected that the statistical properties of the data will change significantly with height. The averaging procedure involved in the eddy correlation analysis or even cross correlation spectra does not readily distinguish between the two regimes described above. Another possibility first described by Shaw (1985) in a discussion of boundary layer momentum transfer is to recognize that both the vertical wind fluctuation w' and the concentration fluctuation ρ can be positive or negative. In this case the plumes would be distinguished by w' being positive and ρ being negative. In fact there are clearly four possibilities + +, + -, - + and - - which could be displayed on a scatter diagram. It was hoped that plotting <u>instantaneous</u> values of w' and ρ on such a scatter diagram might give some evidence of the intermittent plumes in the boundary layer. If the two variables are assumed to be normally distributed with zero correlation then the distributions of any pair of values would be represented by a series of concentric ellipses. If the two variables are perfectly correlated then the distributions would be along a straight line. Thus the shape of the distributions gives an indication of the correlation between the two variables as well as their amplitude distributions.

The carbon dioxide concentration fluctuation and vertical wind data from the St. Hyacinthe flights are shown on such a scatter diagram in figs. 1a to d. These were constructed from more than 10,000 pairs of data from runs at each of four different heights. The carbon dioxide and wind fluctuations are not highly correlated. There is, however, an interesting contrast between the concentration fluctuations, which show relatively normal distributions of values about the mean with relatively few outliers, and the wind fluctuations which show a large number of zero values but also a much larger number of extremely large values particularly in the positive (upwards) direction. It is apparent from fig. 1 that these extreme upward velocity fluctuations are predominantly associated with negative concentration fluctuations which would coincide with the 'hot plumes' discussed above. They are seen to be carrying low carbon dioxide concentration air away from the surface. The compensating high carbon dioxide downflow is at a slower speed over a larger area. This behaviour is seen at the three lower heights where the velocity fluctuations are found to be getting larger and the concentration fluctuations smaller as the altitude increases. In marked contrast is the scatter plot for the highest altitude where the velocity fluctuations have become much smaller while the concentration fluctuations are correspondingly larger. This behaviour is consistent with the aircraft having risen above the turbulent surface layer.

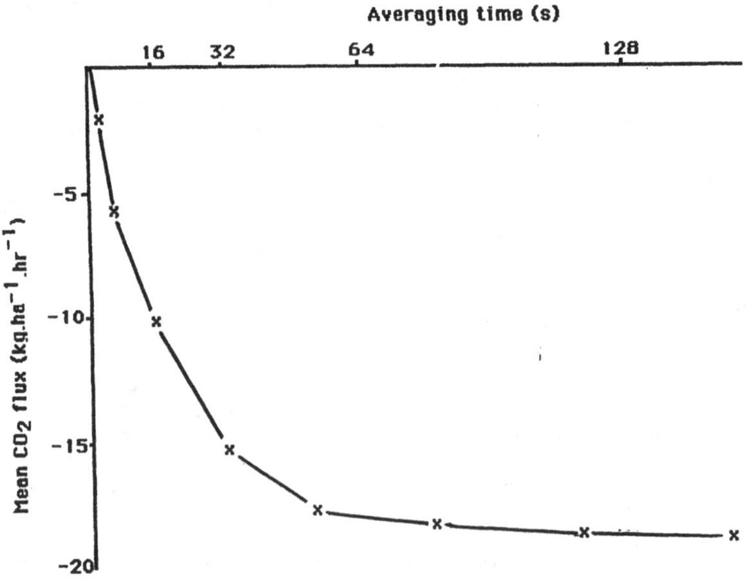

Fig. 2: Change in estimated flux due to the change in time over which the mean CO_2 concentration is taken.

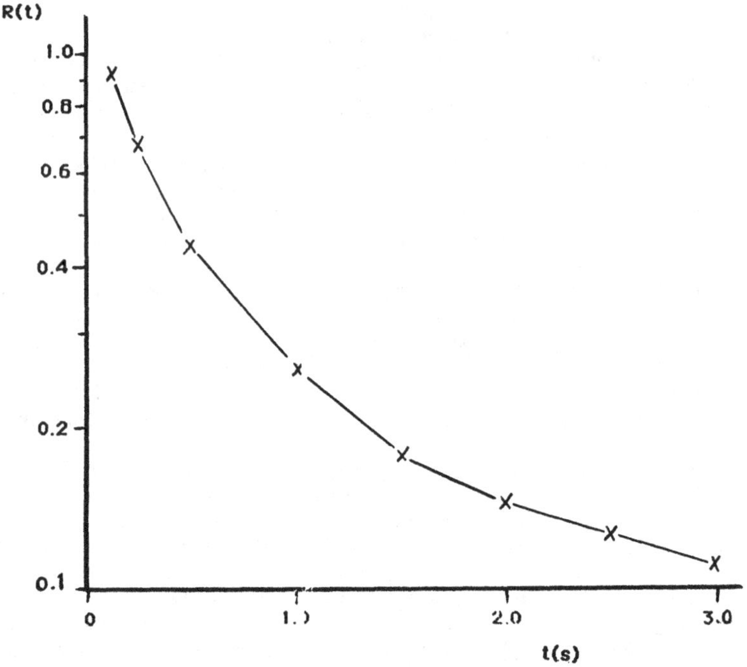

Fig. 3: Auto correlation function of vertical wind for various line lags (time displacements).

4. ANALYSIS

While the techniques described above illustrate a method of studying the structure of the turbulent flow, in the end, the application central to this study requires the estimation of average fluxes. The determination of the fluctuations depends on the subtraction of a mean value. If the time taken for the average is very short, then the "mean" will follow the fluctuating signal resulting in lower values for both w' and ρ. If, on the other hand, they are averaged for too long a period, then the possibly non-stationary values of w' and ρ due either to change in overall meteorological conditions (e.g. sun angle) or drift, will contaminate the results. The hope then is that as we increase the averaging period, t, the fluctuations and consequently the flux will increase in value and eventually become a constant. Carbon dioxide fluxes were calculated using the fluctuation of the concentration from the calculated mean value for different amounts of averaging. The result of doing this is shown in fig. 2. It is evident that provided t is greater than 30 seconds then approximately stable results are obtained.

A more difficult question is how much of this flux data has to be averaged in order to obtain a stable result. This question has been addressed by Wyngaard, (1983) in some detail using ideas and mathematics he attributes to Monin and Yaglom (1975) and Lumley and Panofsky (1964). He considers the averaging of a variable f, which could be flux, for the purpose of the present discussion. The original discussion is in terms of a spatial average; here the analysis is presented in terms of a temporal average to match the data processing methodology to be used.

If f(t) is a stationary random function of t with time average $\overline{f}$ then the ensemble-average variance between the time average $\overline{f}$ and the ensemble average $<f>$ is given by

$$\sigma^2 = \frac{2t_s}{T} <(\overline{f} - <f>)^2> \qquad (1)$$

where T is the duration of the averaging and t_s is the integral time scale and defined as

$$t_s = \int_0^\infty R(t)\, dt \qquad (2)$$

where R(t) is the autocorrelation function,

$$\frac{(f_i)\,(f_{i+t})}{f_i^2}$$

It could be argued, however, that this introduction of a scale is suspect, since the rate at which R(t) falls off depends significantly on the amount of smoothing the data recording system uses.

For the particular case $f = w'\rho$ we obtain

$$\frac{T}{t_s} = \frac{2 <(w'\rho - <w'\rho>)2>}{\sigma^2} \qquad (3)$$

If we further follow the Wyngaard analysis and expand the numerator in eq. (3) and use the result

$$<w'^2\rho^2> = <w'^2><\rho^2> + 2 <w'\rho>^2$$

which makes the serious, and not necessarily correct, assumption that w' and ρ are jointly normal. Then

$$\frac{T}{t_s} = \frac{2\,(1 + r^2)}{r^2} \frac{<f>^2}{\sigma^2} \qquad (4)$$

where the correlation coefficient r is given by

$$r = \frac{<w'\rho>}{<w'^2>^{1/2} <\rho^2>^{1/2}} \qquad (5)$$

Analysis of the St. Hyacinthe and prairie data yields values of the correlation coefficient of r between -0.1 and -0.3. The autocorrelation function for the vertical wind for various lags is shown in fig. 3, from which it may be estimated that the integral time scale computed using the definition eq. 2 is about 4.4 seconds which corresponds to a space scale of 220m.

Substituting r=-0.1 (and -0.3) into eq. 4 yields the result that to obtain an accuracy of estimate ($\sigma/<f>$) within a range of 10% of the true value then T/t_s = 20000 (2,200), so that for t_s = 4.4sec. T = 88000sec (1000sec) which corresponds to an average distance of about 4400km (500 km) for an airplane flown at 50m.sec^{-1}. For an accuracy of within 100% T/t_s can be about 200 (22) which for t_s = 4.4 sec gives a value of T of 880sec (100sec) and hence and averaging distance of 44km (5km). These results show that considerable averaging is required in using these data and how little reliance can be placed on the estimation of carbon dioxide fluxes from limited data sets.

These results have even more serious implications for the trade-off between the accuracy of the flux meaurements, the amount of data which has to be averaged to obtain reasonably accurate measurements and the spatial resolution of the resulting images. If reasonable imagery is to be obtained, either the number of intensity levels has to be reduced or the aircraft has to make many passes over the same region to obtain a large data set, the average of which will constitute accurate flux measurements.

5. VARIABILITY AND EXTREME VALUES

It would seem appropriate to comment on the impact of recent advances in the study of intermitency.

There are two main reasons why theoretical work of this type is relevent to the work in hand.

1) The work of Schertzer and Lovejoy (1985) in particular questions the validity of the introduction of space and time scales into the classical analyses of the type presented earlier.

2) The data collected by the NAE aircraft is entirely appropriate for performing analyses of the type suggested in Schertzer and Lovejoy to demonstrate the existence or non-existence of self similarity of turbulent eddies over very large space and time scales.

Sufficent for our purposes here is to note that:

1) The plotting of log cumulative curves of atmospheric variables for different amounts of averaging, or for differences in the values at different spacing, will yield straight lines for large fluctuations only if the variable is distributed hyperbolically.

2) The slope of the line produced in such a log cumulative plot is significant in that the lower the value of the slope the more variable and intermittent the parameter is and the more extreme values it has.

Fig. 4 shows the result of log cumulative plots of probabilities that a flux estimate (f) based on increasing averaging periods of flux data will exceed a given threshold f_0. There is a difficulty in plotting positive and negative fluxes on the same graph. As may be seen there is evidence of the characteristic straight line behaviour at large fluctuations characteristic of extreme variability. It is clear that a definitive distinction between the hyperbolic and other distributions is at best subjective as is the estimation of the slope of tail of the distribution. However fig. 4 show the least squares fitting of a straight line to the tail of the negative flux values. If the fluxes had normal distributions, then the lines of fig. 4 would have curved down and become completely vertical for extreme values of f_0. These curves show that there are significantly more extreme values than would be present in a normal distribution. These extreme values may well have an important physical role in the flux transfer process and some at least may be associated with the "hot plumes" discussed earlier.

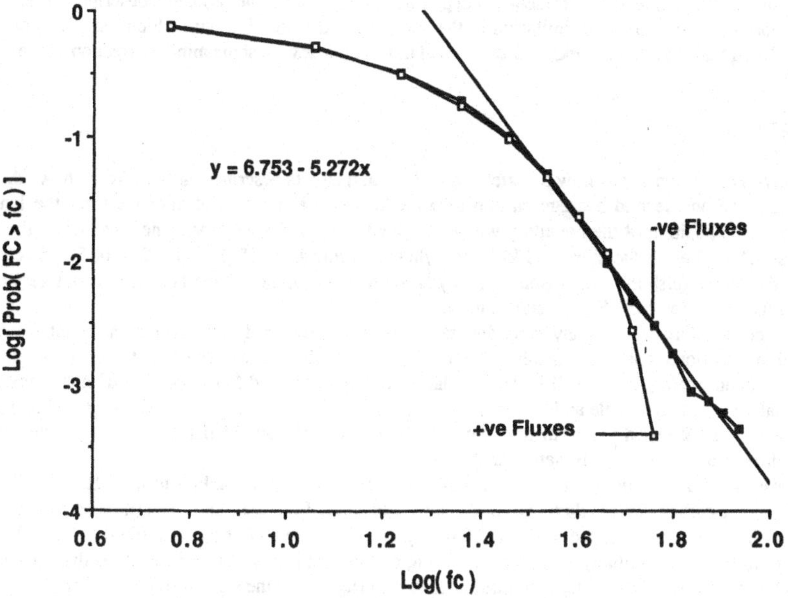

Fig. 4: (a) Negative carbon dioxide fluxes. (b) Positive carbon dioxide fluxes.

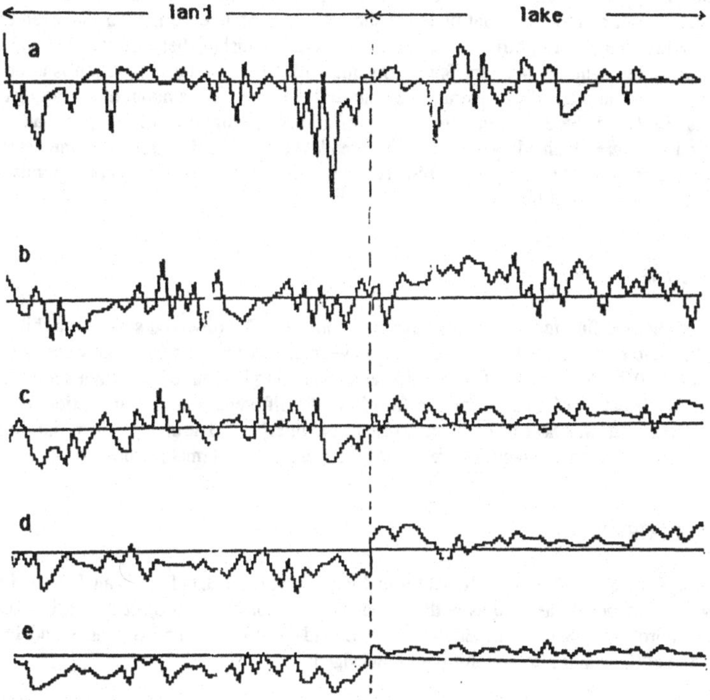

Fig. 5: (a) Onene pass of the aircraft; (b-e) 8, 16, 32, 64 simulated passes.

It is probable that this behaviour is not due to lack of instrument quality, poor data collection or processing but fundamental to turbulent atmospheric motions. Thus the problems described here, to some extent represent a fundamental limitation in the accuracy and resolution with which the eddy correlation technique can be expected to work. The errors will not disappear with some minor correction of the data.

6. RESULTS

The analysis of section 3 may now be extended to the real data sets described earlier. Fig. 5 shows the result of plotting the one second average carbon dioxide fluxes as a function of distance from the land/lake interface. The position of the boundary was established using the video tape. The mean of all the "land" data was -18 kg/hectare/hour and -11 kg/hectare/hour. A threshold of (18 + 11)/2 = 14.5 kg/hectar/hour was used to distinguish the two regions. It may be seen from fig. 5a that the boundary is not very easy to locate from the carbon dioxide flux data alone.

In order to address how many more data points would be required to produce an acceptable land/lake distinction, additional data were obtained by a simulation method. Two data sets, the first containing all the one second average carbon dioxide flux data that were obtained 25m over "land" areas was stored sequentially in a computer file and all the data obtained from "lake" areas, was similarly stored in a second set. These were then used as two distinct data sets. For the "land" set 288 data points were obtained while for the "lake" set 357 data points were stored.

For each of 50 locations on the flight path over the land, corresponding to a length of 2.5km, one second average carbon dioxide fluxes were selected sequentially from the "land" data file starting at a random location. If the sequence started less than 50 data points from the end of the data file then the subsequent values were taken sequentially from the beginning of the data set. A similar set of 50 data points were selected from the "lake" data file. To simulate a second flight over the same terrain, another 50 sequential flux values were selected from the appropriate data file starting at a random point in the file. For each of the 50 locations, the two flux values were averaged. This procedure was repeated to simulate 8, 16, 32 and 64 passes of the aircraft and the resultant plots are shown in figs. 5b to e. It is clear that even the average of 16 flights would still make a significant number of errors distinguishing between the land and the lake.

An average of 32 flights shows very few errors and with 64 flights the distinction is almost perfect! Naturally an alternative approach is to average a single flight in time. The amount of averaging required to achieve results similar to the 32 flight case would, of course, be significantly longer than 32 seconds because these curves were obtained by averaging independent data. The decorrelation time was previously 4.4 sec so that it is probable that 4.4 x 32 or 140 seconds would be required which corresponds to 7km for each of the land and lake legs of the run.

7. CONCLUSIONS

Airborne carbon dioxide flux measurements have been studied. The observed strong variability has been analysed and found to be plausibly hyperbolic and to have relatively low correlation between wind and CO_2 concentrations (0.1-0.3). This means that considerable averaging is required to obtain stable flux values. The results for the detection of a land/lake interface from the differences in carbon dioxide flux show that about 7 km of flight path was necessary on each side of the interface. The changes in variability observed as a function of aircraft height are plausibly related to the presence of thermal plumes.

8. ACKNOWLEDGEMENTS

We thank I. Macpherson and D. Carter (NAE) for making available the data for this analysis and for helpful discussions with these people as well as with M. Duncan. Support of the project by Agriculture Canada, Atmospheric Environment Service of Environment Canada and the Natural Sciences and Engineering Research Council of Canada is also gratefully acknowledged.

9. REFERENCES

Austin, Lydia, B., P. H. Schuepp, and R. L. Desjardins, 1987: The feasibility of using airborne CO_2 flux measurements for the imagery of the rate of biomass production. Ag. and Forest Meteorology, 39, 13-23.

Alvo, P., R. L. Desjardins, P. H. Schuepp, and J.I. Macpherson, 1984: Aircraft measurement of CO_2 exchange over various ecosystems. Boundary Layer Met., 29, 167-183.

Brach, E. J., R. L. Desjardins, and G.T. St. Amour, 1981 : Open path CO_2 analyser. J. Phys E. Sci. Instrum., 14, 1415-1419.

Desjardins, R. L., E. J. Brach, P. Alvo, and P.H. Schuepp, 1982: Aircraft monitoring of surface CO_2 exchange. Science, 216, 732-735.

Desjardins, R. L., J. L. Macpherson, P. Alvo, and P. H. Schuepp, 1985: Measurements of turbulent heat and CO_2 exchange over forest from aircraft. In The Forest-Atmosphere Interaction, pp645-658, D. Reidel, Dordrecht.

Lumley, J. C. and H. A. Panofsky, 1964: The structure of Atmospheric Turbulence, John Wiley and Sons, New York, 239.

Macpherson, J. I. and G. A. Isaac, 1977: Turbulent characteristic of some cumulus clouds. J. Appl. Met., 16, 81-87.

Macpherson, J. I., J. M. Morgan, and K. Lum, 1981: The NAE Twin Otter atmospheric research aircraft. NAE Lab. Tech. rep. LTR-FR-80, 21.

Monin, A. S., and A. M. Yaglom, 1975: Statistical hydromechanics: The mechanics of turbulence, Part 1. Trans-joint Pub. Research Service Washington, 37, 763.

Riley, J., and R. Metcalfe, 1980: Direct numerical simulation of perturbed turbulent mixing layer AIAA 80-0274, 18th Aerospace Sciences Meeting, Pasadena, California.

Schertzer, D., and S. Lovejoy, 1985: The dimension and intermittency of atmospheric dynamics. Turbulent shear flow, 7-33, B. Launder Ed., Springer-Verlag, NY.

Shaw, R. H., 1985: On diffusive and dispersive fluxes in forest canopies. In The Forest-Atmosphere Interaction, pp407-419, D. Reidel.

Swinbank, W. C., 1951: The measurement of vertical transport of heat and water vapour by eddies in the lower atmosphere. J. Meteor., 6, 135-145.

A FRACTAL STUDY OF DIELECTRIC BREAKDOWN IN THE ATMOSPHERE

A. A. Tsonis
Department of Geosciences
The University of Wisconsin-Milwaukee
Milwaukee, WI 53201

ABSTRACT. Analysis of photographs of lightning indicates that lightning has fractal geometry associated with a reproducible fractal dimension of about 1.34. Following this analysis a nonequilibrium model is presented which generates structures which are qualitatively similar to lightning observed in the atmosphere and which exhibit a fractal dimension of 1.37. The result of the model are also used to demonstrate that the observed lightning structures are most probable events.

1. INTRODUCTION

Lightning is the result of dielectric breakdown of gases which occurs when some region of the atmosphere attains a sufficiently large electric charge. Basically, a strong concentration of negative charge within the cloud base produces electric fields which cause some negative charge to be propelled towards the ground. This cloud-to-ground discharge is called the stepped leader because it appears to move downward in steps. When the stepped leader has lowered high negative potential near the ground the resulting high electric field at the ground is sufficient to cause an upward-moving discharge which carries ground potential up the path previously forged by the stepped leader. By doing so, the return discharge illuminates and drains the branches formed by the stepped leader. This return discharge is called the return-stroke. Both the stepped leader and following return stroke are, therefore, usually strongly branched downward.

No two lightnings are alike. Lightning comes in an extraordinary variety of structures which appear random. This is the main reason that lightning has defied any quantitative characterization. Yet anybody can distinguish lightning from any other growth form.

Visual examination of lightning photographs reveals a striking presence of structure at many different length scales. Every branch, for example, looks itself like a lightning and so does every branch of a branch. It seemed, therefore, appropriate to attempt to examine the structure of lightning using the concept of fractals or self-similarity (Mandelbrot, 1983). A fractal (or a scale invariant structure) is an object whose statistical properties are unchanged under a change of spatial length scale. In other words, two pieces of a fractal, one of size A and the other of a size A' (A' < A), are statistically equivalent over some wide range of intermediate lengths, as long as the smaller piece is magnified by a factor A/A'. In this paper we will present evidence for fractal properties of lightning by analyzing photographs and by the study of a simple theoretical stochastic model of dielectric breakdown.

2. THE FRACTAL GEOMETRY OF LIGHTNING

For Euclidean structures the total amount of mass, M, scales with some characteristic length, l, as:

$$M(l) \propto l^d \tag{1}$$

D. Schertzer and S. Lovejoy (eds.), Non-Linear Variability in Geophysics, 167–174.

where d is equal to the spatial or Euclidean dimension of the space in which the structure exists (3 for a sphere, 2 for a plane, 1 for a straight line). For many non-Euclidean objects in nature (clouds, for example) eq. (1) is preserved but the exponent is no longer equal to the Euclidean dimension of the space in which the structure is embedded. In these cases $M(l) \propto l^D$ where D<d and need not be an integer. These objects are called fractals and D is called the fractal dimension (Mandelbrot, 1983). For self-similar structures, the fractal dimension does not depend on l and it is the same for all the structures of a common origin (such as clouds). This is an important result because it implies an order in structures that appear random.

Accordingly, if a lightning is a fractal structure, the relation between the total length, L, of all branches inside a circle (or square) of radius (or half side) l, and l itself should be a power law with noninteger D (Mandelbrot, 1983):

$$L(l) \propto l^D \tag{2}$$

In order to estimate the fractal dimension of lightning we have applied the above principle to twenty photographs. Most of the photographs were taken from Salanave, 1980. The selected photographs exhibit some degree of branching and in our analysis it was assumed that the thickness of the branches is zero and that lightning exists on a plane. The photos were enlarged or reduced to a common size of about 10 x 10 cm. For each lightning we placed randomly on the photograph a square of a side $2l = 1$ cm (i.e., $l = 0.5$ cm) and we measured the total length (using a step-length of 2 mm) of all branches inside that square. The square was repositioned several times to obtain a mean estimated value for the length, $\langle L \rangle$, for the above value of l. We then repeated the above procedure for a range of increasing l values against log l. This procedure results in a graph which for the above range of l values is almost linear with slope D. The quantity D is the fractal dimension of the structure under examination (Morse et al., 1985; Niemeyer et al., 1984).

Our analysis indicated that $D = 1.34 \pm 0.05$. The mean of the twenty estimated fractal dimensions is 1.34, and 0.05 is the standard deviation. One other widely-used technique to estimate the fractal dimension of an object is the following: Take a big square of side set to 1 which includes the object. Then pave it with subeddies of side $r=1/2$, and find the number of squares, N, which intersect the object. Repeat the above process with subsubeddies of side r^2. Continue as far as feasible. The number N scales as a function of r according to the relation $N \propto r^{-D}$ where D is an estimation of the fractal dimension of the object (Mandelbrot, 1983). When we applied the above technique to the photos we obtained the same result: $D = 1.34 \pm 0.05$. Therefore, lightning is a fractal with a reproducible fractal dimension of about 4/3. Such a result provides for the first time a quantitative characterization of lightning. Having such a characterization, we can now present a nonequilibrium model which simulates fractal structures which are quantitatively and qualitatively similar to lightning. A model is called nonequilibrium when randomizing effects dominate stabilizing (deterministic) effects.

3. A MODEL FOR THE SIMULATION OF LIGHTNING

The model employed here for simulating lightning is a modification of the nonequilibrium model proposed by Niemeyer et al. (1984) for the modeling of two-dimensional radial discharge commonly referred to as Lichtenberg Figures. The details of the model used here are as follows: the simulation is a stepwise procedure carried out on a two-dimensional lattice (fig. 1) in which the potential (ϕ) of the top and bottom row is fixed at a value of $\phi = 0$ and $\phi = 1$ respectively. Periodic boundary conditions are assumed at the sides of the lattice. Only the middle point of the top row (A_1) is capable of growth. Given these initial conditions the potential at every point of the lattice is obtained by solving the Laplace equation $\nabla^2 \phi = 0$. On a two-dimensional lattice this is obtained by iterating the following equation using successive over relaxation (SOR):

$$\phi_{i,j} = \frac{1}{4}\left(\phi_{i+1,j} + \phi_{i-1,j} + \phi_{i,j+1} + \phi_{i,j-1} \right) \tag{3}$$

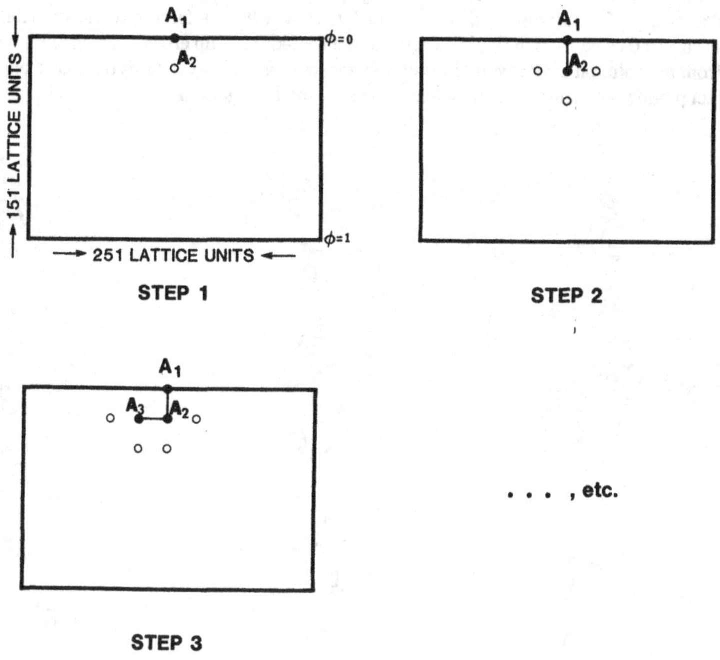

Fig. 1: Illustration of the model used to simulate lightning. The discharge is indicated by the black dots, connected with solid lines and it is considered equipotential. The open circles indicate the possible growth sites. The probability of each one of these sites is proportional to the local potential field (see text for details).

All the immediate non-zero neighbours to point A_1 are then considered as possible candidates, one of which will be added to the evolving discharge pattern. In fig. 1 the candidates are indicated by open circles and the evolving pattern by black dots. In Step 1 there is only one possible candidate. Therefore point A_2 will be added to the discharge pattern, which is considered equipotential ($\phi = 0$). In Step 2 one solves again the Laplace equation taking into account that the boundary conditions should include the discharge pattern. The possible candidates in Step 2 are three and each one of them is assumed to be associated with a probability P (commonly called "growth" probability) which is defined as:

$$P_i = \phi_i^2 / \sum_{i=1}^{N} \phi_i^2 \tag{4}$$

where i = 1, ..., N and N is equal to the number of the possible candidates. In Step 2, apparently N = 3. *The growth probability depends, therefore, on the local field determined by the equipotential discharge pattern.* At each step a probability distribution is defined. Given this probability distribution, a point is randomly selected and added to the evolving pattern. The above procedure is then repeated until the first point of the bottom row is added to the discharge pattern. Figures 2, 3 and 4 show three examples of computer-generated lightning. In order to obtain good convergence when iterating eq. (3) we performed 50 interactions at each step. Early experimentation indicated that after 50 iterations the potential field on each possible growth site changes by around 0.1% per iteration which indicates very good convergence. The problem, however, is that the program becomes very laborious and expensive. Each lightning simulation

(~700 points) takes a UNIVAC 1100 computer about 3.5 hours in CPU time! The above model reproduces very effectively the influence of a given discharge pattern on the growth probability of each candidate. For example, the tip of the line (indicated by A in fig. 2) will have a larger growth probability than points inside a cage (indicated by B in fig. 2). These are the well-known "tip effect" and "Faraday screening" which result from the solution of the field equation (Niemeyer et al., 1984) and it is obvious from the results that the model produces structures which unquestionably look like lightning.

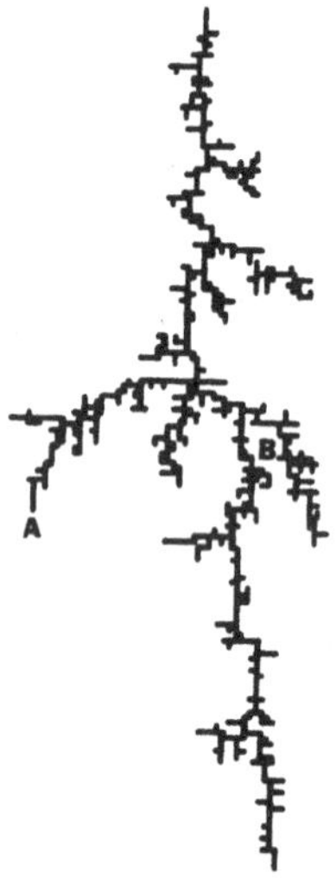

Fig. 2: A first example of a computer generated lightning. This structure is made up from 726 points.

Fig. 3: A second example of a computer generated lightning. This structure is made up from 666 points.

Fig. 4: A third example of a computer generated lightning. This structure is made up from 592 points.

Apart from the fact that the model produces structures qualitatively similar to lightning, it is necessary to verify whether or not these structures are also quantitatively similar to lightning. In order to do this we have calculated the fractal dimension of the five computer generated lightnings. A widely used method to determine the fractal dimension of computer simulated nonequilibrium growth structures is the following: for each point of a given structure we determined the number, $n(l)$ of all the other points within a square of half side l (l is now measured in lattice units) and its average $\langle n(l) \rangle$ over all the points of that structure, for varying l. We then plotted log $\langle n(l) \rangle$ against log l. The structure is a fractal if for a relatively wide range of scales the graph is approximately linear with a slope D. The quantity D is called the correlation dimension (Witten and Sander, 1981; Hentschel and Procaccia, 1983; Lovejoy et al., 1986) and again it provides an estimation of the fractal dimension of that structure. Figure 5 shows log $\langle n(l) \rangle$ versus log l for the structure shown in fig. 2. It can be observed that for a considerably wide range of l values (from 2 to about 50), log $\langle n(l) \rangle$ varies linearly with log l. The $l^{1.36}$ function is also shown for comparison. One may, therefore, estimate the fractal dimension of the structure in fig. 2 as about 1.36. From the five simulations we obtained that $D = 1.37 \pm 0.02$. This is very close to the value of 1.34 ± 0.06 derived from the photographs. The larger standard deviation is probably due to the finite size and resolution of the photos which do not allow a higher accuracy in determining D. In addition to be above described method there are other techniques which are commonly used in order to estimate the fractal dimensionality of computer simulated nonequilibrium growth structures (Meakin, 1986). For example, the mass (number of occupied lattice sites), M(r), contained within the distance r measured from the initial growth site, scales with r according to the relation $M(r) \propto r^{D'}$ where D' is an estimation of the fractal dimension of the structure. According to this method we obtained that D' = 1.36 ± 0.02.

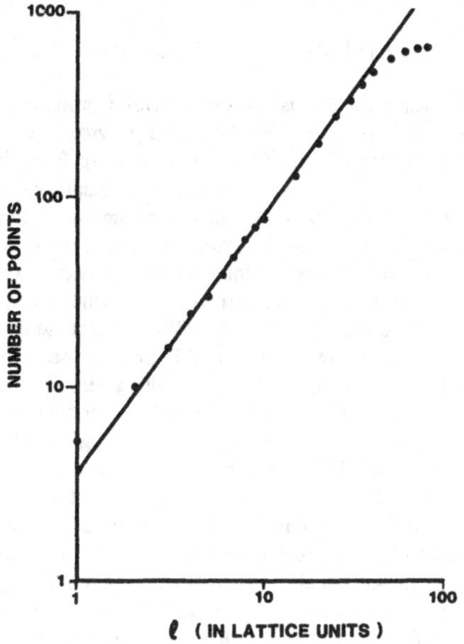

Fig. 5: For the structure in fig. 2, this graph shows the average number of points <n(*l*)> within a square of increasing half side *l*. The $l^{1.36}$ function is shown for comparison. It can be observed that for the range of *l* values between 2 and 50 the variation of <n(*l*)> with *l* is almost linear with a slope D = 1.36. Therefore the estimated fractal dimension of the structure in fig. 3 is 1.36.

The above results indicate that in addition to producing structures that are qualitatively similar to lighning, the model employed produces structures that are also quantitatively similar to lightning. It should be mentioned at this point that the above model simulates the stepped leader process. The photographs on the other hand most likely represent return strokes because only these are luminous enough. The stepped leader takes some milliseconds to reach the ground whereas the return stroke propagates upward in a fraction of a millisecond. However, since the return stroke propagates upward on the path previously forged by the stepped leader, we expect the geometry of the stepped leader to be identical to the geometry of the return stroke.

One may wonder about the choice of the exponent in eq. (4). This exponent may vary from 0 to d, where d is the Eudidean dimension of the space in which the growth process is embedded. It has been demonstrated by Niemeyer et al. (1984) that the fractal dimension of the generated structures depends on the exponent. For larger values of the exponent the generated fractal structures tend to be more "linear" and the fractal dimension is smaller. What justifies the choice of the exponent in this study is the fact that for a value of the exponent equal to two, we simulate structures that are qualitatively (look like) and quantitatively (same D) similar to observed lightning.

One final note. In our simulations we have used a rectangular lattice. This is only a reflection of our feeling that lightning takes place in a three dimensional space where the z direction is smaller compared to the x or y direction. Preliminary experimentation indicates that the consideration of a square lattice or a different rectangular lattice will not significantly affect the results, although differential stratification as investigated in Lovejoy, Schertzer and Tsonis (1987) could make a difference.

4. SOME PROBABILISTIC ASPECTS OF THE OBSERVED FRACTAL GEOMETRY

Let us assume that at the nth step of a simulation the generated structure is made up from n points denoted as $A_1, A_2 \ldots A_n$ which are selected in that order. The probability, P(n), of this structure is given by:

$$P(n) = P(A_1, A_2, \ldots A_n) = P(A_1)\, P(A_2/A_1)\, P(A_3/A_1A_2) \ldots P(A_n/A_1A_2 \ldots A_{n-1}) \quad (5)$$

As was mentioned above, at a given step each candidate is associated with a probability which is determined by the observed pattern in the previous step. Therefore the model provides the information needed to evaluate equation (5) as a function of n. By applying eq. (5) to structure in fig. 2 we find that the probability of the whole structure is of the order of $10^{-1,075}$. This means that the probability of generating a certain structure is extraordinarily small and explains why no two lightnings are the same. In a perfectly uniform atmosphere one will expect the breakdown to spread out in a straight line (which is not a fractal structure). As a matter of fact according to eq. (5) a straight line lightning simulation is the most probable one with a probability of the order of 10^{-80}. Because there is always some noise in the system, a growth instability will occur and irregularities will appear (Nittmann and Stanley, 1986). In our case this noise can be temperature or humidity or density fluctuations. This noise is reproduced in the proposed model via the random selection procedure at each step. There is always a chance that a site with a very small probability will be selected. Thus, the evolving structure can soon become very irregular. Under such conditions is there a significance in the apparent fractal geometry of dielectric breakdown? Are the generated structures the most probable ones? If fractal structures are most probable events, then one would expect the chance of getting a nonfractal discharge pattern out of the model to be much, much smaller. A theoretical way to prove this does not exist and to demonstrate this by simulating structures until a nonfractal one is generated, is hopeless due to the probabilities involved. The best alternative is to "cook up" a nonfractal discharge pattern (i.e., a structure with a fractal dimension equal to 1) and "force" the model to generate it. This means that if at each step the model does not select one of the candidates that fit the nonfractal structure the selection procedure is repeated until a point that fits that structure is selected. Such a nonfractal discharge pattern is shown in fig. 6. It consists of approximately equal number of "points" as the fractal structure in fig. 2 and its probability as a function of the step is shown in fig. 7 by the dashed line. The solid line corresponds to the probability of the structure in fig. 2 as a function of the step. As it can be observed in fig. 7 the probability of the nonfractal structure is much, much smaller than that of the fractal structure and it becomes even smaller as the number of steps increases. After only 170 steps the probability of the fractal structure is greater by a factor of about 10^{40}. After 200 steps this factor has become 10^{500}.

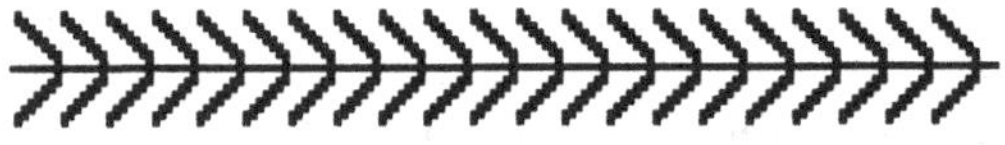

Fig. 6: A hypothetical nonfractal discharge pattern. This structure consists of 738 points. The vertical extent of this structure is 150 lattice units. The mass (number of occupied lattice sites) (M) of this structure scales with distance (r) from the origin as $M(r) \propto r^D$ where $D = 1$. Therefore this structure is not a fractal.

In view of the above results the following interesting questions may be posed: are those differences observed because the generation of the nonfractal structure was "forced"? Will those differences be observed if the model is forced to generate a given fractal structure (such as the one shown in fig. 2)? These questions can be answered if we answer a more general question: obviously, a given structure may be generated in many different ways (the one we force will be one of the possible ways). Does the total probability, P(n), of a given structure depend on the order in which it is built up? Experimentation with the structures in figs. 2 and 6 indicates that when a structure is "young" ($n \leq 30$) significant differences in P(n) may be observed. These differences diminish rapidly and for a given structure and a given $n > 70$, P(n) has a reproducible value which does not depend on the order in which the structure is built up. Therefore, the probability of the fractal structure will always be infinitely higher than the probability of the nonfractal structure no matter how these structures are generated.

Apparently, with n points one may produce a great number of different structures. Some of them will be more probable than others. From the more probable ones the most probable will most likely be generated. The above results demonstrate that the "chosen" structure will be a fractal structure.

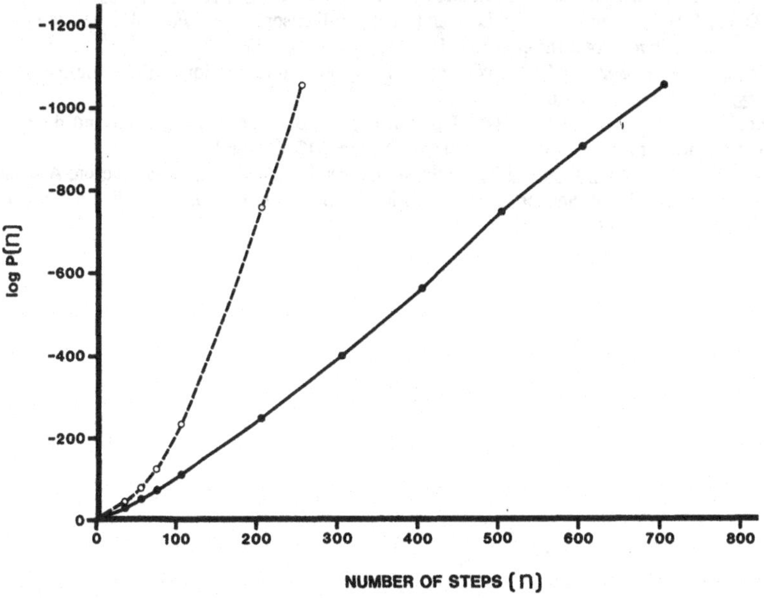

Fig. 7: The probability of an evolving structure as a function of the step. The solid line refers to the structure in fig. 2 and the dashed line refers to the structure in fig. 6.

5. CONCLUSIONS

We have examined the fractal properties of lightning from photographs and have presented results from a dielectric breakdown model that generates structures which are qualitatively and quantitatively similar to lightning. The major conclusions from this study are two: 1) lightning is fractal with a reproducible fractal dimension of about 1.37, and 2) lightning can be simulated assuming a growth probability which depends on the local potential field that is determined by the equipotential discharge pattern. Such an assumption naturally leads to fractal structures which are consistent with photographs of observed lightning. It is speculated that in the real atmosphere these fractal structures may be most probable outcomes.

We already know that cloud-to-ground discharge pattern is a stepwise procedure. The assumptions of the model suggest that at each step the evolving pattern modifies the potential field, thus affecting the possibilities of its evolution constantly. The success of the model employed may indicate that its basic assumptions capture the essence and the principles behind the underlying processes that produce lightning in the atmosphere.

6. REFERENCES

Hentschel, H. G. E., and I. Procaccia, 1983: The infinite number of generalized dimensions of fractals and strange attractors.Physica, 8D, 435-444.

Lovejoy, S., D. Schertzer, and A. A. Tsonis, 1987: Functional Box Counting and Multiple Elliptical Dimensions in Rain. Science, 235, 1036-1038.

Lovejoy, S., D. Schertzer, and P. Ladoy, 1986: Fractal characterization of inhomogeneous geophysical measuring networks. Nature, 319, 43-44.

Mandelbrot, B., 1983: The Fractal Geometry of Nature. Freeman and Company, New York, 461 pp.

Meakin, P., 1986: A new model for biological pattern formation. J. Theor. Biol, 118, 101-113.

Morse, D. R., J. H. Lawton, M.M. Dodson, and M.H. Williamson, 1984: Fractal dimension of vegetation and the distribution of arthropod body lengths, Nature, 314, 731-733.

Niemeyer, L., L. Pietronero, and H. J. Wiesmann, 1984: Fractal dimension of dielectric breakdown. Phys. Rev. Lett., 52, 1033-1036.

Nittmann, J., and E. H. Stanley, 1986: Tip splitting without interfacial tension and dendritic growth patterns arising from molecular anisotropy. Nature, 312, 663-668.

Salanave, L. E., 1980: Lightning and its spectrum. University of Arizona Press, Tucson, Arizona. 136 pp.

Witten, T. A., Jr., and L. M. Sander, 1981: Diffusion-Limited Aggregation, a kinetic critical phenomenon. Phys. Rev. Lett., 47, 1400-1403.

ON LOGNORMALITY AND SCALING IN SPATIAL RAINFALL AVERAGES?

V. K. Gupta*
C.I.R.E.S.
Campus Box 449,
University of Colorado,
Boulder, Colorado 80309

E. Waymire
Department of Mathematics
Oregon State University
Corvallis, Oregon 97331

ABSTRACT. Empirically-motivated assumptions of lognormality and statistical self-similarity (or scaling) of fluctuations of spatially-averaged rainfall are examined theoretically. Various basic properties of self-similarity are described in an effort to keep the exposition self-contained. It is shown that if the rainfall probability distributions are assumed to be lognormal, then the fluctuation field can be scaling in at most a second order sense; but not in the sense of the full distribution. Therefore, an application of this framework to rainfall requires some modification of the notion of scaling or the assumption of lognormality, or both.

1. INTRODUCTION

The primary aim of this paper is an attempt to present in the clearest possible terms some of the basic theoretical and empirical issues pertaining to lognormality and scaling in the current models for rainfall fluctuations. The problem being addressed here has persisted since Kolmogorov's (1962) introduction of lognormality into turbulence theory.

Let $\{r(\underline{x}):\underline{x} \in \Re^d\}$ be a d-dimensional random field (d = 2 or 3) describing the rain rates at points $\underline{x}$ in space. Fix time and consider the spatial distribution of rain rates. The <u>fluctuation field</u> is the random field $\{\xi(\underline{x}) = r(\underline{x}) - \langle r(\underline{x})\rangle\}$ describing the fluctuations of rain rates about their mean levels $\langle r(\underline{x})\rangle$, $\underline{x} \in \Re^d$. We shall focus on the properties of the fluctuation field. Throughout this article it is assumed that the fluctuation field $\{\xi(\underline{x})\}$ is <u>homogeneous</u>, i.e., has a translation invariant probability distribution in the sense that,

$$\{\xi(\underline{x}+\underline{z})\} \overset{\text{dist}}{=} \{\xi(\underline{x})\} \text{ for any (fixed) } \underline{z} \in \Re^d, \tag{1.1}$$

where $\overset{\text{dist}}{=}$ denotes equality in distribution. Furthermore, it is assumed that $\{\xi(\underline{x})\}$ is <u>second order</u>, i.e., $\xi(\underline{x})$ has finite second moments,

$$\langle \xi^2(\underline{x})\rangle < \infty, \underline{x} \in \Re^d \tag{1.2}$$

* Also at Utah Water Research Laboratory, Utah State University, Logan, UT.

D. Schertzer and S. Lovejoy (eds.), Non-Linear Variability in Geophysics, 175–183.

From the definition of $\xi(\underline{x})$ as $r(\underline{x}) - \langle r(\underline{x})\rangle$, it is obvious that $\langle\xi(\underline{x})\rangle = 0, \underline{x} \in \Re^d$, so that (1.2) gives the mean-square fluctuation (or variance) parameter,

$$\sigma^2 = \mathrm{Var}\,(r(\underline{x})) = \langle\xi^2(\underline{x})\rangle, \underline{x} \in \Re^d. \tag{1.3}$$

The parameter σ^2 does not depend on spatial location $\underline{x}$ under the homogeneity condition (1.1).

A familiar example of a fluctuation field which occurs in diverse areas of science and engineering is the Gaussian white noise (GWN). The GWN $\{\xi(\underline{x})\}$ refers to a random field of (jointly) Gaussian variables $\xi(\underline{x}), \underline{x} \in \Re^d$, having mean $\langle\xi(\underline{x})\rangle = 0$ and covariances given by

$$\mathrm{Cov}(\xi(\underline{x}), \xi(\underline{y})) = \sigma^2 \cdot \delta(\underline{x} - \underline{y})\, , \underline{x}, \underline{y} \in \Re^d, \tag{1.4}$$

where $\delta(\underline{x} - \underline{y})$ is the Dirac delta function. According to (1.4), the values are uncorrelated. Therefore they are also independent, since $\{\xi(\underline{x})\}$ is Gaussian.

To describe the GWN rigorously requires the use of generalized functions, e.g., the Dirac delta function in (1.4), and moreover, $\{\xi(\underline{x})\}$ is a generalized random field. From a 'practical' point of view, this means that 'integral averages' i.e., $\xi(f) = \int_{\Re^d} f(\underline{x})\xi(\underline{x})d\underline{x}$, for a test function f, are being described instead of the 'point values' $\xi(\underline{x})$ at $\underline{x} \in \Re^d$. We shall return to this matter at the close of this section. For the present it is enough to consider the especially simple test functions given by

$$f(\underline{x}) = \begin{cases} 1 \text{ if } \underline{x} \in U \\ 0 \text{ if } \underline{x} \notin U \end{cases} \tag{1.5}$$

where U is a closed or bounded region in $\Re^d$. For f of this form we have,

$$\xi(U) = \xi(f) = \int_{\Re^d} f(\underline{x})\xi(\underline{x})d\underline{x} = \int_U \xi(\underline{x})d\underline{x} \tag{1.6}$$

Let

$$\bar{\xi}(U) = \frac{1}{|U|}\xi(U) = \frac{1}{|U|}\int_U \xi(\underline{x})d\underline{x}\, , \tag{1.7}$$

where $|U|$ is the d-dimensional volume of U. It now follows from our earlier assumptions (1.1) and (1.2) that,

$$\langle\xi(U)\rangle = \int_U \langle\xi(\underline{x})\rangle d\underline{x} = 0 \tag{1.8}$$

$$\sigma^2(U) = \langle\xi^2(U)\rangle = \int_{U\times U} \mathrm{Cov}(\xi(\underline{x}),\xi(\underline{y}))d\underline{x}d\underline{y}$$

$$= \int_{U\times U} \sigma^2\delta(|\underline{x} - \underline{y}|)d\underline{x}d\underline{y}$$

$$= \sigma^2\, |U| \tag{1.9}$$

Let

$$\bar{\sigma}^2(U) = \langle\bar{\xi}^2(U)\rangle = |U|^{-2}\langle\xi^2(U)\rangle = |U|^{-2}\,\sigma^2(U) = \sigma^2\,|U|^{-1}$$

We now assume that the fluctuation field $\{\xi(\underline{x})\}$ is scaling or statistically self-similar. In particular, this means that if the spatial coordinate $\underline{x}$ is scaled by a parameter $\lambda > 0$, then the distribution of $\{\xi(\underline{x})\}$ scales as follows,

$$\{\xi(\lambda\underline{x})\} \overset{\text{dist}}{=} \lambda^{-\theta}\{\xi(\underline{x})\} \tag{1.10}$$

where $\theta > 0$, is called the scaling exponent. Since

$$\begin{aligned}\xi(\lambda U) &= \int_{\lambda U} \xi(\underline{x})d\underline{x} \\ &= \lambda^{d}\int_{U} \xi(\lambda\underline{y})d\underline{y} \\ &\overset{\text{dist}}{=} \lambda^{d}\int_{U} \lambda^{-\theta}\,\xi(\underline{y})d\underline{y} \\ &= \lambda^{d-\theta}\,\xi(U)\end{aligned} \tag{1.11}$$

the following scaling property is obtained for $\{\xi(\underline{x})\}$,

$$\{\xi(\lambda U)\} \overset{\text{dist}}{=} \lambda^{d-\theta}\{\xi(U)\} \tag{1.12}$$

Noting that $|\lambda U| = \lambda^{d}|U|$, it follows from (1.7) and (1.12) that

$$\{\bar{\xi}(\lambda U)\} \overset{\text{dist}}{=} \lambda^{-\theta}\{\bar{\xi}(U)\} \tag{1.13}$$

In the GWN example it is easy to check that $\{\xi(\underline{x})\}$ is scaling with exponent $\theta = d/2$, since $\xi(\lambda U)$ and $\lambda^{d/2}\,\xi(U)$ are both Gaussian with mean 0 and variance $\sigma^2\,|\lambda U| = \lambda^{d}\,|U|\,\sigma^2$. Likewise, $(\xi(\lambda U_1),\ldots,\xi(\lambda U_k))$ is Gaussian with mean vector (0,0,.......,0) and covariance $\sigma^2\,|\lambda U_i \cap \lambda U_j| = \sigma^2\,\lambda^{d}\,|U_i \cap U_j|$ as is the random vector $(\lambda^{d/2}\xi(\lambda U_1),\ldots\ldots,\lambda^{d/2}\xi(\lambda U_k))$ for any choice of regions $U_1,\ldots,U_k$.

Other Gaussian random fields which are scaling can be obtained by specifying power law correlations, see Taqqu (1987). However, as our main purpose here is to examine recent proposals for non-Gaussian theories, the Gaussian case shall not be discussed any further.

In closing this section, we wish to make note of an interesting observation of Dobrushin (1980) with regard to the use of generalized random fields in the formulation of scaling theories. In particular, Dobrushin (1980) observed that if $\{\xi(\underline{x})\}$ is an ordinary random field whose sample paths are continuous in probability (i.e., $\xi(\underline{x})$ is likely to be close to $\xi(\underline{y})$ if $\underline{x}$ is taken sufficiently close to $\underline{y}$), then homogeneity and scaling imply that with probability one $\xi(\underline{x}) \equiv \xi$ is a random constant, i.e., is a trivial homogeneous and scaling random field. The argument is quite simple so it is instructive to sketch it here. Note that, by homogeneity, for any $\underline{x}$ the random variable $\xi(\underline{x})$ has the same probability distribution as $\xi(\underline{x}+\underline{x}) = \xi(2\underline{x})$.But by self-similarity, $\xi(2\underline{x})$ is distributed as $2^{-\theta}\,\xi(\underline{x})$. Therefore, $\xi(\underline{x}) \overset{\text{dist}}{=} 2^{-\theta}\,\xi(\underline{x})$. This implies that $\theta = 0$. It now follows that

$$\xi(\underline{x}) - \xi(\underline{y}) \overset{\text{dist}}{=} \lambda^{0}\,[\xi(\lambda\underline{x}) - \xi(\lambda\underline{y})] = \xi(\lambda\underline{x}) - \xi(\lambda\underline{y}) \tag{1.14}$$

for any fixed $\underline{x}, \underline{y}$. Letting $\lambda \to 0$ so that $|\lambda\underline{x} - \lambda\underline{y}| = \lambda|\underline{x} - \underline{y}| \to 0$, it follows from continuity in probability that $\xi(\underline{x}) - \xi(\underline{y}) \equiv 0$ for any $\underline{x}, \underline{y}$, i.e., $\xi(\underline{x})$ is a random constant.

2. SOME FURTHER CONSIDERATIONS ON SCALING

There are various ways in which one can test the plausibility of the scaling hypothesis for rain rates. In this section are recorded a few simple observations with regard to scaling of the second moments. Obviously, if the full probability distribution of the fluctuations are scaling, then the mean square fluctuations will scale as well. In view of the proposition below, if one calculates an exponent $\theta > d/2$ then there must be negative correlations present. That is,

Proposition (2.1). Suppose that $\{\xi(\underline{x})\}$ is homogeneous, second order and scaling. If $\mathrm{Cov}(\xi(U),\xi(V)) \geq 0$ for all regions U and V of $\Re^d$ then $\theta \leq d/2$.

Proof. Let I be a unit cube in $\Re^d$. For any whole number $\lambda > 0$ partition λI as

$$\lambda I = I_1 \cup \ldots \cup I_{\lambda^d}$$

where $I_1, I_2, \ldots, I_{\lambda^d}$ are λ^d unit cubes partitioning the region λI.

Then,

$$\lambda^{2(d-\theta)}\sigma^2(I) = \sigma^2(\lambda I) = \left\langle \left(\int_{\lambda I} \xi(\underline{x})d\underline{x} \right)^2 \right\rangle$$

$$= \left\langle \int_{\lambda I}\int_{\lambda I} \xi(\underline{x})\xi(\underline{y})d\underline{x}d\underline{y} \right\rangle$$

$$= \sum_{r,s=1}^{\lambda^d} \int_{I_r}\int_{I_s} \langle\xi(\underline{x})\xi(\underline{y})\rangle d\underline{x}d\underline{y}$$

$$= \sum_{r=s}^{\lambda^d} \int_{I_r}\int_{I_s} \langle\xi(\underline{x})\xi(\underline{y})\rangle d\underline{x}d\underline{y} + \sum_{r\neq s}^{\lambda^d} \int_{I_r}\int_{I_s} \langle\xi(\underline{x})\xi(\underline{y})\rangle d\underline{x}d\underline{y}$$

$$= \lambda^d \sigma^2(I) + \sum_{r\neq s} \mathrm{Cov}(\xi(I_r),\xi(I_s))$$

$$\geq \lambda^d \sigma^2(I)$$

Therefore, dividing by $\lambda^d \sigma^2(I) > 0$, we have

$$\lambda^{d-2\theta} \geq 1$$

Since $\lambda \geq 1$ it follows that $d - 2\theta \geq 0$.

Note that the extreme case $\theta = d/2$, for non-negative correlations, is attained in the GWN example. For random fields satisfying the FKG inequalities from statistical physics, Newman (1980) has shown that the GWN is the only second order homogeneous and isotropic scaling case with $\theta = d/2$.

As noted above to test the plausibility of the scaling hypothesis for the probability distribution of the rain rate fluctuations, we can consider first the root mean square fluctuations. First observe that,

$$\sigma(\lambda U) = (<\xi^2(\lambda U)> - <\xi(\lambda U)>^2)^{1/2}$$

$$= (<\xi^2(\lambda U)>)^{1/2}$$

$$= (<\lambda^{2(d-\theta)}\, \xi^2(U)>)^{1/2}$$

$$= \lambda^{d-\theta}\, (<\xi^2(U)>)^{1/2}$$

$$= \lambda^{d-\theta}\, \sigma(U) \tag{2.1}$$

In particular, it follows from (2.1) that

$$\log \sigma(\lambda U) = (d - \theta) \log (\lambda) + \log \sigma(U) \tag{2.2}$$

Take $U = I$ as a unit square ($d = 2$), or a unit cube ($d = 3$). For convenience, let

$$\xi(\lambda) = \xi(\lambda I),\ \sigma(\lambda) = \sigma(\lambda I),\ \bar{\sigma}(\lambda) = \bar{\sigma}(\lambda I) \tag{2.3}$$

with this notation, we may also write $\sigma^2(1) = \sigma^2(I)$.

In view of (2.3), (2.2) can be expressed as a linear relationship as a function of length scale on a log-log plot,

$$\log \sigma(\lambda) = (d - \theta) \log (\lambda) + \log \sigma(1) \tag{2.4}$$

i.e., the root mean square fluctuations in $\{\xi(\underline{x})\}$ scale as,

$$\sigma(\lambda) = \lambda^{d-\theta}\, \sigma(1) \tag{2.5}$$

Equivalently,

$$\log \bar{\sigma}(\lambda) = -\theta \log(\lambda) + \log \bar{\sigma}(1) \tag{2.6}$$

The fit of (2.6) is remarkably good when applied to the Phase I GATE data areal average rain rates in the form tabulated in Laughlin (1981) with $d = 2$, and slope $-\theta = -1/3$ [see Bell, 1987]. In particular,

$$\theta = \frac{1}{3} \tag{2.7}$$

In spite of the evidence in favor of scaling with $\theta = 1/3$ implied by the Laughlin computations, the above empirical evidence is only for the root mean square fluctuations. This we will call second order scaling. However, by definition (1.10), scaling is a property of the full probability distribution. While in the case of Gaussian random fields it is enough to check second order scaling, in general it is insufficient. This fact plays an important role in the next section.

3. THE ISSUE OF LOGNORMALITY

In the present section we consider the construction of a non-Gaussian theory of rain rate fluctuations based on the lognormal distribution for rain rates. We shall show that such a theory is possible under second order scaling of fluctuations, but breaks down under scaling of the full distribution of fluctuations. The proofs are based on the simple arguments sketched by Mandelbrot (1972, 1974) in the context of turbulence theory. The details are provided here to make issues pertaining to rainfall data on (2.5)-(2.7) and its extrapolations as transparent as possible.

Suppose that the rain rate $r(\lambda)$ over a rectangular region λI of side lengths λ is lognormal. That is

$$r(\lambda) \equiv r(\lambda I) = e^{Z(\lambda)} \tag{3.1}$$

where $Z(\lambda)$ is Gaussian with mean $m(\lambda)$ and variance $s^2(\lambda)$.

The moments of all orders n of $r(\lambda)$ are easily computed from the moment generating function or from the Fourier transform of the Gaussian kernel. That is,

$$<r^n(\lambda)> = <e^{nZ(\lambda)}> = \exp\left\{ nm(\lambda) + \frac{1}{2} n^2 s^2(\lambda) \right\}, \quad n = 1, 2, \ldots \tag{3.2}$$

Let $\xi(\lambda) = r(\lambda) - <r(\lambda)>$, $\lambda > 0$. In view of homogeneity, we have

$$<r(\lambda)> = \mu \cdot \lambda^d \tag{3.3}$$

where μ is a constant. Therefore from (3.2), taking $n = 1$, we have

$$m(\lambda) + \frac{1}{2} s^2(\lambda) = \log \mu + d \log(\lambda) \tag{3.4}$$

Without loss of generality, since $r(\lambda)$ is non-negative, $\mu > 0$, so that we may take $\mu = 1$, i.e., if necessary rescale $r(\lambda)$ as $\mu^{-1} r(\lambda)$. Then (3.4) becomes

$$m(\lambda) + \frac{1}{2} s^2(\lambda) = d \log(\lambda) \tag{3.5}$$

Next observe that under second order scaling, defined by (2.5), we have

$$\begin{aligned} \lambda^{2(d-\theta)} \sigma^2(1) &= \sigma^2(\lambda) \\ &= <\xi^2(\lambda)> \\ &= <r^2(\lambda)> - <r(\lambda)>^2 \\ &= e^{2m(\lambda)+2s^2(\lambda)} - e^{2m(\lambda)+s^2(\lambda)} \\ &= <r(\lambda)>^2 (e^{s^2(\lambda)} - 1) \\ &= \lambda^{2d} (e^{s^2(\lambda)} - 1) \end{aligned} \tag{3.6}$$

Therefore,

$$e^{s^2(\lambda)} = \lambda^{-2\theta} \sigma^2(1) + 1 \tag{3.7}$$

The equations (3.5) and (3.7) can be solved simultaneously for $m(\lambda)$ and $s^2(\lambda)$ to get

$$s^2(\lambda) = \log(1 + \sigma^2(1) \lambda^{-2\theta}), \quad \lambda > 0 \tag{3.8}$$

$$m(\lambda) = \log\left(\frac{\lambda^{d}}{(1+\sigma^{2}(1)\,\lambda^{-2\theta})^{1/2}}\right),\ \lambda>0 \tag{3.9}$$

So, in particular, with this choice of $s^2(\lambda)$, and $m(\lambda)$, $\lambda > 0$, the rain rate described in (3.1) is lognormal, homogeneous, and possesses fluctuations $\xi(\lambda) = r(\lambda) - \langle r(\lambda)\rangle$ which are second order scaling.

To see that the fluctuations are not scaling in the sense of distribution, we suppose that they are and we then arrive at a contradiction as shown below. If $\xi(\lambda)$ is scaling (see, (1.12)) then in particular we have, writing $\mu_3(1) = \langle\xi^3(1)\rangle$,

$$\begin{aligned}\lambda^{3(d-\theta)}\,\mu_3(1) &= \langle\xi^3(\lambda)\rangle \\ &= \langle[r(\lambda) - \langle r(\lambda)\rangle]^3\rangle \\ &= \langle r^3(\lambda)\rangle - 3\lambda^{d}\langle r^2(\lambda)\rangle + 2\lambda^{3d}\end{aligned} \tag{3.10}$$

Now apply (3.2), (3.8) and (3.9) to get

$$\langle r^3(\lambda)\rangle = e^{3m(\lambda) + (9/2)\,s^2(\lambda)} = \lambda^{3d}\,(1+\sigma^2(1)\,\lambda^{-2\theta}) \tag{3.11}$$

$$\langle r^2(\lambda)\rangle = e^{2m(\lambda) + 2s^2(\lambda)} = \lambda^{2d}\,(1+\sigma^2(1)\,\lambda^{-2\theta}) \tag{3.12}$$

In particular, taking $\lambda = 1$ in (3.11) and (3.12), then

$$\begin{aligned}\mu_3(1) &= \langle\xi^3(1)\rangle = \langle r^3(1)\rangle - 3\langle r^2(1)\rangle + 2 \\ &= (1+\sigma^2(1))^3 - 3(1+\sigma^2(1)) + 2 \\ &= \sigma^4(1)\,(\sigma^2(1)+3)\end{aligned} \tag{3.13}$$

Substituting (3.11) and (3.12) into the right hand side of (3.10),

$$\begin{aligned}\langle\xi^3(\lambda)\rangle &= \lambda^{3d}(1+\sigma^2(1)\,\lambda^{-2\theta})^3 - 3\lambda^{3d}(1+\sigma^2(1)\,\lambda^{-2\theta}) + 2\lambda^{3d} \\ &= \lambda^{3(d-\theta)}\sigma^4(1)\left(3\lambda^{-\theta} + \lambda^{-3\theta}\sigma^2(1)\right)\end{aligned} \tag{3.14}$$

Therefore (3.14) and (3.13) show that (3.10) does not hold, i.e.,

$$\langle\xi^3(\lambda)\rangle \neq \lambda^{3(d-\theta)}\mu_3(1) \tag{3.15}$$

thus contradicting the definition of scaling.

4. FINAL REMARKS

The lognormal distribution is a natural candidate (Bell, 1987) for rainfall fluctuation distributions with near 1/f-spectrum. This means that large fluctuations occur less frequently than small ones, but in such a way that doubling the size halves the frequency, while still retaining the finiteness of the second order moments (see Montroll and Schlesinger, 1982). On the basis of the available spatial rainfall data from Phase I of the GATE experiment this latter property seems desirable and, as noted earlier, it is at this level, i.e., mean square fluctuations, where scaling is observed empirically. However, as has been demonstrated here, the proper theoretical framework does not extend beyond this without some modification of the basic notion of statistical self-similarity or the assumption of lognormality or both. An interesting notion connecting scaling and lognormality is discussed by Montroll [1987, eq. (29)]. Similarly, various issues involving scaling as well as considerations of tail probabilities are being entertained in the recent literature on rainfall which are not addressed here [see, e.g., Waymire 1985; Gupta and Waymire, 1987; Zawadzki, 1987; Lovejoy and Mandelbrot,1985; Schertzer and Lovejoy, 1987].

We will close with a brief discussion of a question which arose at the McGill NVAG1 conference on scaling, fractals, and non-linear variability in geophysics with regard to the definition of scaling itself. The question is whether or not one can generalize the form of (1.10), i.e., select scaling functions $g(\lambda)$ different from the power law form $\lambda^{-\theta}$? To answer this question we suppose that $\lambda^{-\theta}$ is replaced by a scaling function $g(\lambda)$ in (1.10), such that g is continuous and non-negative, say. Then for any positive numbers λ_1 and λ_2 apply (1.10) iteratively in two ways as follows:

$$g(\lambda_1\lambda_2)\{\xi(\underline{x})\} \overset{\text{dist}}{=} \{\xi(\lambda_1\lambda_2\underline{x})\} \overset{\text{dist}}{=} g(\lambda_1)\{\xi(\lambda_2\underline{x})\} \overset{\text{dist}}{=} g(\lambda_1)g(\lambda_2)\{\xi(\underline{x})\} \qquad (4.1)$$

Therefore,

$$g(\lambda_1\lambda_2) = g(\lambda_1)g(\lambda_2) \qquad (4.2)$$

The power law form of $g(\lambda)$ appearing in (1.10) now follows as the general solution of (4.2).

5. ACKNOWLEDGEMENTS

We acknowledge discussions with D. Short, T. Bell, Long Chiu, J. North, A. McConnell and B. Kedem of NASA during our visit at Goddard Space Flight Center, sponsored by ASEE-NASA Summer Faculty Fellowships. This research was supported in part by grant 21078-GS from the Army Research Office. The authors are grateful to Utah Water Research Laboratory at Utah State University for its support during the academic year 1985-86. The first author also wishes to acknowledge support of the NSF grant CEE830384 and of the Institute for Mathematics and Its Applications, University of Minnesota during the Spring of 1986. Detailed comments by Shaun Lovejoy and Daniel Schertzer on the original manuscript resulted in a number of improvements.

6. REFERENCES

Bell, T. L., 1987. A space-time stochastic model of rainfall for satellite remote-sensing studies. J. Geophys. Res. (Atmosphere), 92 (D8), 9631-9643.

Dobrushin, R. L., 1980. Automodel generalized random fields and their reform-group. In: Multicomponent Random Systems, (ed) R.L. Dopbrushin and Y.G. Sinai, Advances in Probability and Related Topics, Vol. 6, Marcel Dekker, Inc., New York, pp. 153-198.

Gupta, V. K., and E. Waymire, 1987. On Taylor's hypothesis and dissipation in rainfall. J. Geophys. Res. (Atmosphere), 92 (D8), 9657-9660..

Kolmogorov, A., 1962. A refinement of previous hypothesis concerning the local structure of turbulence in viscous incompressible fluid at high Reynolds number. J. Fluid Mech., 13, 82-85.

Laughlin, C. R., 1981. On the effect of temporal sampling on the observation of mean rainfall, Precipitation Measurements from Space, NASA report, (ed) D. Atlas and O.W. Thiele, Goddard Space Flight Center, Greenbelt, MD 20771.

Lovejoy, S. and B. B. Mandelbrot, 1985. Fractal properties of rain, and a fractal model, Tellus, 37A, 209-232.

Mandelbrot, B. B., 1972. Possible refinement of the lognormal hypothesis concerning the distribution of energy dissipation in intermittent turbulence. In: Statistical Models and Turbulence, Rosenblatt and Van Atta (ed.), Springer.

Mandelbrot, B. B., 1974. Intermittent turbulence in self-similar cascades: Divergence of moments and dimension of carrier. J. Fluid Mech., 62, 331-358.

Montroll, E. W. and M. F. Shlesinger, 1982. On 1/f noise and other distributions with long tails, Proc. Nat. Acad. Sci. 79, 3380-3383.

Montroll, E. W., 1987. On the dynamics and evolution of some sociotechnical systems, (ed) B.J. West, Bul. Am. Math. Soc. (New Series), 16(1), 1-46.

Newman, C. M., 1980. Self-similar random fields in mathematical physics, Proc. of Dekalb Measure Theory Conference, Northern Illinois University, Dekalb, Illinois.

Schertzer, D., and S. Lovejoy, 1987. Physical modeling and analysis of rain and clouds by anisotropic scaling multiplicative processes. J. Geophys. Res. (Atmosphere) 92 (D8), 9693-9714.

Taqqu, M. S., 1987. Random processes with long-range dependence and high variability, J. Geophys. Res., (Atmosphere), 92(D8), 9683-9686.

Waymire, E., 1985. Scaling limits and self-similarity in precipitation fields. Water Resour. Res., 21(8), 1271-1281.

Zawadzki, I., 1987. Fractal structure and exponential decorrelation in rain. J. Geophys. Res. (Atmosphere) 92 (D8), 9586-9590.

CONTINUOUS MULTIPLICATIVE CASCADE MODELS OF RAIN AND CLOUDS

J. Wilson, D. Schertzer*, S. Lovejoy
Physics Department,
McGill University,
3600 University st., Montréal, Québec
Canada, H3A 2T8,

ABSTRACT. Early scaling stochastic models of cloud and rain fields were designed to respect an unrealistically simple scale invariant symmetry called "simple scaling" or "scaling of the increments", involving a single fractal dimension. These monofractal processes were produced by summing a large number of random structures: pulses and/or wave packets (depending on whether the model was based in real or Fourier space) and in spite of this extreme simplification yielded simulations of cloud fields with some realistic features including texture, clustering, bands and intermittency. The linear (additive) nature of these processes is however intrinsically related to the fact that they have a single fractal dimension and is in sharp contrast with the non-linear nature of the true dynamical processes and with the observed multiple fractal dimensions of the fields. From this perspective, even the relatively simplified case of stochastic modeling of passive clouds - i.e. passively advected by a turbulent velocity field - is already beyond the scope of linear stochastic approaches, because of the statistical implications of the highly non-linear distortion of the concentration field by the turbulent velocity field. Focusing on this problem, we show how to build up models corresponding to coupled cascade processes, non linearly conserving the fluxes of energy and concentration variance. These cascades, have generally been based on a series of highly artificial discrete steps, are here taken to their continuous limit. Continuous cascades are not only far more realistic than their discrete counterparts, they also admit universality classes characterized by 2 parameter generators. This means that a simple (dynamical) generator describes the entire multifractal spectrum rather than an infinite number of (geometric) dimension parameters. We also show how to numerically simulate such processes with both Gaussian and Lévy generators as well as how to perform rescaling iteration on the results which is scale invariant "zooming" procedure.

1. INTRODUCTION

In recent years, many efforts have been made to tackle the problem of extreme variability and intermittency in turbulent atmospheric fields. Interest in this problem comes from both geophysical observations and from the emerging dynamical understanding of the importance of chaotic behaviour. From the empirical point of view, our knowledge of the relationship between the various fields obtained from different remotely sensed or in-situ measurements is somewhat limited as shown for instance, by the problem of infering rain or cloud liquid water fields from radar reflectivities or satellite radiances (e.g. radar calibration by rain gauge networks, the "albedo-paradox", etc.; see various papers in this volume). Even the relationship between similar sets of data spatially averaged at different resolutions, is not at all trivial, especially since the different fields are generally non-linearly related.

From the dynamical point of view, the understanding of the ("meso-scale") inner limit in numerical weather prediction (NWP) models is often the aspect of the extreme variability that is most seriously considered. However, understanding the resolution dependence of atmospheric fields is important well beyond the problem of NWP modelling. Even if we had complete information about the 10^{27} - odd

*EERM/CRMD, Météorologie Nationale, 2 Av. Rapp, Paris 75007, France

D. Schertzer and S. Lovejoy (eds.), Non-Linear Variability in Geophysics, 185–207.

velocity, pressure and other meteorological elements that are required to specify the state of the atmosphere down to the dissipation scale (which is roughly one millimeter), we would still not know what to do with it, i.e. we would still not have "understood" the weather, nor be in a position to meaningfully describe the atmosphere - we would still require appropriate macroscopic (statistical) concepts. Even within the context of the existing relatively limited number of degrees of freedom in NWP models (typically, about 10^6) sophisticated initialization procedures are required, which smooth out (if not literally truncate), small scale atmospheric structures. In actual fact the physically relevant number of degrees of freedom is much smaller since most of the dynamics occur in small fraction of the total available space i.e. the processes occuring in only a small number of grid points will be decisive for the evolution of the system. This suggest that NPW models would have much to gain by being able to "zoom" into active regions (this idea is already somewhat exploited in "nested" models in which high resolution regional models are placed inside larger global models). The systematic rescaling procedure we study here provides the proper theoretical framework for these problems and suggests possible operational procedures in the near future.

Furthermore, NWP-modelers use ad hoc pseudo-viscosities to dissipate energy at the inner scale to avoid a constant accumulation of energy (and hence "explosion"). These models can therefore not deal directly with processes in which energy is transfered from smaller physical scales up to grid scales. A related problem is that of sensitive dependence on initial conditions which is a general feature of atmospheric models, and presumably also of the atmosphere. Because of these (and other) limits to the deterministic dynamical approach it is of interest to study other types of models which are primarily aimed at producing realizations with realistic statistical properties. These models are phenomenological in the sense that while they may respect many of the symmetries (conservation laws) of the deterministic dynamical equations, but they are not (yet!) linked to them in a more fundamental way. They have an important role to play in better understanding the statistical properties of the deterministic equations, and perhaps by forming the starting point for a new dynamical-stochastic approach to the problem (e.g. "flux dynamics" in Schertzer and Lovejoy (this volume)).

We start from the idea that atmospheric fields featuring strong variability over wide ranges in scale should not be artificially split apart into large and small scales. We seek rather to study their overall statistical nature over a wide range of scales in relation with their dynamics. For the moment, one has to use all available means, including the investigation of their dimensions, symmetry properties, and topology, their mathematical nature (the singular limit of their support differentiability) their statistical properties (multiple scaling exponents, extreme variability, stationarity) and their general links to the atmosphere dynamics.

The most promising stochastic models are those physically based on the cascade phenomenology of the atmosphere and its governing equations. This makes possible the development of stationary processes applicable to fields conserving their flux densities. In the present paper we limit ourselves to continuous multiplicative cascade models, applied to passively advected clouds. This velocity field is related to the energy field obtained by a scale invariant multiplicative cascade, which is the continuous generalization of the now familiar discrete cascade models[1] (see e.g. Lavallée et al., this volume). Before giving the details of the construction steps, we will investigate the physics of passive scalar fields. Finally, we give the details of the numerical implementation of the latter process.

2. PHYSICAL BASIS OF MULTIFRACTAL PASSIVE SCALAR CLOUD SIMULATIONS

We consider the simplest type of clouds which result from the passive advection of water concentration (ρ) by a velocity field ($\underline{v}$):

$$\frac{\partial \underline{v}}{\partial t} + (1-P(\nabla))\, \underline{v}\cdot\nabla\, \underline{v} = f + \nu\nabla^2\, \underline{v} \qquad (2.1)$$

$$\frac{\partial \rho}{\partial t} + \underline{v}\cdot\nabla\, \rho = f' + \kappa\nabla^2\rho$$

[1] It is worth noting that although Mandelbrot (1972) apparently used Fourier series in certain numerical simulations of cascades, he did not investigate the continuous limit (Fourier transforms). Since the universal properties of multiplicative cascades appear only in this limit, this explains his recent (Mandelbrot, 1989) strong statement that there are no universal dimension functions (or, equivalently, generators) for such cascades.

Where[1] $P(\nabla) = \nabla^{-2}\nabla_i\nabla_j$. First we note that the nonlinear terms $((1-P(\nabla))\ \underline{v}\cdot\nabla\ \underline{v}\ ,\ \underline{v}\cdot\nabla\ \rho)$ of both the incompressible Navier-Stokes equations and the equation of (passive) advection dynamically conserve the fluxes of energy and scalar variance (having densities ε and χ respectively). This conservation is simply expressed in terms of the fluxes of energy and scalar variance:

$$\varepsilon = -\ \partial\langle v^2\rangle/\partial t = \text{constant}$$
$$\chi = -\ \partial\langle \rho^2\rangle/\partial t = \text{constant} \tag{2.2}$$

Furthermore, in Fourier space the non-linear terms are heavily weighted to interactions involving neighboring wavenumbers- thus the energy flux is mainly transfered from one scale to a neighboring scale, hence the Richardson- Kolmogorov idea of energy cascading from large to small scales. From here, various phenomenological arguments (which boil down to dimensional analysis) leads to the well-known Kolmogorov scaling (power law) spectra. These all rely on the invariance of the equations under the dilation transformation $\underline{x}\rightarrow\underline{x}/\lambda$, $\underline{v}\rightarrow\underline{v}/\lambda^{H}$ and $\rho\rightarrow\rho/\lambda^{H'}$ (here and below, the prime refers to the passive scalar cascade). This scaling leads to the following relations for the "fluctuations" at scale l of the fields $\underline{v}$ and ρ (these fluctuations can be characterized for example by the standard deviations of the differences, or increments in v and ρ at points separated by distance l) denoted $\Delta v(l)$ and $\Delta\rho(l)$. Furthermore, dimensional analysis applied to χ and ε implies H=H'=1/3, or:

$$\Delta v(l) \approx \varepsilon^{1/3}\ l^{1/3}$$
$$\Delta \rho(l) \approx \varphi^{1/3}\ l^{1/3} \tag{2.3}$$

This invariance leads to power law spectra $E_v(k) \sim k^{-\beta}$ and $E_\rho(k) \sim k^{-\beta'}$ which are respectively the power spectra for the velocity and passive scalar fields (depending on k, the wave number: $k\approx 2\pi/l$) and β=2H+1 and β'=2H'+1 since the power spectrum is the Fourier transform of the autocorrelation function:

$$E_v(k) \approx \varepsilon^{2/3}\ k^{-5/3}$$
$$E_\rho(k) \approx \varphi^{2/3}\ k^{-5/3} \tag{2.4}$$

where $\varphi = \chi^{3/2}\ \varepsilon^{-1/2}$ is the flux resulting from the nonlinear interactions of the velocity and water. Note that the above are scale invariant since their forms are conserved under the dilation $k\rightarrow k\lambda$).

The values H=H'=1/3 and $\beta=\beta'$=5/3 were obtained only because the first (h=1) moment of the flux is conserved. So, the question arises as to what should be the scaling exponents for the orders $h\neq 1$. And as far as scale invariance of equations 2.1 is concerned, a different value is possible for every (integer or non integer order) moment hence the terminology of "multiple scaling". This can be expressed for local density values of the field, called (which, at homogeneity scale l) with the following scaling laws:

$$\langle\varepsilon_l^{\,h}\rangle \sim l^{\ -(h-1)C(h)}$$
$$\langle\chi_l^{\,h}\rangle \sim l^{\ -(h-1)C'(h)} \tag{2.5}$$

It can be shown that in general the functions C(h)=d-D(h) and C'(h)=d-D'(h) are codimension functions[2], where D(h), D'(h) are the general multifractal dimensions of the various moments of the process. In the simplest case of a process having a unique fractal codimension C(1), all the various moments of the physical quantities have scaling exponents which are linearly related (differing only by the factor (h-1)), and in this restrictive situation all moments scale the same way. Aside from this, when the functions C(h) and

---

[1]This form of the equations is obtained after the pressure term is eliminated under the assumption of incompressibility.

[2]A co-dimension is the difference between the dimension d of the embedding space and the fractal dimension of the set of interest.

C'(h) are nonlinear in h we obtain a full spectrum of scaling exponents which are all consistent with the scale invariance of the leading dynamical equations. Since higher order moments single out and enhance the more intense regions of the field, multiple scaling simply expresses the fact that the most intense regions of the field will scale differently than the weak regions. So, in the case of multiple scaling (which was first proposed by Kolmogorov (1962) and Obukhov (1962)) who this singular statistical behaviour was later proposed by Parisi and Frisch (1985), analyzed this in term of singularities of various order γ in the local energy-flux density field: $\varepsilon_l \geq l^{-\gamma}$. These singularities are indeed consistent with the multiple scaling behaviour and they are the expressions of the minimal rate of divergence of ε_l as l tends to zero, this will be discussed in more details in section 3.2 a spectrum of scaling exponents for the moments implies a hierarchy of singularities.

3. CONTINUOUS CASCADE MODELS

3.1. General outline

In a series of attempts to take into account the highly variable (intermittent) nature of the ε, χ fields, a considerable literature on cascade processes has developed. Nearly without exception, the cascades discussed in the literature have been discrete, i.e. eddies divided into integral numbers of sub-eddies, each an integral fraction of size of the parent (see appendix A for an introductory discussion). Discrete cascades are unsatisfactory models of physical phenomena for many reasons, perhaps the most obvious being the multitude of artefacts such as square remnants and ghostly lines due to the explicit splitting by an (integer) factor λ of the basic rectangularly shaped eddies.

It is helpful to recall some elements of probability theory. For a random variable g, denote by $\varphi_g(h)$ the "second Laplacian characteristic function of g";

$$\langle e^{hg} \rangle = e^{\varphi_g(h)} \tag{3.1}$$

then, if $g_1, g_2, \ldots g_n$ are independent random variables, then:

$$\varphi_{sg_i}(h) = \sum_i^n \varphi_{g_i}(h) \tag{3.2}$$

In particular, consider g to be a sinusoid, wavenumber $\underline{k}$, amplitude $f(\underline{k})u_{\underline{k}}$, phase $\phi_{\underline{k}}$, where u is a Gaussian random variable with $\langle u^2 \rangle = \sigma^2$, $\langle u \rangle = 0$, $\phi_{\underline{k}}$ is a random variable uniform of $[0,2\pi]$. In section 3.5 and in appendix B (see also appendix A of Schertzer and Lovejoy, this volume) we consider the more mathematically involved case where u is an infinite variance Lévy random variable. We obtain:

$$g_{\underline{k}} = e^{i\underline{k}\cdot\underline{x}} f(\underline{k}) u_{\underline{k}} e^{i\phi_{\underline{k}}} \tag{3.3}$$

$$\varphi_{g_{\underline{k}}}(h) = -\frac{\sigma^2}{2} h^2 |f(\underline{k})|^2$$

Using the addition property (eq. 3.2, generalized from sums to integrals), we obtain:

$$\varphi_{\int g_{\underline{k}} d\underline{k}}(h) = -\frac{\sigma^2}{2} h^2 \int |f(\underline{k})|^2 d\underline{k} \tag{3.4}$$

Eq. 3.4 makes it clear why the choice of a Gaussian is judicious - the central limit theorem shows that , under the infinite sum limit above a Gaussian limit will be obtained (as long as the random variables are chosen independently and have a finite variance). In order to model a multiple scaling (multifractal) field, recall that we seek a field ε_λ with resolution (finest scale) λ, (c.f. eq. 2.5) satisfying:

$$\langle \varepsilon_\lambda{}^h \rangle = \lambda^{K(h)} \tag{3.5}$$

Rewriting this slightly:

$$\langle e^{h\ln\varepsilon_\lambda} \rangle = e^{K(h)\ln\lambda} \tag{3.6}$$

Where K(h) is the exponent characterizing the scaling of the hth moment. It is therefore obvious that $K(h)\ln\lambda$ is the second Laplacian characteristic function of $\ln\varepsilon_\lambda$. We therefore construct a stochastic model for $\ln\varepsilon_\lambda$ by using eq. 3.4 and adding random sinusoids with wavenumbers between 1 and λ with $f(\underline{k})$ chosen so that:

$$\int_1^\lambda |f(\underline{k})|^2 d^d\underline{k} = n_d \int_1^\lambda |f(\underline{k})|^2 |\underline{k}|^{d-1} d|\underline{k}| = \ln\lambda \tag{3.7}$$

Where n_d is a constant depending on the dimension of space that results from the angular integration (=1, 2π, 4π in d=1,2,3 respectively). Since $n_d|f(\underline{k})|^2|\underline{k}|^{d-1}$ is simply the spectral energy E(k), the above is equivalent to stating that the energy spectrum of the logarithm of a multifractal field must be a "1/f noise" (i.e. have a wavenumber spectrum $E(k) \propto |\underline{k}|^{-1}$). The final ingredient in a continuous multifractal cascade is the determination of the proportionality constant in eq. (3.7) by imposing the conservation condition $\langle\varepsilon_\lambda\rangle=1$. Using eq. (3.5) this implies K(1)=0, which can be achieved by dividing an unnormalized ε_λ by $\langle\varepsilon_\lambda\rangle$. If $K_u(h)$ is unnormalized, then the corresponding normalized second characteristic function is $K(h)\ln\lambda$ with $K(h)=K_u(h)-hK_u(1)$.

Putting together all the above elements using the normalization as indicated, we can now give the following method for modelling a continuous cascade in d dimensional space. We choose:

$$f(\underline{k}) = |\underline{k}|^{-d/2} n_d^{-1/2} \tag{3.8}$$

with u, Gaussian: $\langle u^2\rangle = 2C_1$ (i.e. $C_1 = \sigma^2/2$) and $\langle u\rangle=0$.

These choices, followed by Fourier transformation, and exponentiation will lead to a field that requires division by the normalization factor $\lambda^{\sigma^2/2}$. The resulting normalized field will have $K(h)=C_1(h^2-h)$. The codimension function corresponding to h^{th} order moments is $C(h)=K(h)/(h-1)= C_1h$, hence the value C_1 is the codimension of the mean field (i.e. for h=1).

It should be noted at this point that the requirement that the Laplacian characteristic function (eq. 3.6) must converge restricts the possible probability distributions of the random variables. This is not surprising since if the random variable $\ln\varepsilon_\lambda$ has sufficiently frequent large positive values, the left hand side of eq. 3.6 may not converge for any $h\geq0$. In the Gaussian case discussed here, the extreme values are very rare and there is no difficulty, however, in infinite variance Lévy cases, the extreme tails of the probability distributions are generally algebraic (exponent $-\alpha$) and eq. 3.6 will diverge. The only exceptions are the "extremal" (maximally asymmetric) Lévy distributions which can be chosen so that their positive values have weak (exponentially decaying) fluctuations with only their negative values being algebraic. These are the only Lévy distributions with $\alpha<2$ that yield convergence and are discussed in more detail in appendix B.

3.2. Numerical simulation of two dimensional ε_λ fields with Gaussian generator (α=2)

One-dimensional simulations are sufficient for the study of many of the statistical properties of the cascade. However, if we are interested in the ("coherent") structures of the multifractal field, it is clearly more interesting to produce simulations in two or more dimensions. Ultimately, four dimensional (space-time) simulations would provide information on temporal evolution and hence on the predictability of the fields.

In the Gaussian case, the implementation is straightforward. We use Gaussian white noise (mean zero, variance given below) distributed over a rectangular grid in 2D Fourier space up to the maximum excited wave number $2\pi\lambda/l_0$ which corresponds to the largest scale of homogeneity (=smallest scale of variability) l_0/λ. The variance $\langle u^2\rangle = 2C_1$ is chosen so that the codimension function is:

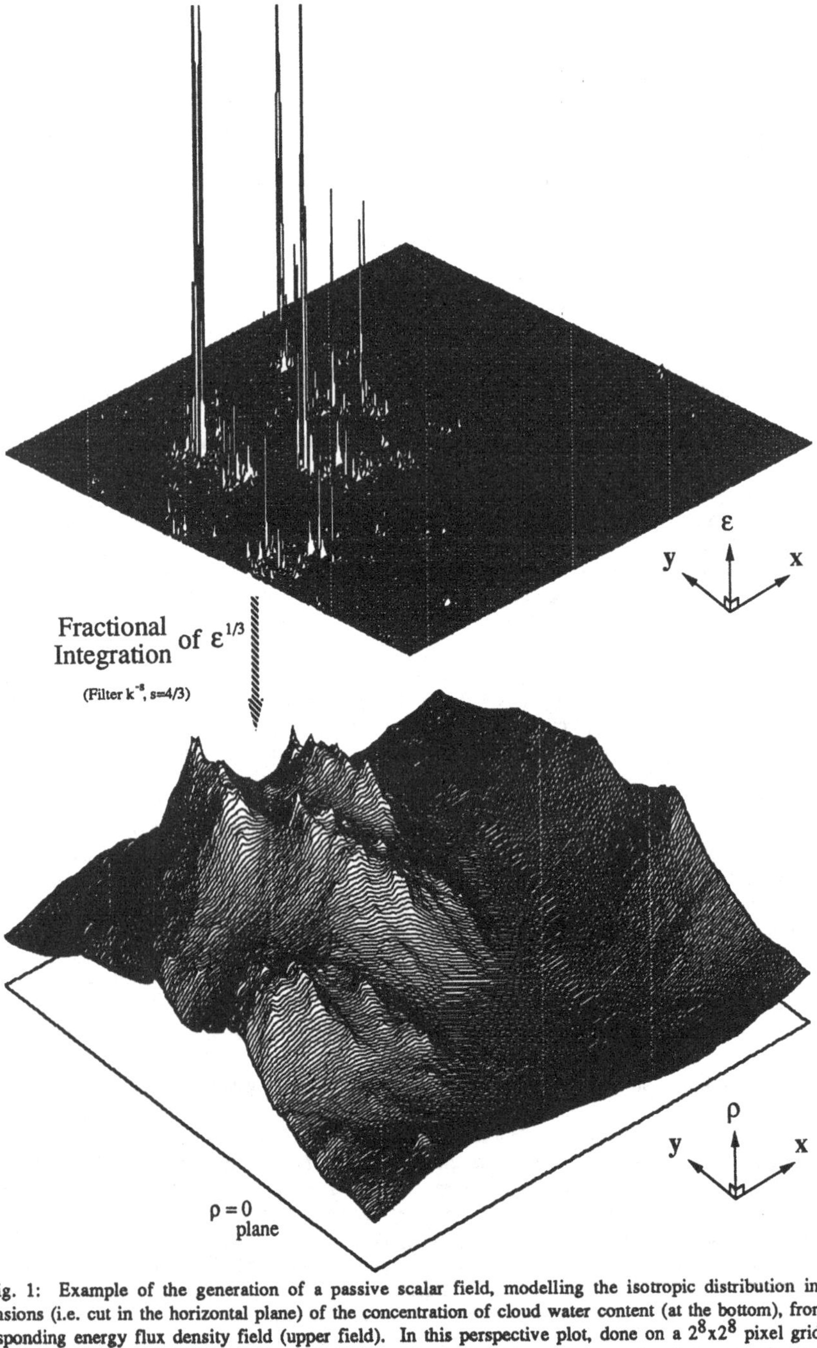

Fig. 1: Example of the generation of a passive scalar field, modelling the isotropic distribution in two dimensions (i.e. cut in the horizontal plane) of the concentration of cloud water content (at the bottom), from the corresponding energy flux density field (upper field). In this perspective plot, done on a $2^8 x 2^8$ pixel grid, the height (in the vertical) of the structures represents respectively the values of energy ε and concentration ρ. The transformation from the former to the latter field is insured by a scale invariant smoothing operation called the fractional integration applied to the one third power of the energy field. The energy flux density field is obtained by a continuous multiplicative cascade using a Gaussian generator (α=2) and has a fractal dimension of D_1=1.2 associated with the first order moment (C_1=2-D[h=1]=0.8).

$$C(h) = C_1 h \tag{3.13}$$

Finally, we exponentiate the Fourier transform (FFT) of the field. The normalization of the resulting ε_λ field must take into account the fact that the continuous integral is replaced by a discrete sum, and this implies a correction in the form of the Euler constant γ_e=0.57..., since

$$\left(- \int_1^n \frac{dx}{x} + \sum_{i=1}^n \frac{1}{i} \right) \xrightarrow[n \to \infty]{} \gamma_e \tag{3.14}$$

The upper part of figure 1 shows a two dimensional energy field simulation having a (1/f) Gaussian generator (α=2) and resulting from a continuous multiplicative cascade done over a $2^8 x 2^8$ pixel grid. As an illustration, an example of the needed Gaussian subgenerator field is provided at the top of figure 8 (which is simply a symmetric unitary Gaussian white noise). So, we have chosen in the simulation $C_1 = 0.4$, such that fractal dimension associated with the mean of the field is d-C_1 = D(h=1) = 1.6.'

3.3. Simulation of two dimensional cuts passive scalar clouds- fractional integration

To model passive scalar density field ρ, we require both ε, χ.which will generally be coupled (dependent) cascade quantities. ρ is then determined from $\Delta\rho \approx \varphi^{1/3} l^{1/3}$ where $\varphi = \chi^{3/2}\varepsilon^{-1/2}$ (see section 2).

The simplest (albeit extreme) procedure for obtaining a φ field is to assume that the energy and passive scalar variance fluxes are completely correlated, ($\varepsilon \propto \chi$) hence $\Delta\rho \approx \chi^{1/3} l^{1/3}$ or alternatively, $\Delta\rho^3 \approx \chi l$. The factor l represents an integration of order one which can easily be accomplished in Fourier space by dividing by $|\underline{k}|$. Hence, in order to produce a field with the correct scaling properties, we Fourier transform χ, divide by $|\underline{k}|$, transform back into real space, and then take the cube root. These operations act only on the amplitudes of the field, leaving the phases entirely unchanged, and has the main mathematical effect of a scale invariant smoothing operation which establishes the required spectrum for the field of concentration ρ.

The only numerically delicate part of this scheme is the need for some care in the memory management. Since the procedure uses purely real physical fields, one should take advantage of this when performing the fast Fourier transform (FFT) which is the time limiting step in the simulation. Large scale simulations are easily performed. The example given in figure 1 shows the resulting concentration field over a $2^8 x 2^8$ pixel grid simulated on a personal computer. The generation of a passive scalar field proceeds in the same way for any other value of α (α<2) once the energy field is provided (see below).

Numerical implementation of the process is straightforward without any special difficulty except for the use of discrete fast Fourier transforms which have their own limitations. Possible alternative methods might involve use of "wavelet" transforms, together with windowing techniques. Even without zooms (see below) very large simulations are possible (perhaps up to 512^3 on Cray 2 supercomputers). If we assume that the homogeneity scale of the process corresponds to the viscous scale =1 pixel, then $Re_1 = 1$ and using $Re_l = v l/\nu \approx \varepsilon^{1/3} l^{4/3}/\nu$ where ν is the viscosity. Hence:

$$Re_l \approx \left(\frac{\varepsilon^{1/3} l^{4/3}}{\nu}\right) \left(\frac{\varepsilon^{1/3}}{\nu}\right)^{-1} Re_1 \approx l^{4/3} \tag{3.15}$$

Hence, using l =512, we have $Re_l \approx 4000$ which compares very favorably, with the largest direct numerical simulation of Navier-Stokes, for which $Re_l \approx 100$ (on a 128X128X512 grid). However the real advantage of these models is the possibility of zooming outlined below which allows us to simulate over effectively unlimited range scale.

Examining figure 1, one notices the periodic character due to the Fourier treatment. The latter is more evident in the case of the concentration field rather than for the energy field since the smoothing by fractional integration forces the spectral amplitude of a given structure of the energy field to slowly extend its influence over the neighboring points of the grid. If the structure is near the border the corresponding effect appears on the other side. To avoid this artifact, one may choose to eliminate a small strip near each border or use one of the numerous window filtering techniques with a view to minimizing the effect on the spectrum (which is biased anyway at high wave numbers due to both aliasing and the periodicity of the Fourier transform).

3.4. Normalized zoom sequences

Although unrealistic, discrete cascades have the interesting property that by a simple iteration process we multiplicatively insert more and more small scale detail: we "zoom" into smaller and smaller regions. In this sub-section we show how to perform such zooms on continuous cascades (see Wilson et al. (1986) for more discussion on zooming).

The main principle in the actual implementation of zooming is first, to retain a small square window from the original two-dimensional energy field, and second, to map this new field on the original grid by enlarging by a factor λ_z, the values retained being repeated the appropriate number of times on the neighboring pixels. For the sake of simplicity we will use in the following a zoom factor $\lambda_z=2$, so that when we double, each pixel becomes a 2x2 field element. The last step is simply the multiplicative modulation of smaller structures to refine the details of the field by "reactivating" all the newly available degrees of freedom at the new resolution. For simulations done on a $\lambda x \lambda$ pixel grid, since we retain $1/\lambda_z$ of the former field, this additional information represents then $\lambda x \lambda(1-1/\lambda_z)$ degrees of freedom, or, for $\lambda_z=2$, exact one half.

The insertion of the high wavenumber (small scale) details is performed in a multiplicative way with the appropriate normalization. To do this, a new field is created in the same way as previously except that the range in scales is the exact complement of that of the doubled field. In Fourier space a cascade is performed in a spectral band spanning $2\pi\lambda_z/l_0 \leq |\underline{k}| \leq 2\pi/l_0$. This partial field is then transformed back to physical space and exponentiated with a normalization corrected this time by the zoom factor in such a way that all points of this new field act as independent multiplicative increments. Since both fields have the same spatial extent (the same number of degrees of freedom), the final modulation is performed by direct point by point multiplication. Thus, the zooming technique applied here, applied multiplicatively in discrete steps yields the same continuous cascade as before, except that we now we have separately modeled the two contiguous ranges of scale related by the zoom factor. A concentration field can be obtained from each of the newly zoomed χ fields by taking integration of order one as before and then the cube root.

Figures 2 and 3 illustrate three successive zooms with $\lambda_z=2$ on a 256x256 pixel concentration field with a Gaussian generator. The first figure shows the technique applied to the center 128x128 window whose corresponding χ field is singled out, doubled and then modulated with finer scale structures before being reconverted to a zoomed concentration field. Repeating the process two more times in figure 3, we obtain an enlarged 256x256 view of the former 32x32 centered pixels of the first field, or equivalently after a global factor 8 zoom (each step of process being repeated recursively) the center window looks as if it were taken from a 2048x2048 primitive field. One can see the signature of the Fourier transforms by the periodicities introduced in the transformation relating χ to ρ which is an artifact of the finite Fourier transforms.

This can also be seen with more details in bigger "zooming" simulations (done over a 512x512 grid scale): as in the case of figures 4 and 5 featuring a topographical representation of passive scalar fields with a one iteration zoom; and also: in the case of figures 6 and 7 where we simulate cloud like representations with a three iterations zoom based on a different realization of a normalized Gaussian generator (see legends for details). The imaging procedures used here (ray tracing or logarithmic grey scale) are standard ones applied on multiplicative cascade fields.

The scale invariant zooming technique can be used as a diagnostic tool to check our premises about the correct normalizations involved in all parts of the process both theoretically and numerically. As mentioned earlier, since we are dealing with highly intermittent fields which are multiplicating one each other, any small numerical or other errors lead quickly to numeric overflows.

3.5. Simulation of ε_λ fields and passive scalar clouds with Lévy-stable generator ($\alpha<2$)

In this subsection, we describe the modelling of the (non-Gaussian) universality classes involving Lévy distributions with $\alpha<2$. Recall that Lévy distributions are the only limiting distributions for sums of independent, identically distributed (i.i.d.) random variables, and hence are the generalizations of the Gaussian (which corresponds to $\alpha=2$). They are the limit of (appropriately normalized) sums of i.i.d. random variables whose variances are infinite. It is due to this "attractive" property that they form the basis of the universality classes for multiplicative processes (see Schertzer and Lovejoy (this volume) for

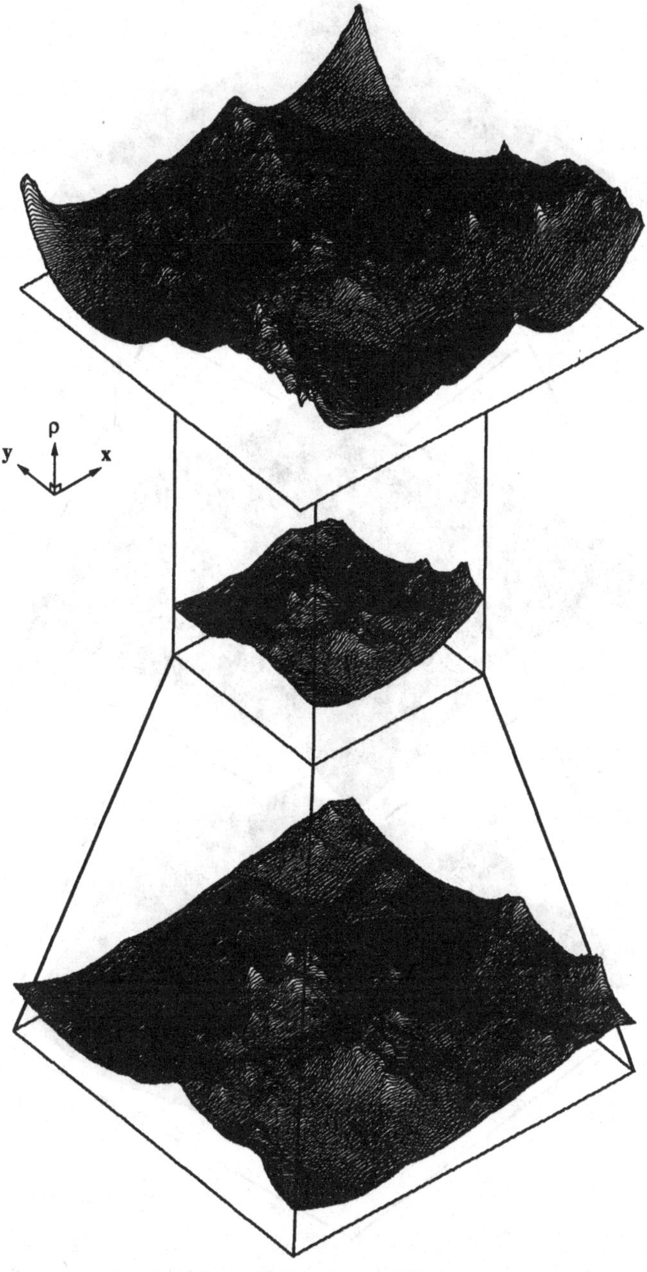

Fig. 2: Illustration of the "zooming" technique applied specifically in the case of the continuous generalization of multiplicative cascades. Here we are zooming by a factor 2 on a 256x256 concentration field issued from a Gaussian generated energy field. The 128x128 window chosen exactly in the center is mapped onto the former resolution where all the newly available degrees of freedom are reactivated. The adjunction of small scale details is done in a multiplicative way which is perfectly coherent with the cascade principles (see text). This illustrates the equivalent discrete step of the former discrete cascades.

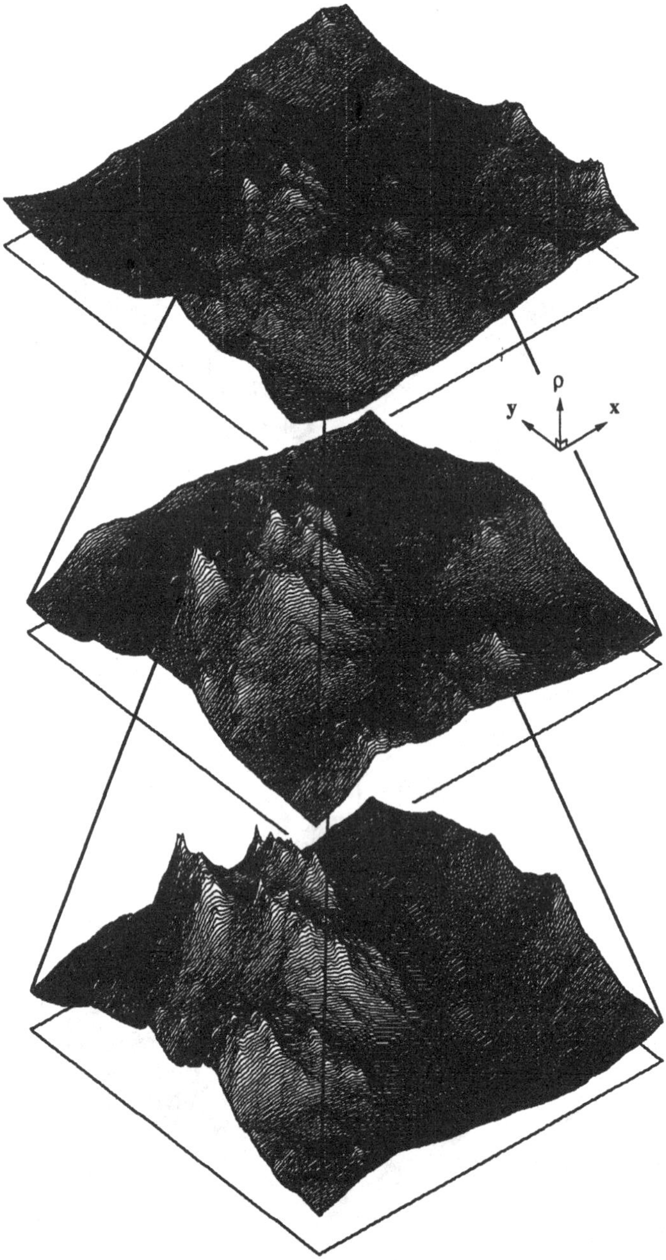

Fig. 3: Examples of generation of two other daughters concentration fields as in figure 2. Repeating here the process of "zooming" in the cascade, by pushing down the inner scale limit two more times with factor 2 zooms, we obtain then a widened 256x256 view of the former 32x32 centered pixels of the first concentration field of figure 2. Equivalently after a global factor 8 zoom (each step of process being repeated recursively) the center window looks like if it were taken from a 2048x2048 primitive field.

Fig. 4: This figure and the next one illustrate the "zooming" technique applied specifically in the case of the continuous generalization of multiplicative cascades. From this figure to the next one we are zooming by a factor 2 on a 512x512 concentration field issued from a gaussian generated energy field. The 256x256 window chosen exactly in the center is mapped onto the former resolution where all the newly available degrees of freedom are reactivated. The adjunction of small scale details is done in a multiplicative way which is perfectly coherent with the cascade principles. The topographic imaging is done by the application of a ray tracing technique directly to the passive scalar field.

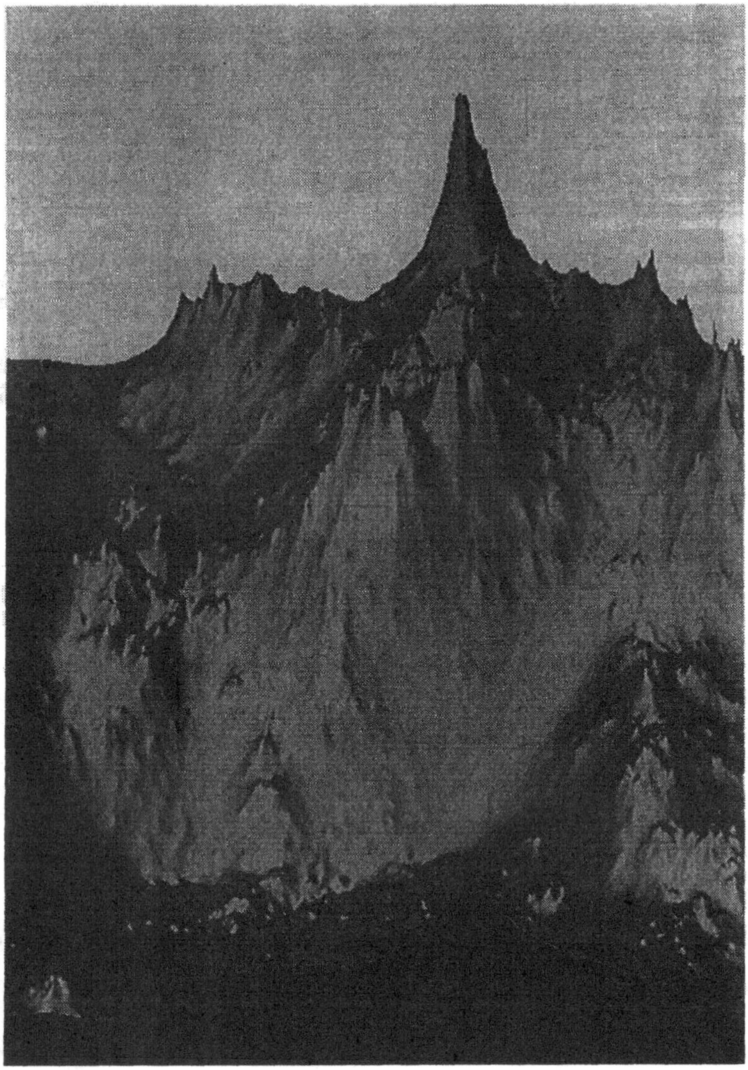

Fig. 5: In this figure we apply the "zooming" procedure to figure 4. The topographic imaging is done again by the application of a ray tracing technique directly to the passive scalar daughter field. Both fields are obtained from a gaussian generator and do feature a fractal dimension of $D_1=1.2$ associated with the first order moment ($C[h=1]=2-D_1=0.8$) .The synthetic topographical field, shown here, is the daughter of the previous one since only small scale details where added in a multiplicative way. Such scale invariant zooms can be iterate easily downward to smaller scales.

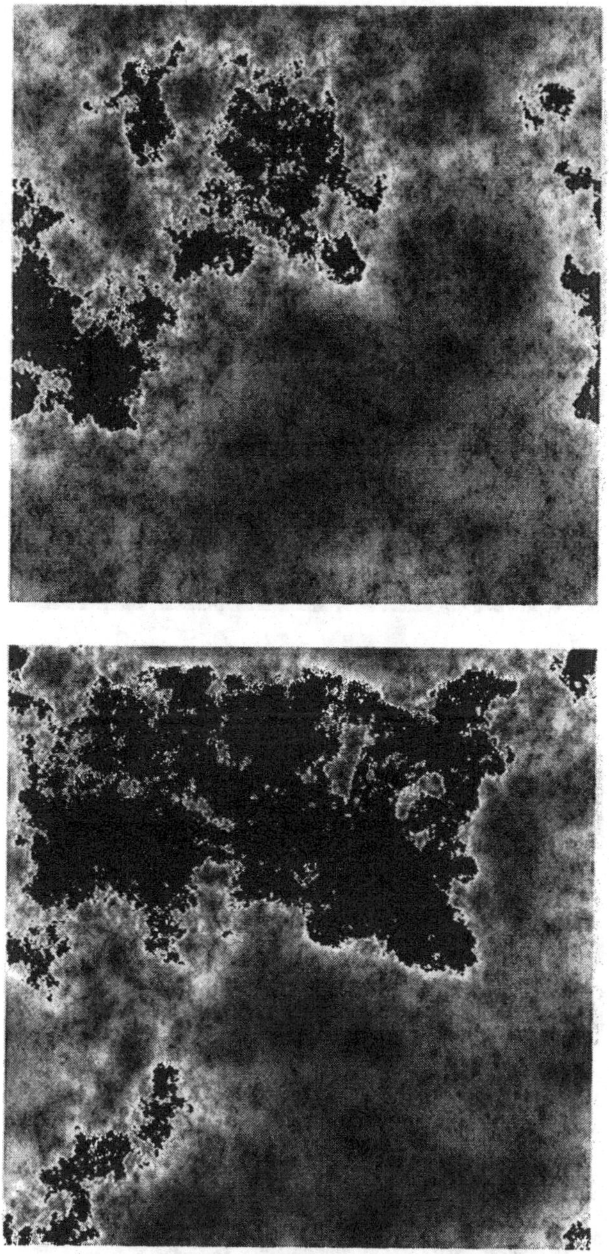

Fig. 6: From top (level 0) to bottom (level 1) we illustrate the "zooming" technique on synthetic cloud fields We are zooming by a factor 2 on a 512x512 concentration field issued from a Gaussian generated energy field which has a fractal dimension of D_1=1.2 associated with the first order moment (C[h=1]=2-D_1=0.8). The 256x256 window chosen exactly in the center is mapped onto the former resolution where all the newly available degrees of freedom are reactivated. The adjunction of small scale details is done in a multiplicative way which is perfectly coherent with the cascade principles (see text). The cloud imaging effect is obtained by applying a linear grey scale to the logarithm of the field intensities to simulate the action of optical thickness in the cloud field.

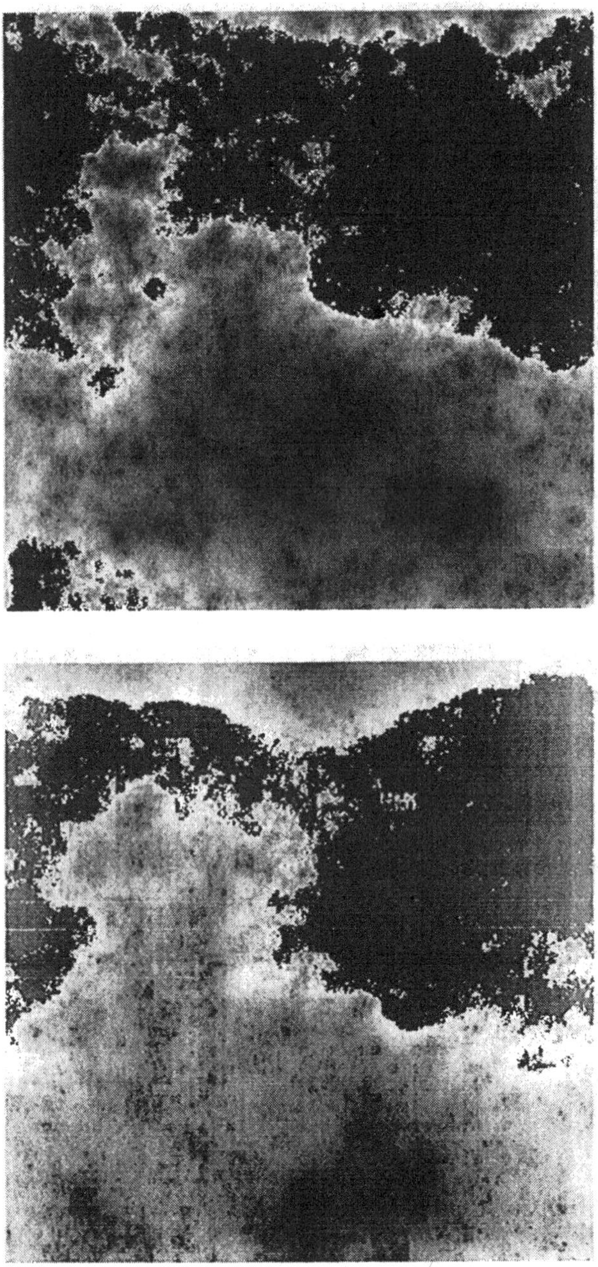

Fig. 7: From top (level 2) to bottom (level 3) we continue the "zooming" procedure of figure.6 The synthetic cloud fields obtained are daughters of the previous ones. The "cloud imaging" is again obtained by applying a linear grey scale to the logarithm of the field intensities to add and simulate the effect of optical thickness to the cloud fields. At each iteration of the "zooming", a threshold under which the sky appears, is determined according to a given level of singularities $\gamma = -.44$ which is chosen to be the same for all fields (level 0 to 3). Such scale invariant zooms can be iterate easily downward to smaller scales.

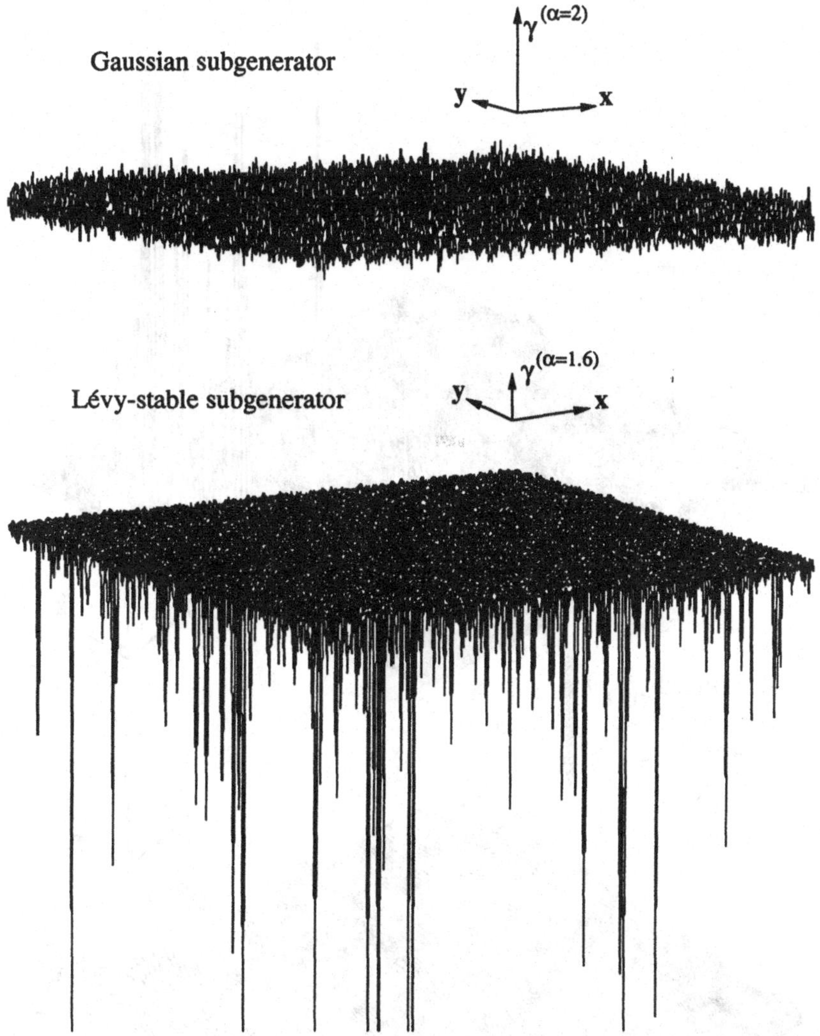

Fig. 8: Two dimensional illustration of two examples of subgenerator field used in continuous multiplicative cascades. The field at the top (a) is made of unitary (mean=0, variance=1) symmetric gaussian (α=2) white noises. And, at the bottom (b) the field is obtained by independent sums of 50 centered asymptotically hyperbolic variables (mean=0, α_h=1.6) converging towards asymmetrically negative Lévy-stable white noises (α=1.6). The perspective plots of the field are depicted on the same physical space, using a 2^8x2^8 pixel grid. The height (in the vertical) of the structures represents respectively the values of the associated random variables, here uniformly independently distributed (the lengths of the appearing vertical axes provide a reference unit). The transformation from this subgenerator to the appropriate 1/f generator field of the cascade requires a fractional integration insured by a $k^{-d/\alpha'}$ (d=2, $1/\alpha'=1-1/\alpha$) filter in fourier space.

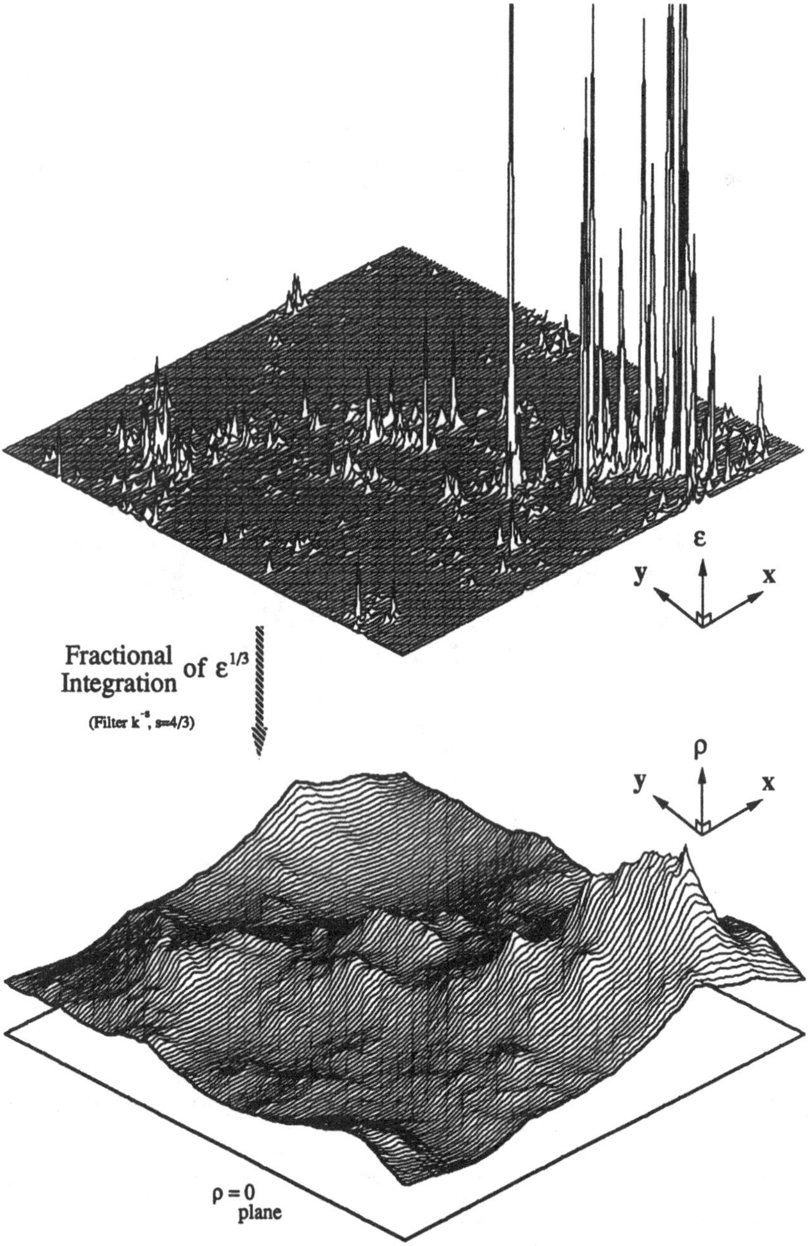

Fig. 9a: Example of the generation of a passive scalar field, modelling the isotropic distribution in two dimensions (i.e. cut in the horizontal plane) of the concentration of cloud water content (at the bottom), from the corresponding energy flux density field (upper field). Here the energy flux density field is obtained by a continuous multiplicative cascade using a 1/f Lévy-stable generator (α=1.6) and has a fractal dimension of D_1=1.2 associated with the first order moment (C[h=1]=2-D_1=0.8). In this perspective plot, done on a 2^7x2^7 pixel grid, the height (in the vertical) of the structures represents respectively the values of energy ε and concentration ρ. The transformation from the former to the latter field requires to do a fractional integration on $\varepsilon^{1/3}$.

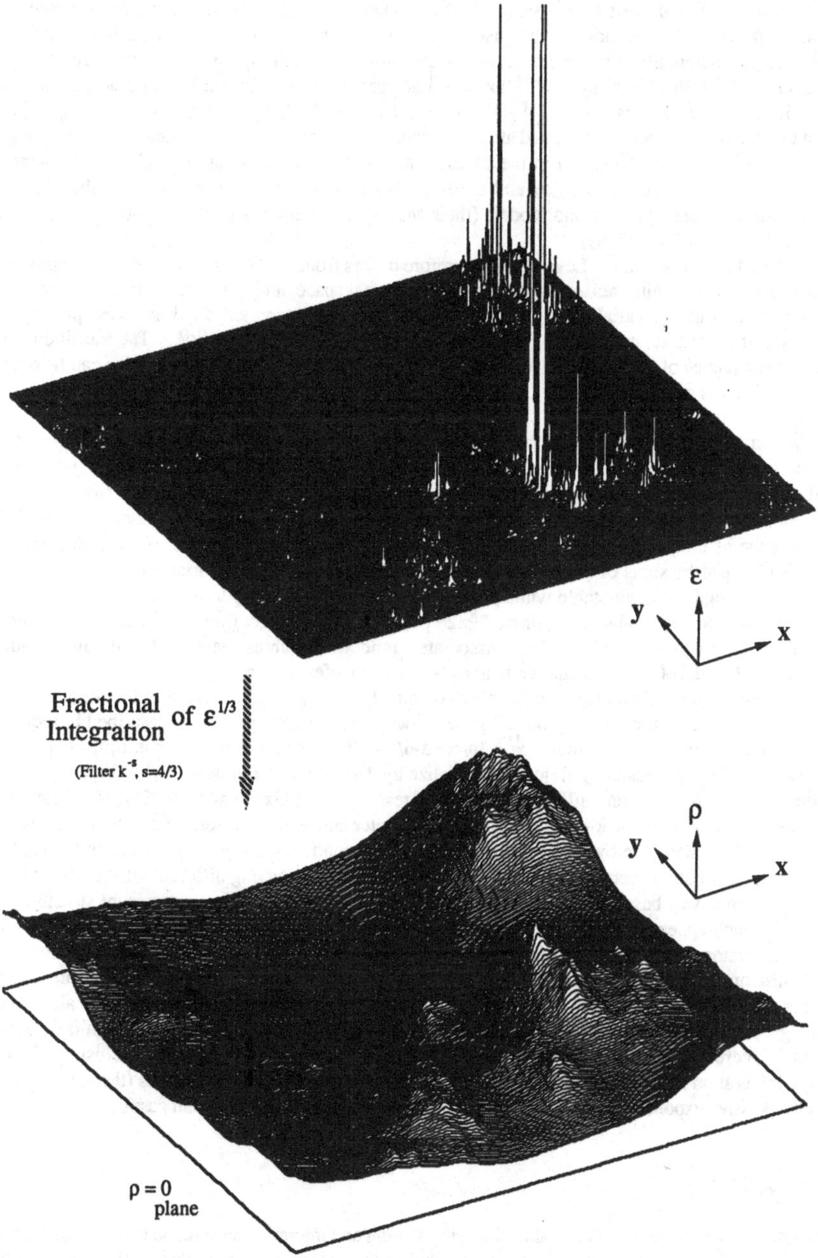

Fig. 9b: Another example of the generation of a passive scalar field, representing also the concentration of cloud water content (at the bottom), from the corresponding energy flux density field (upper field). This latter field, having a codimension associated with the first order moment $C[h=1]=2-D_1=0.8$, comes from the exponentiation of a 1/f generator which is obtained by effecting a fractional integration (filter $k^{-0.75}$) of the Lévy-stable ($\alpha=1.6$) subgenerator field depicted at the bottom of fig.8. In this perspective plot, done here on a $2^8 x 2^8$ pixel grid, the height (in the vertical) of the structures represents also the values of energy ε and concentration ρ. The transformation from the former to the latter field needs another fractional integration (filter $k^{-1.33}$).

discussion). To model such processes, the obvious method to proceed is by replacing the unit Gaussian $u_{\underline{k}}$ in eq. 3.3 by a unit Lévy denoted $u^{(\alpha)}$). However, since the model is constructed by exponentiating these noises in real space, and must satisfy $<\varepsilon_\lambda>=1$, the probability distribution of positive real space noises $u_{\underline{x}}^{(\alpha)}$ must decay at least exponentially fast so that we obtain finite moments of positive order. This requires use of "extremally" asymmetric Lévy distributions. In such distributions, the extreme negative values decrease very slowly as $|u_{\underline{x}}^{(\alpha)}|^{-\alpha}$ however, the probability distribution for extreme positive values is quite different, it decays as $\exp(-|u_{\underline{x}}^{(\alpha)}|^{\alpha'})$ where $1/\alpha+1/\alpha'=1$. Large negative values of $u_{\underline{x}}^{(\alpha)}$ are so frequent that the random variable obtained after exponentiation $(\exp(u_{\underline{x}}^{(\alpha)}))$ is very nearly 0, and all negative moments diverge. We shall see that in the cascade model this has the effect of producing frequent deep "holes" or very weak regions; Lovejoy and Schertzer (this volume) suggest on this basis that this makes these simulations promising as cloud models (their analysis of satellite data yields α~0.63,1.7 for visible ansd infrared channels respectively).

The procedure for producing Lévy clouds therefore differs from the Gaussian case in that (appropriately normalized) extremal (white) noises must be produced in real space, and transformed into Fourier space to yield appropriate random variables $u_{\underline{k}}^{(\alpha)}$. Once this Lévy "sub-generator" field has been produced, we proceed as in the Gaussian case, except that we now require a filter $f(\underline{k}) = |\underline{k}|^{-d/\alpha'}$. The transition from α to α' is a consequence of passing from real to Fourier space. The extremal Lévy variables can be obtained by convergent sums of a large number of negative hyperbolic variables (see appendix B for details on the correct normalization).

The physical interpretation of the result of using such variables is straightforward: we are still multiplicatively creating Gaussian "mountains" but now they are pierced by strong Lévy-stable "holes" created by the exponentiation of occasional large negative values. Figure 8 shows a two dimensional illustration of two examples of subgenerator field used in continuous multiplicative cascades. The field at the top is made of unitary (mean=0, variance=1) symmetric Gaussian (α=2) white noises. At the bottom the field is obtained by sums of centered asymptotically hyperbolic random variables converging towards asymmetrically negative Lévy-stable white noises (mean=0, α=1.6). The perspective plots of the field are depicted on the same physical space, using a 2^8x2^8 pixel grid. The height (in the vertical) of the structures represents respectively the values of the associated random variables, here uniformly independently distributed (the lengths of the appearing vertical axes provide a reference unit).

Once the (real space) subgenerator field is obtained as in fig. 8 (bottom) using the normalization in appendix B, we produce the 1/f Lévy noise by effecting a fractional integration as described in section 3.3 except that we now use here the filter $k^{-d/\alpha'}$ ($d/\alpha'=d-d/\alpha=0.75$, if d=2 and α=1.6 as in figures 9a and 9b). We then exponentiate the resulting field and normalize by dividing by c_N from eq. B.14. Figures 9a and 9b illustrates two such fields having different resolutions: respectively 128x128 and 256x256. Both fields have a first order moment codimension C_1=0.8. The cloud water concentration fields are obtained in the same way as for the Gaussian case by taking 1/3 powers and a (second) fractional integration. In figure 9a, the passive scalar distribution (bottom) features a hierarchy of structures having different intensities, and figure 9b (bottom), which was based on the subgenerator field of fig. 8 (bottom), has a dominant structure in the back corner which appears to be the random result of a low clustering in that region of the negative values of the subgenerator, so that their broad attenuating effect is almost absent even if the multiplicative aspect of small structures "growing" on this "mountain" is still clearly evident. The clustering influence is not surprising since the passage from the subgenerator to the generator invokes, as well, a scale invariant smoothing operation (fractional integration) which do not alter the phases of the well localized negative Lévy-stable increments of the subgenerator but allows to group their influence in clusters since the fractional integration scatters the effect of given increments onto their neighbourhoods (the same argument applies for positive exponential increments or for symmetric values in the Gaussian case).

4. CONCLUSIONS

We have shown how passive scalar clouds, featuring scaling and intermittency, can be modeled by fractional integration over appropriate powers of conserved but highly intermittent multifractal fluxes. In order to produce multifractal (multiple scaling) fields, we show that it is sufficient to create a field whose logarithm is a "1/f noise". Such fields can be readily produced in Fourier space using either Gaussian or Lévy generators (white noises with Gaussian or (extremal) Lévy distributions with random phases and appropriate power law filters). Just as the latter are the basic universality classes for the addition of random variables, the mulifractal fields they generate form the universality classes of multiplicative processes. Unlike the cascade processes most often discussed in the literature, the cascades produced here are "canonical" in the sense that energy flux is conserved only on average rather than for each realization and at each scale[1]. Aside from being physically more realistic (since it treats eddies at each scale as open rather than closed systems), canonical cascades have the advantage of lending themselves relatively easily to a continuous Fourier treatment. In contrast, the rigid constraints placed on the cascade at each step (and everywhere in space) in the microcanonical case cannot easily be imposed in Fourier space, hence the of construction of continuous microcanical cascades is far more difficult. All parts of the simulations described here were reentrant and recursive allowing an unlimited number of downward and upward zooms. The physical basis is provided by the multiplicative character of the cascade simulating the breaking of eddies due to nonlinear interactions or internal instabilities, and also by scaling and intermittency, which is of broadly the same sort as that specified by the dynamical (nonlinear, partial differential) equations. Because of the possibility of "zooming", these models are effectively capable of examining turbulent fields at arbitrarily high resolution and may be expected to be indispensable in modelling fully developed turbulence.

These models have many possible applications; particularly in remote sensing, hydrology and meteorology since explicit stochastic models of broadly this type are required for solving basic problems in measurement, calibration and forecasting. Even in the (unusual) case where measuring devices directly sense signals linearly proportional to the quantity of interest, the sub temporal/spatial resolution fluctuations will not be easy to take into account because of the fundamental distinction between the "bare" and "dressed" cascade quantities. In the more usual case (such as radar or satellite remote sensing of rain and clouds), the instruments measure a (radiation) field which is non-linearly related to that of interest, hence explicit subresolution modelling is imperative. In both cases, the multifractal nature of the underlying field will lead to resolution dependent measurements which are of little intrinsic value since they depend mainly on the sensor rather than the characteristics of the field under study. We view the development of multifractal stochastic models as an essential component in developing scale invariant approaches two the problem of remote sensing; see Davis et al. (this volume).

As far as numerical weather prediction is concerned, one may divide applications into two categories. First there are a number of applications which involve evaluating the "stochastic coherence" (Schertzer et al., 1983) of the models i.e. do any of the steps proceeding from analysis, initialization to numerical integration artificially break the scaling and/or attenuate extreme fluctuations (such as meteorological fronts)? This question is directly related to the types of ("coherent") structures produced in the NWP models and in multifractal cascades. A further, related problem that can be studied with the help of cascade processes is the problem of predictability and its limits as well as developing procedures for stochastic forecasting.

A second category of applications is more theoretical - how can the stochastic cascade generator be related to the dynamical equations? This is likely to be the key problem in any clean "renormalization" (i.e. sub-grid parametrisation of the model equations. Since the addition of other fields (such as radiation, humidity, etc.) coupled non-linearly with the basic dynamical equations will introduce new conserved fluxes, we may expect the true system of non-linear partial differential equation governing the atmosphere to be described by coupled cascade processes and to share many of the basic properties (multiple scaling, intermittency) discussed here. In this regard, the universality classes predicted by continuous cascade models provides promise for great simplifications.

APPENDIX A: SUMMARY OF DISCRETE MULTIPLICATIVE CASCADES: THE α- AND β-MODELS

The basic idea of dynamical cascade models is that the structures at neighboring scales modulate each other in a multiplicative way, simulating the breaking of eddies (and consequent transfer of energy fluxes to smaller scales) due to nonlinear interactions and internal instabilities. The original idea, goes back to

---

[1]This is "microcanonical" case which in the small scale limit involves the extraordinarily restrictive property of conservation of flux at each point in space; see Schertzer and Lovejoy (this volume).

Richardson, but has evolved considerably especially since the impetus of Kolmogorov's (1962) suggestion that intermittency involves log-normally distributed fluxes.

The cascade is built with respect to the energy-flux and passive scalar-flux conservation requirements: $<\varepsilon>=1$ and $<\chi>=1$. The basic procedure for producing discrete cascades is as follows: starting from the outer scale l_0 with unit density of energy flux, representing a primitive uniform eddy, we pass to smaller scale sub-eddies (λ times smaller, where in discrete cascades, λ must be an integer >1) by breaking the eddy, transfering the energy flux to the λ^d sub-eddies (d is the dimension of space in which the process occurs).

Various energy flux redistribution mechanisms have been proposed in the literature, the interesting cases being stochastic. The simplest of these is the "β model" in which energy fluxes are modulated by random factors $\mu\varepsilon$ chosen from a binomial process such that for each sub-eddy:

$$\Pr(\mu\varepsilon=\lambda^C) = \lambda^{-C} \qquad \text{(A.1)}$$
$$\Pr(\mu\varepsilon=0) = 1- \lambda^{-C}$$

As the cascade proceeds to smaller and smaller scales, the energy flux is concentrated in smaller and smaller regions while conserving the ensemble average flux $<\varepsilon>=1$, ultimately becoming a singularity of order C over fractal sets (codimension C). Many ways of generalizing this process exist; the simplest is the "α model"[1] in which the above "dead"/"alive" choice is replaced by "active/"weak":

$$\Pr(\mu\varepsilon=\lambda^{C/\alpha}) = \lambda^{-C} \qquad \text{(A.2)}$$
$$\Pr(\mu\varepsilon=\lambda^{-C/\alpha'}) = 1 - \lambda^{-C}$$

The α model yields multifractal rather than monofractal measures (Schertzer and Lovejoy, 1983), and is thus far more interesting. More recently, "random β models" in which the parameter C is chosen randomly (Benzi et al., 1984), and with certain (complex) correlations (Siebsema et al., 1988) have been proposed. The latter models were also subjected to an additional (very restrictive) constraint that the energy flux be exactly conserved everywhere in space and at each scale, i.e.:

$$\sum_{i=1}^{\lambda^d} \mu\varepsilon_i = \lambda^d \qquad \text{(A.3)}$$

for the sum of the random factors $\mu\varepsilon_i$ over each of the λ^d subeddies at each step. In analogy with the statistical physics of energy fluctuations (rather than energy *flux* fluctuations), cascades subjected to this constraint (which induces subtle correlations between the previously independent $\mu\varepsilon_i$) are called "microcanonical" (see below), whereas the α and β models are canonical"- they only conserve the fluxes on average. Recently, Meneveau and Sreenivasan (1987) baptized the microcanonical α model the "p model". To produce discrete versions of the universal continuous cascades discussed in the text, $\mu\varepsilon$ must be taken as either log-normal (α=2), or log-Lévy (α<2).

One of the distinguishing features of microcanonical cascades is that unlike its canonical counterpart, averages of moments over completed cascades (dressed moments, see below and Lavallée et al., this volume), over sets with dimension d, always converge. Although this seems to be one of the reasons for the popularity of microcanonical models, it is not compelling, particularly since averages over sets with dimension arbitrarily close to d (but less) feature divergence and most of the other properties of the canonical models.

APPENDIX B: PRODUCING EXTREMAL LÉVY VARIABLES

Lévy variables generally have probability distributions of the form $\Pr(>x) \sim |x|^{-\alpha}$ (α<2) for sufficiently large absolute values of the random variable x. Extremal Lévy variables are special cases in which the algebraic tail is confined to either the positive or negative extremes only - the other tail will be generally $\exp(-|x|^{\alpha'})$ where $1/\alpha+1/\alpha'=1$. Extremal Lévy variables are required for convergence of the Laplace

[1]Note that this α is quite different from α in the Lévy distribution used in the body of this paper.

characteristic function (eq. 3.4) and hence for generating cascades with $\alpha<2$. Below, we describe how to produce such cascades including the correct normalization.

Consider w as a uniform random variable in the interval [0,1] (efficient numerical routines are readily available for generating these). The variable $y' = w^{-1/\alpha}$ will therefore have the density:

$$p(y') = \alpha y'^{-1-\alpha} \qquad y' \geq 1$$
$$= 0 \qquad y' < 1 \tag{B.1}$$

The central limit theorem (see in Schertzer and Lovejoy (this volume) for discussion) ensures that summing n independent variables p(y) will approach a Lévy limit distribution parameter α when n tends to infinity.

We now seek the correct centering and normalization constants μ and ξ such that an extremal Lévy will result in the large n limit. Consider the normalized and centered variable:

$$y = \xi(\mu - y') \tag{B.2}$$

Its second Laplacian characteristic function $\varphi_y(h)$ is given by;

$$e^{\varphi_y(h)} = \langle e^{yh} \rangle = e^{\xi\mu h}\alpha \int_1^\infty e^{-\xi h y'} y'^{-1-\alpha} dy' \tag{B.3}$$

$$e^{\varphi_y(h)} = \alpha e^{\xi\mu h} \xi^{\alpha} \Gamma(-\alpha, \xi h) \tag{B.4}$$

where Γ is the incomplete gamma function:

$$\Gamma(v,x) = \int_x^\infty e^{-t}\, t^{v-1} dt \tag{B.5}$$

and furthermore:

$$\Gamma(-\alpha,x) \approx \Gamma(-\alpha) + \frac{x^{-\alpha}}{\alpha} + \frac{x^{-\alpha+1}}{1-\alpha} - \ldots \tag{B.6}$$

Hence:

$$e^{\varphi_y(h)} = (1+\xi\mu h + \frac{(\xi\mu h)^2}{2!} + \ldots)(1 + \frac{\alpha\xi h}{1-\alpha} + \alpha\Gamma(-\alpha)(\xi h)^{\alpha} + \ldots) \tag{B.7}$$

$$e^{\varphi_y(h)} = 1 + \xi h(\mu - \frac{\alpha}{\alpha-1}) + \alpha\Gamma(-\alpha)(\xi h)^{\alpha} + \ldots$$

Hence, taking:

$$\mu = \frac{\alpha}{\alpha-1} \tag{B.8}$$

we obtain:

$$e^{\varphi_y(h)} = \langle e^{yh} \rangle = 1 + \alpha\Gamma(-\alpha)(\xi h)^{\alpha} + \ldots \tag{B.9}$$

Hence:

$$\langle \exp\{\frac{h}{n^{1/\alpha}} \sum_{i=1}^{n} y_i\} \rangle = (1 + \alpha\Gamma(-\alpha)\frac{(\xi h)^{\alpha}}{n} + ..)^n \rightarrow \exp(\alpha\Gamma(-\alpha)(\xi h)^{\alpha}) \tag{B.10}$$

Thus for large n, the random variable:

$$Y = \frac{1}{n^{1/\alpha}} \sum_{i=1}^{n} \xi\left(\frac{\alpha}{\alpha-1} - w_i^{-1/\alpha}\right) \tag{B.11}$$

has (Laplacian) characteristic function:

$$\langle e^{Yh} \rangle = \exp(\alpha\Gamma(-\alpha)(\xi h)^{\alpha}) \tag{B.12}$$

Hence, since the (unnormalized) scaling exponent $K_u(h) = C_1 h^{\alpha}\alpha'/\alpha$, (subscript "u" for unnormalized) we must choose ξ so that:

$$\xi^{\alpha} = \frac{C_1\alpha'}{\alpha^2\Gamma(-\alpha)} \tag{B.13}$$

or, using the identity $\Gamma(2-\alpha)=\Gamma(-\alpha)(-\alpha)(1-\alpha)$ and the fact that $1/\alpha'+1/\alpha=1$, we obtain:

$$\xi = \left(\frac{C_1}{\Gamma(2-\alpha)}\right)^{1/\alpha} \tag{B.14}$$

furthermore, the resulting field must be normalized by dividing by $\lambda^{K_u(1)}$, i.e.

$$c_N = \lambda^{C_1\alpha'/\alpha} \tag{B.15}$$

yielding a normalized field whose scaling exponent is:

$$K(h) = K_u(h) - hK_u(1) = \frac{C_1\alpha'}{\alpha}(h^{\alpha}-h) \tag{B.16}$$

In summary, to obtain extremal Lévy variables which will generate cascades with parameters α, C_1, we produce a large number[1] (n) of uniform random variables w, and hyperbolic variables $y' = w^{-1/\alpha}$. We then transform this into a new (centered, normalized) random variable y using equation B.2 with centering and norming constants given by B.8, B.14. We sum n of these, and normalize with $n^{1/\alpha}$ as in eq. B.11. The result is an extremal Lévy random variable with the required properties.

ACKNOWLEDGMENTS

We acknowledge fruitful discussions with J.P. Kahane, P. Ladoy, A. Davis, P. Gabriel, A. Saucier, G. Sarma, Y. Tessier, P. Brenier and R. Viswanathan.

REFERENCES

Benzi , R., G. Paladin, G. Parisi, A. Vulpiani, 1984: J. Phys., A17, 3521.

Davis, A., S. Lovejoy, D. Schertzer, 1990: Radiative transfer in multifractal clouds. (this volume).

Kolmogorov, A. N., 1962: A refinement of previous hypothesis concerning the local structure of turbulence in viscous incompressible fluid at high Reynolds number. J. Fluid Mech., 13, 82-85

Lavallée, D., D. Schertzer, S. Lovejoy, 1990: On the determination of the co-dimension function. (this volume).

---

[1]The number required for convergence can be quite large, and increases as α approaches 2. For example, for $\alpha=1.5$, it was found that $n \approx 30$ was required.

Lovejoy, S., D. Schertzer, 1990: Multifractal analysis techniques and the rain and cloud fields from 10^{-3} to 10^{-6}m. (this volume).

Mandelbrot, B, 1972: in Statistical models of turbulence, Lecture notes in physics, 12, eds. M. Rosenblatt and C. Van Atta, Springer Verlag, p. 333.

Mandelbrot, B., 1988: An introduction to multifractal distribution functions. Fluctuations and pattern formation, eds. H.E. Stanley and N. Ostrowsky, Kluwer.

Meneveau, C., K.R. Sreenivasan, 1987: Simple multifractal cascade model for fully developed turbulence, Phy. Rev. Lett., 59, 1424-1427.

Obukhov, A., 1962: Some specific features of atmospheric turbulence. J. Geophys. Res, 67, 3011-3014.

Parisi, G., U. Frisch, 1985: A multifractal model of intermittency, Turbulence and predictability in geophysical fluid dynamics and climate dynamics, 84-88, Eds. Ghil, Benzi, Parisi, North-Holland.

Schertzer, D., S. Lovejoy, G. Therry, J. Coiffier, Y Ernie, and J. Clochard, 1983: Are current NWP system stochastically coherent? Preprints, IAMAP/WMO Symp. Maintenance of the Quasi-Stationary Components of the Flow in the Atmosphere and in Atmospheric Models, Paris. WMO, Genrva, 325-328.

Schertzer, D., S. Lovejoy, 1987a: Singularités anisotropes, et divergence de moments en cascades multiplicatifs. Annales Math. du Qué., 11, 139-181.

Schertzer, D., S. Lovejoy, 1987b: Physically based rain and cloud modeling by anisotropic, multiplicative turbulent cascades. J. Geophys. Res., 92, 9692-9714.

Schertzer, D., S. Lovejoy, 1990: Scaling nonlinear variability in geodynamics: Multiple singularities, observables and universality classes. (this volume).

Siebsema, A.P., R. R. Tremblay, A. Erzan, 1988: Multifractal cascades with interactions. (Preprint).

Wilson, J., S. Lovejoy, D. Schertzer, 1986: An intermittent wave packet model of rain and clouds. 2nd conf. on satellite meteor. and remote sensing, AMS, Boston, 233-236.

FRACTAL CHARACTERIZATION OF INTERTROPICAL PRECIPITATIONS VARIABILITY AND ANISOTROPY

P. Hubert
URA-CNRS 1367
C.I.G., Ecole des Mines de Paris
35 rue Saint-Honoré, 77305 Fontainebleau, France

J.P. Carbonnel
URA-CNRS 1367
Laboratoire de Géologie Appliquée, Université P. et M. Curie,
4 Place Jussieu, 75252 Paris Cedex 05, France

ABSTRACT. Burkina Faso rainfall study (1986 June 12-13) conducted over 10,000 km^2 thanks to 111 raingauges. Multifractal nature of the precipitation is shown. Variability and anisotropy of the rainfield is characterized by fractal parameters.

INTRODUCTION

Fractal techniques (Mandelbrot, 1975, 1977) have already been applied in numerous applications relating to the shape of cloud systems and rain areas or to rainfall intensity (Lovejoy, 1983 ; Schertzer and Lovejoy, 1984 ; Lovejoy and Mandelbrot, 1985 ; Lovejoy and Schertzer, 1986). The applications we present here are related to rainfall data associated to a single shower. The small amount of these informations, compared to the huge amount coming from satellites and radars, is one peculiarity of our study.

The shower under study was observed during the 1986 12 to 13 June night in the vicinity of Ouagadougou (Burkina Faso). It came from a squall line which originated in the evening of the 12 in the East of Burkina Faso and disappeared during the afternoon of the 13 above the South-West of Mali. This squall line went across the square degree of Ouagadougou (11-12 degrees North, 1-2 degrees West) from June 12 at 23 h to June 13 at 7 h along an East North-East to West South-West direction at a speed of around 60 km/h (Hubert and al., 1987).

1. CHARACTERISTICS OF THE NETWORK

The shower associated with the squall line crossing was observed thanks to a network of gauging stations scattered in and around the square degree of Ouagadougou. This network operated during the whole 1986 summer. For the episode we are speaking about, 111 stations were active. From a geometrical point of view, this network can be looked at as a set of points distributed in a two dimension space. We assign to this network a dimension Dm, the value of which is bounded by 0 and 2.

We computed the correlation dimension of the network (Lovejoy, Schertzer and Ladoy, 1986). For each network point, we determined the number of network points n (L) included in an L radius circle centered on this point, then $< n(L) >$ the mean of these numbers for all the network points. Finally, we

D. Schertzer and S. Lovejoy (eds.), Non-Linear Variability in Geophysics, 209–213.

displayed $< n(L) >$ against L on a Log-Log diagram (figure 1). It can be seen that for L values between 1 and 80 km :

$$< n(L) > \approx L^{Dm}$$

where the representative line slope is equal to 1.55 ± 0.05. This network fractal dimension imposes some detection limits. Indeed, a phenomenon coming out on a fractal set of dimension Dp, would be detected by the network, which is already a fractal set of dimension Dm, under the condition that the intersection of the two sets is not empty. These two sets being included in a two dimension space, this condition can be expressed from the intersection theorem (Schertzer and Lovejoy, 1987 a, 1987 b), by the following inequality :

$$(Dm + Dp) > 2$$

In our network case, a phenomenon has to come out on a set the dimension of which has to be at least 0.45 to be detected.

2. RAINFALL HEIGHT DISTRIBUTION

The 111 rainfall measurements extend from 0 (there are two null values) to 73.2 mm. The mean value is 27.3 mm and the asymmetry is slightly positive ($C_s = 0.56$). We assigned to the 109 observed positive values an empirical probability according to the Weibull formula. These probabilities are displayed against corrresponding heights on a Log-Log diagram (figure 2). For rainfall greater than 35 mm a hyperbolic behaviour can be admitted :

$$\text{Prob}(H > h) \approx h^{-\alpha}$$

From our diagram α can be estimated at about 4.4. This shows a moderate intermittency as compared to some other meteorological phenomena (Lovejoy and Schertzer, 1986) but nevertheless implies divergency of moments of order greater than 4.

3. RAINFALL SPATIAL FIELD STUDY

The definition of a rainfall threshold delimits a region where rainfall has been greater or equal to this threshold. The dimension D (the value of which is between 0 and 2) of this region can be computed for example by applying the "functional box counting" method (Lovejoy, Schertzer and Tsonis, 1987). This technique supposes that the number of L-sided boxes (here squares) necessary to cover entirely a set, the fractal dimension of which is D, is a function of L :

$$N(L) \approx L^{-D}$$

To apply this method, the Ouagadougou square degree has been divided into 1024 (32 x 32) elementary squares. The rainfall height on each elementary square has been estimated by the Thiessen method (Thiessen, 1911), using the network measurements (figure 3). Setting the elementary square side equal to 1, and using successive grids divided into square meshes with sides equal to 1, 2, 4, 8 and 16, we computed the fractal dimension of the regions associated to different rainfall thresholds by computing the slope of the line joining the representative points of N (L) against L plotted on a Log-Log diagram for each threshold (figure 4). We stopped at the 40 mm threshold in order to be able to rely upon a sufficient number of measurements. Results gathered in table I show that the fractal dimension of threshold-delimited regions is a decreasing function of the threshold, confering a multifractal structure on the rainfall field.

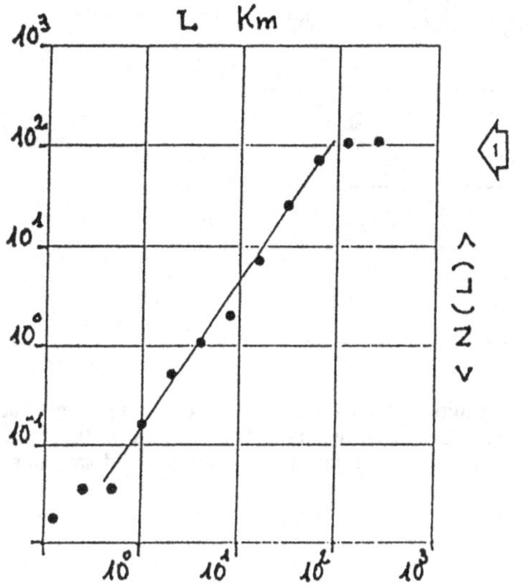

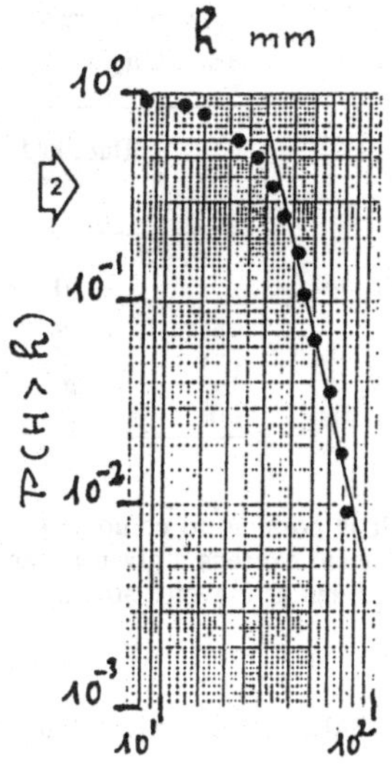

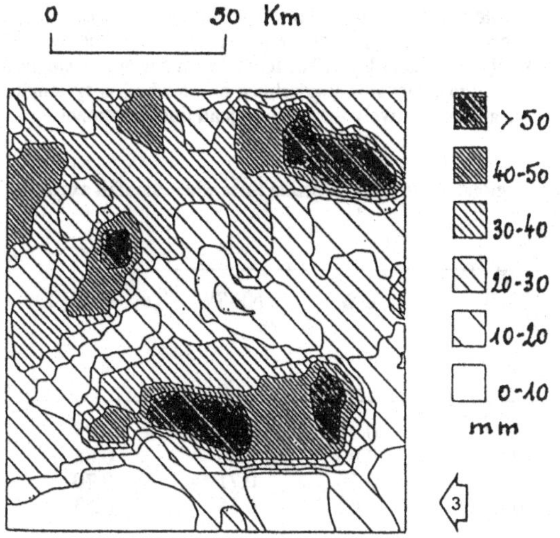

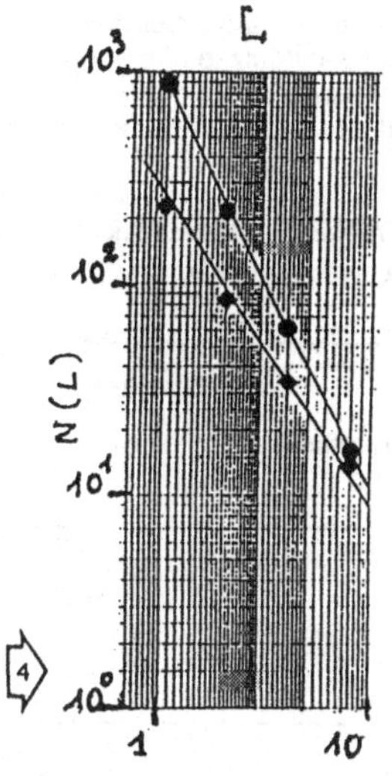

Figure 1. Fractal dimension of the gauging network.
Figure 2. Empirical exceeding probability of rainfall depths.
Figure 3. Rainfall isohyetal map.
Figure 4. Fractal dimension of sets delimited by (from top to bottom) 10, 40 mm thresholds.

TABLE I. Fractal dimension D_2 of regions where rainfall exceeds a given threshold.

Threshold mm	Dimension
10	1.94
20	1.86
30	1.68
40	1.44

It is possible to compute directly the correlation dimension of the set of points where the measurement exceeds a given threshold. This procedure yields results almost identical to those just described, but as it avoids an estimation step, it prevents us from tackling the rainfall field anisotropy problem.

4. RAINFALL FIELD ANISOTROPY

The functional box counting method we just applied in a two dimension space can be applied in an one dimension space. It is sufficient to substitute squares of sides 16, 8, 4, 2 and 1 for rectangles of sides 16 x 1, 8 x 1, 4 x 1, 2 x 1 and 1 x 1 oriented along the same direction (Schertzer and Lovejoy, 1987 a). We computed the dimension (bounded between 0 and 1) associated to regions delimited by different thresholds according to four directions (table II). In spite of limited differences, an anisotropy can be seen, which increases with the threshold. The maximum dimension (underlined in the table) is systematically obtained for the East-West direction which is approximatively that of the squall line movement.

TABLE II. Fractal dimension D_1 of regions of different direction lines where rainfall exceeds a given threshold.

Direction Threshold mm	SE-NW	E-W	NE-SW	N-S
10	0.94	0.96	0.94	0.95
20	0.86	0.90	0.86	0.86
30	0.74	0.76	0.71	0.70
40	0.58	0.62	0.55	0.53

CONCLUSIONS

These first attempts to apply fractal techniques to a single shower rainfall field call for the following comments.

- The multifractal structure of a single shower rainfall field was shown at mesoscale, from a small data set.
- The network spatial outline imposes a detection limit characterized by a fractal dimension associated to a threshold. It would be useless to study the spatial arrangement of possible observations exceeding this threshold.
- A hyperbolic distribution, always associated with intermittent phenomena leading to fractal structures, can be fitted to rainfall heights. This fact should lead hydrologists to ask themselves about the systematic use of exponential distributions.
- It becomes workable to quantify variability and anisotropy of rainfall fields. In our climatic context, anisotropy seems to be controlled by the direction of movement of meteorological perturbations. We can then hope to be able to compare diffferent rainfall fields.
- It would be useful to study just up to which spatial and temporal level these conclusions could be extended.
- It is necessary to coordinate such a spatial approach with temporal ones (Hubert and Carbonnel, 1989).

REFERENCES

Hubert P., Carbonnel J.P., 1989. 'Dimensions fractales de l'occurrence de pluie en climat soudano-sahélien", *Hydrologie Continentale*, **4**, 3-10.

Hubert P., Thiao W., Cadet D., Carbonnel J.P., Desbois M., 1987. 'Comparaison des données satellitaires infrarouge et des mesures au sol concernant l'épisode pluvieux du 12 au 13 juin 1986 dans la région de Ouagadougou (Burkina Faso)', *Veille Climatique Satellitaire*, **20**, 31-36.

Lovejoy S., 1982. 'Area-Perimeter relation for Rain and Clouds areas', *Science*, **216**, 185-187.

Lovejoy S., 1983. 'La géométrie fractale des nuages et des régions de pluie et les simulations aléatoires', *La Houille Blanche*, **516**, 431-436.

Lovejoy S., Mandelbrot B.B., 1985. 'Fractal properties of rain, and a fractal model', *Tellus*, **37 A**, 209-232.

Lovejoy S., Schertzer D., 1986. 'Scale invariance, symmetries, fractals, and stochastic simulations of atmospheric phenomena', *Bull. Amer. Meteor. Soc.*, **67**, 1, 21-32.

Lovejoy S., Schertzer D., Ladoy P., 1986. 'Fractal characterization of inhomogeneous geophysical measuring networks', *Nature*, **319**, 6048, 43-44.

Lovejoy S., Schertzer D., Tsonis A.A., 1987. 'Functional box-counting and multiple elliptical dimensions in rain', *Science*, **235**, 1036-1038.

Mandelbrot B.B., 1975. *Les objets fractals, forme, hasard et dimension*, Paris, 192 p.

Mandelbrot B.B., 1977. *The fractal geometry of nature*, San Francisco, 461 p.

Schertzer D., Lovejoy S., 1984. 'Des fractales dans l'atmosphère', *Sciences et Techniques*, **5**, 17-19.

Schertzer D., Lovejoy S., 1987 a. 'Physical modeling and analysis of rain and clouds by anisotropic scaling multiplicative processes', *J. Geophys. Res.*, **92**, D8, 9693-9714.

Schertzer D., Lovejoy S., 1987 b. 'Singularités anisotropes, divergence des moments en turbulence : Invariance d'échelle généralisée et processus multiplicatifs', *Ann. Sc. Math. Québec*, **11**, 1, 139-181.

Thiessen A.M., 1911. 'Precipitation averages for large areas', *Monthly Weather Review*, **39**, 1082-1084.

4 - MODELING AND ANALYSIS OF THE CLIMATE, OCEANS AND IN SOLID EARTH GEOPHYSICS

FRACTAL NATURE OF SURFACE GEOMETRY IN A DEVELOPED SEA

Roman E. Glazman
Jet Propulsion Laboratory
California Institute of Technology
4800 Oak Grove Drive, Pasadena, CA 91109

ABSTRACT. The spectral density function for spatial variations of the sea surface elevation field is characterized by a power law in the high-frequency range. Commonly used spectral models, the Pierson-Moskovitz or the JONSWAP spectrum, are based on the Phillips law that gives ω^{-q} in the frequency domain and k^{-p} in the wavenumber domain, with $q = 5$ and $p = 4$. Other power laws of spectrum decay, for instance ω^{-4} and $k^{-7/2}$, have also been proposed in the literature. The surface geometry corresponding to such laws possesses many features of a fractal: the field of its slope is discontinuous in the mean-square and the surface itself exhibits a cascade pattern with an infinite number of superimposed wavelets of decreasing size, etc. A new approach to analyzing such broad-band fields is presently proposed, as based on the decomposition of a surface into the narrow-band (averaged) part $\bar{\zeta}$ and the broad-band ("microscopic") part ζ'. The latter component is considered in the "fractal" limit of a small surface patch. Employing the Karhunen-Loève expansion, the microscopic field is expanded in series of "universal" orthogonal functions whose elementary properties are investigated numerically, and simplifications yielded by this representation are discussed.

1. INTRODUCTION

The geometry of a wind-disturbed sea surface is extremely complex, even for highly idealized wind conditions. However, in many cases a statistical characterization of only certain, simpler features of a surface is sufficient. In the present work we consider a small patch of the sea surface, such that the pertinent wave number range in the power spectrum for the surface elevation field is described by a power law. This allows one to concentrate on a cascade pattern appearing in the surface topography within this range.

There are several reasons for this study. The geometrical cascade results in an increased surface density of wave crest occurrence whose statistics are of great interest both in the general context of wave studies and in connection with microwave remote sensing of the ocean (i.e., satellite altimetry, scatterometry, microwave radiometry, and radar imagery). Namely, the wavelets whose steepness becomes too high are unstable and tend to break, thus providing a mechanism of wave energy dissipation and of substance transport across the air-sea interface through whitecapping. The wavelets that are rather steep but not yet breaking tend to have sharp crests, which dramatically affects scattering of centimeter and decimeter electromagnetic and acoustic waves. A brief review of statistical treatments on the breaking wave problem can be found in Glazman (1985) and Glazman and Weichman (1989) and the role of steep wavelets in microwave signatures of the sea surface has been highlighted by Kwoh and Lake (1984, 1985), by Lyzenga et al. (1983), by Glazman et al. (1988), as well as by various other studies. The knowledge of the surface density and of the surface flux density of the sharp-crested wavelets' occurrence would be most useful with respect to the problems mentioned above and to many other problems, indeed.

However, traditional mathematical techniques (that assume the surface to be sufficiently narrow-banded in order to satisfy certain conditions of differentiability) are inappropriate for this analysis. In the present work, possible advantages of representing the surface as $\bar{\zeta} + \zeta'$, where $\bar{\zeta}$ is a narrow banded, "macroscopic"

D. Schertzer and S. Lovejoy (eds.), Non-Linear Variability in Geophysics, 217–226.

component and ζ' is a broad-banded "microscopic" component approximated as a fractal, are explored. The attention is focused on ζ'. Further discussion is provided in a recent paper by Glazman and Weichman (1989).

2. CASCADE PATTERN IN GEOMETRY OF A WELL-DEVELOPED SEA: REVIEW

Wave observations in field as well as in laboratory conditions reveal a considerable range of frequencies (up to two decades) where the spectral density function of a stationary or near stationary wave field can be approximated by a power law. As a result, most empirical wave spectra have the form

$$S(\omega) = \alpha H(\omega/\omega_o)\omega^{-q} \tag{1}$$

where $H(\omega/\omega_o)$ is a smeared step-function that yields a maximum of $S(\omega)$ at $\omega \approx \omega_o$, where ω_o is the spectral peak frequency, and α the (dimensional) coefficient of proportionality. The corresponding wave number modulus spectrum is given by

$$F(k) = \beta h(k/k_o)k^{-p} \tag{2}$$

with β and h having a meaning similar to α and H. The exponents p and q vary (e.g., Barenblatt and Leykin, 1981). In a fetch-limited sea, most observations indicate $q = 5$ is a good approximation (Hasselmann et al. 1973). This value, as well as $p = 4$ in eq. (2), expresses the Phillips law for the equilibrium range of wave spectra (Phillips, 1977). In recent years, arguments have been presented in favor of less rapid decay of wave spectra. In particular, power laws ω^{-4} and $k^{-7/2}$ have been proposed by Zakharov and Filonenko (1966) and by various other authors; and even more erratic behaviour of the sea surface has been predicted by the equilibrium range theory of Zakharov et al. (1975, 1982, 1983): $\omega^{-11/3}$ and $k^{-10/3}$. The broadening of the spectrum takes place when the wind energy input is shifted towards high frequencies, and the lower-frequency components of the wave field receive energy from the higher-frequency range through weakly non-linear interactions among the (resonant) wave modes. This case of the inverse Kolmogorov cascade (Zakharov et al. 1975, 1982) is rather typical for the open-ocean conditions, since it is realized when the phase velocity of the spectral-peak wave component exceeds the mean wind speed.

Since the subject of our interest is the surface itself, the question arises as to the implications of the spectrum shape for surface geometry. As a simple measure of a surface's erraticity (the term suggested by Adler (1981)) one can estimate its Hausdorff dimension D_H. If D_H is stricly greater than D, where D is the topological dimension (D = 2), one anticipates appearance of a cascade pattern which will be the more pronounced, the greater the difference between D_H and D. One can easily calculate the Hausdorff dimension of a Gaussian process/surface whose spectral density function obeys a power law at high frequencies. The final result is:

$$D_H^{(1)} = (5-q)/2$$

$$D_H^{(2)} = (8-p)/2 \tag{3}$$

where the superscripts indicate the topological dimension. The second equation presumes that the spectral density function can be separated into a wave-number modulus factor and an angular factor (as in eq. 11), so that the directional properties of the surface are statistically independent of the range of spatial scales under consideration. It is easy to show that the spatial variations of the sea surface are much more erratic than the temporal variations. This is due to the gravity-wave dispersion relationship, $k = \omega^2/g$, that reflects direct proportionality between the first-order spatial derivatives (wave slope) and the second-order temporal derivative (local vertical acceleration) of the surface elevation. The cascade pattern becomes noticeable for wave spectra containing the Phillips saturation range where $p = 4$. This case, yielding $D_H^{(2)} = D^{(2)}$, can be referred to as a marginal fractal (Berry's (1979) terminology).

Assuming Gaussian statistics, the mean number, n_2, of crests per "basic" (dominant) wave and the mean surface density, $v_{2,Y}$ of sharp-crested events have been evaluated for such spectra by Glazman (1986), by studying relationships between level-crossing statistics of the surface elevation field $\zeta(x)$ and those of the wave slope modulus field $Y(x) = \nabla\zeta$. Extending those results to the three-dimensionsl case when the surface flux density for the events of the steep wavelets occurrence is sought, one obtains (employing results on the rate of high-level excursions by a three-dimensional Gaussian field, presented by Adler (1981)) for a time-varying two-dimensional surface:

$$v_{3,Y} = (2\pi)^{-2} \frac{|\Lambda|^{1/2}}{\sigma_Y^3} \frac{U^2}{\sigma_Y^2} \exp(-U^2/2\sigma_Y^2) \qquad (4)$$

Here Λ is the covariance matrix for the field of wave slope modulus $Y(x_1,x_2,x_3)$ with $x_3 = t$, σ_Y is the total variance of the Y-field, and U is the level to be surpassed by Y provided $U >> \sigma_Y$. This effort, as well as any attempt at evaluating statistics of broad-band fields, requires explicit specification of a high-frequency cutoff or, more rigorously, the use of partial averaging for a random field in order to eliminate the contribution of the high-frequency part of the spectrum that describes the dissipation sub-range. Due to the averaging, the resulting excursion rates and other statistics describe only the "macroscopic" component $\bar{\zeta}$ of the field, whereas without the averaging, $|\Lambda|$ and σ_Y tend to infinity. The extent of the "fractal" range in wave spectra is highly important for the rates of wave crest occurrence. Thus, the so-called microscale (i.e. the averaging scale) [Sections 3, 5 and 9 in Glazman (1986) and section 6 in Glazman and Weichman (1989)] becomes a crucial parameter of the problem. It can be shown that for $D_H \geq D$, the surface flux density $v_{3,Y}$ increases as the microscale decreases, tending to infinity as the latter goes to zero. This can be viewed a consequence of a surface's geometrical cascade: each larger-scale wave carries wavelets which in turn serve as carriers for yet smaller-scale wavelets, and so on to infinity. The averaging truncates the infinite cascade by smoothing out sufficiently small oscillations, thus limiting the surface density of wave crest events.

The "microscale" component ζ' becomes very important if the equilibrium range spans several decades. Then, one may inquire whether this component can be viewed as a fractal. The surface shapes at different scales are statistically similar only if they all are sufficiently far away from the dominant wave (k_o) in the frequency space. Thus, the "truly fractal" regime requires $k >> k_o$. In reality, the conditions of invariance under magnification are hardly satisfied: the equilibrium range usually extends over one or two decades and the angular factor $T(\theta)$ in eq. (11) for the wave number spectrum, as is known from field measurements, is (weakly) wave-number dependent but could be explored in the more general framework of Generalise Scale Invariance. Let us explore the simplifications yielded by the "fractal assumption": are they so important as to justify neglecting the "non-fractal" corrections in some cases?

3. FINITE SURFACE PATCH AND THE KARHUNEN-LOÈVE EXPANSION

Due to their special properties, certain random processes are employed as a tool for studies of more complicated random functions. For instance, the Gaussian white noise is used for construction of arbitrary near-Gaussian processes (Meecham and Clever, 1971). Here we demonstrate some simple properties of the fractal functions.

For the purpose of this study, it is natural to consider a finite area A of the supporting plane $z = 0$. Introduce the spatial covariance function $W(x,y)$.

$$W(x,y) = \langle\zeta(x)\zeta(y)\rangle \qquad (5)$$

where x and y are two points on the plane $\{x_1,x_2\}$, within this area. Let us follow the procedure of the Karhunen-Loève expansion (e.g., Adler, 1981). Define a set of real functions $z(x)$ and real numbers λ as solutions of the eigenvalue problem:

$$\lambda z(x) = \int_A W(x,y) z(y) dy \tag{6}$$

For a statistically uniform surface, the kernel of this homogeneous Fredholm equation of the second kind is symmetric and non-negative definite. Its eigenvectors are all real and satisfy the conditions of orthogonality and normalization (for a given interval A):

$$\int_A z_i(x) z_j(x) dx = \delta_{ij} \tag{7}$$

Furthermore, the kernel can be expanded in a series absolutely and uniformly converging within A:

$$W(x,y) = \sum_i \lambda_i z_i(x) z_i(y) \tag{8}$$

In theory of integral equations, this is known as Mercer's expansion.

Finally, assuming field $\zeta(x)$ to be Gaussian leads to the Karhunen-Loève expansion:

$$\zeta(x) = \sum_i \lambda_i^{1/2} z_i(x) \Theta_i \tag{9}$$

where Θ_i is a set of zero-mean, unit-variance, Gaussian variates satisfying the condition of statistical orthogonality $<\Theta_i \Theta_j> = \delta_{ij}$. Indeed, substituting (9) into (5) yields (8).

To investigate the fractal regime, we assume the wave number modulus spectrum F(k) to have the form of eq. (2) and perform sample calculations for a surface patch of decreasing size A. The spatial covariance function is related to the wave number spectrum by

$$W(x) = \int_0^\infty k dk \int_0^{2\pi} G(k,\theta) e^{ikr\cos(\theta-\phi)} d\theta \tag{10}$$

where x has components $x_1 = r\cos\phi$ and $x_2 = r\sin\phi$; and $G(k,\theta)$ is the two-dimensional spectrum in polar co-ordinates. A necessary condition for scale invariance of the surface statistics is:

$$G(k,\theta) = T(\theta) F(k) \tag{11}$$

For simplicity, only one-dimensional cases will be treated in what follows, implying the surface vertical profile $z = \zeta(x)$. With respect to eqs. (10), (11), this means that $T(\theta) = \delta(\theta)$, and the profile is taken along a radial direction. The role of x is then played by $r\cos\phi$. Scaling the wave number by the peak value k_o and introducing a non-dimensional distance $\bar{x} = rk_o\cos\phi$; eq. (6) can be rewritten as

$$\bar{\lambda} \bar{z}(\bar{x}) = \int_0^R w(\bar{x} - \bar{y}) \, \bar{z}(\bar{y}) \, d\bar{y} \tag{12}$$

where $(\bar{x},\bar{y}) = k_o(x,y)$, $\bar{z}(\bar{x}) = z(\bar{x}/k_o)$.

$$\bar{\lambda} = \lambda k_o^{p-1}/M_o \tag{13}$$

where $M_o = W(0)$ is called the zeroth-order spectral moment (total variance) related to $w(\bar{x})$ by: $M_o w(\bar{x}) = W(r\cos\phi)$. The upper limit of integration, R, represents a fractional size of the surface patch A with respect to the "dominant wavelength": $R = k_o A$.

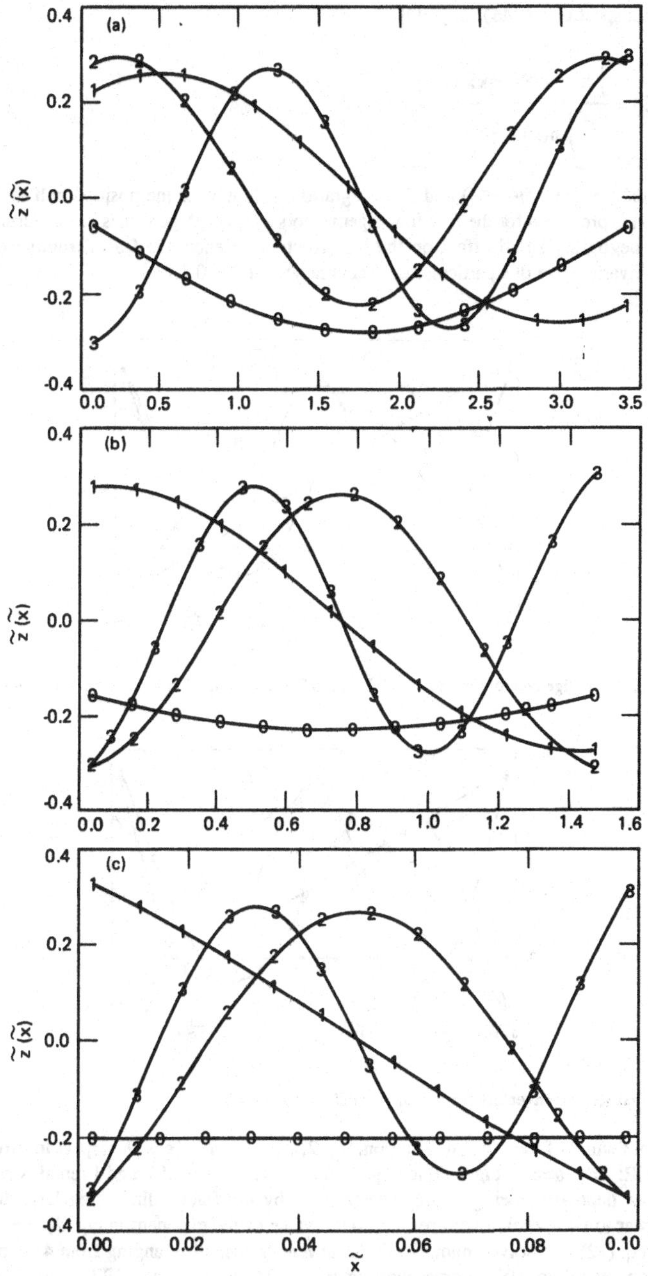

Fig. 1: Dependence of the first four eigenvectors on the relative size R of the surface patch: (a) R = 3.5, (b) R = 1.5, (c) R = 0.1 The spectral density function is given by eq. (2) with p = 4 and $h(k/k_0)$ corresponding to the Pierson-Moskovitz spectrum. Numbers on the curves designate the order of each eigenvector.

Now the fractal regime can be introduced as a limiting case of $\bar{\lambda}$ and $\bar{z}$ when R $\rightarrow 0$. We start with the Pierson-Moskovitz spectrum (PMS). It yields

$$w(\bar{x}) = \frac{\int h(\kappa)\kappa^{-3}\cos\kappa\bar{x}\, d\kappa}{\int h(\kappa)\kappa^{-3} d\kappa} \tag{14}$$

where $\kappa = k/k_o$, $h(\kappa) = \exp(-5/4\ \kappa^{-2})$ and the integration is done over the positive half-axis. In fig. 1 a numerical solution is presented for the first four eigenvectors $z_i(\bar{x})$ for three values of R. Further decrease of R (fig. 2) has a negligibly small effect on the eigenvectors. Hence, the fractal regime (i.e. when the solution is scale-invariant), in this particular case commences at R $\approx$ 0.1.

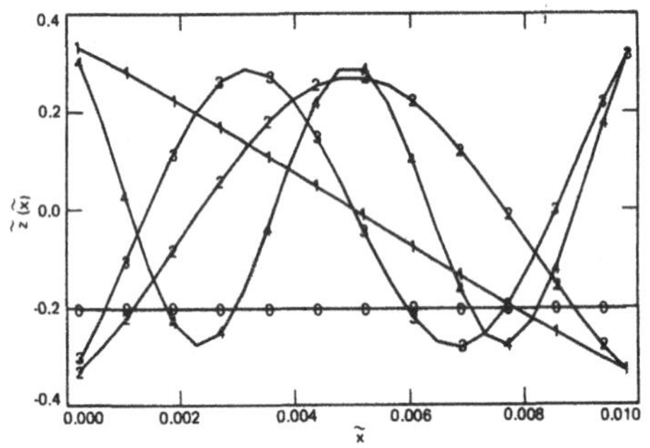

Fig. 2: The first five eigenvectors in the "fractal regime" of R = 0.01. The spectrum is the same as in fig. 1.

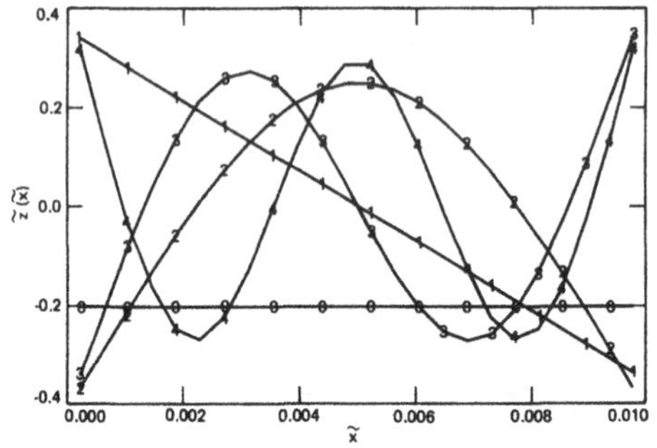

Fig. 3: The first five eigenvectors for a "non-fractal" case of p = 5.

An important feature of the "fractal" solution, fig. 2, is its invariance with respect to further reduction of the patch size R: the ratio of eigenvalues λ_k/λ_l, for any fixed pair of k and l remains constant as the relative patch size decreases, being entirely determined by the fractal dimension D_H. Moreover, the eigenvectors appear to be invariant with respect to the value of the exponent in eq. (2) (i.e., to the fractal dimension D): eq. (12) was solved numerically for several values of p ranging from 4 (corresponding to Philips' saturation range) to 10/3 (corresponding to the Zakharov et al. (1975, 1982) wave number spectrum), yielding virtually the same set of $z_i(\bar{x})$ for sufficiently small R. For the sake of comparison, numerical solutions have also been obtained for several values of p greater than 4 (regular surfaces as opposed to the fractal ones). These "non-fractal" cases yielded eigenvectors (fig. 3) only slightly different from the "fractal" eigenvectors of fig. 2. The differences appear only at the boundaries. Furthermore, it has been found that for $R \to 0$, after appropriate scaling of the dependent and independent variables, the eigenvectors tend to the Legendre polynomials of corresponding order.

TABLE 1

i	0	1	2	3	4	5	6	7	8	9
λ	1.0×10^{-2}	9.6×10^{-7}	3.6×10^{-8}	1.0×10^{-8}	3.8×10^{-9}	1.9×10^{-9}	1.1×10^{-9}	6.6×10^{-10}	4.3×10^{-10}	3.0×10^{-10}

Table 1: Eigenvalues for the case of R = 0.01 in eq. (12) and $w(\bar{x})$ given by eq. (14).

In table 1 the first 10 eigenvalues are presented for the fractal range (R = 0.01) of the PMS. Since the only quantity in the Mercer expansion that depends on the exponent p is λ, it follows that in the fractal range ($R \to 0$) of a Gaussian surface patch, all the information about the surface geometry is contained in a set of eigenvalues λ_i.

Several uses of the orthogonal expansion become apparent immediately. Namely, limited patches of Gaussian surfaces can be simulated, filtered, and analyzed very efficiently by employing the Karhunen-Loève expansion. The relatively fast convergence of the Karhunen-Loève expansion is its major advantage over the traditional harmonic analysis. In general, by virtue of the corresponding variational principle for λ_i's (see any text on integral equations), the Karhunen-Loève expansion gives the smallest possible number of terms among all other orthogonal expansions.

For a surface only slightly deviating from normal, one may achieve significant simplifications when solving kinetic equation for the evolution of the wave spectrum (or of the wave action spectral density, as done by Zakharov et al. 1975, 1982a). This is due to the fact that the third-order and higher moments in the statistical formulation of the weakly non-linear problem for water waves can be expressed via the second-order moments (e.g., Holloway, 1986). Hence, it appears to be possible to reduce the functional equation studied by Zakharov et al. to a set of algebraic equations for eigenvalues.

4. GAUSSIAN SURFACE

The fact that many hydrodynamical fields might be near-Gaussian (though this would not be the true if they were hyperbolic)hence can be constructed based on a Gaussian field as a zero-order approximation, makes the "fractal decomposition" based on the Karhunen-Loève expansion interesting as a possible instrument for turbulence and wind-wave studies. Let me suggest a useful interpretation of its terms.

Figure 4 shows how the sum of the first two terms in the Karhunen-Loève expansion, eq. (9), for a case of eq. (14), changes as the dimensionless size R of the surface patch A decreases (from 9.72 which corresponds to the "dominant wavelength" to 0.01 which corresponds to the fractal regime). In the limit of small R, the zeroth term tends to a constant, the first term to a linear function, the second term to a parabola, etc. Considering the magnitude of the first eigenvalue relative to the subsequent eigenvalues (Table 1), the physical meaning of this term emerges as the following: it represents the (random) elevation of a given small patch, as determined by the underlying, large-scale wave profile. The second term characterizes the (random) slope of the small patch, as determined by the large-scale wave profile.
Its r.m.s. value can be obtained taking into account the scaling relationship, eq. (13). This value depends on the patch size, because the slope here is defined as $(\zeta(x+A) - \zeta(x))/A$. This is not surprising in view of the fact that the r.m.s. slope of the surface whose Hausdorff dimension exceeds the topological dimension is infinitely large. This slope is made finite by averaging the surface over the so-called microscale (Glazman, 1986) or by neglecting all but the first two terms in eq. (9), as is implied in the present interpretation. Apparently, both these methods of low-pass filtering are equivalent, and the second term in eq. (9) along with eq. (13) allow one to appreciate the roles of p and R in statistics of the "macroscopic field" slope $\nabla\bar{\zeta}$.

Finally, in the fractal regime of the small patch, one can introduce

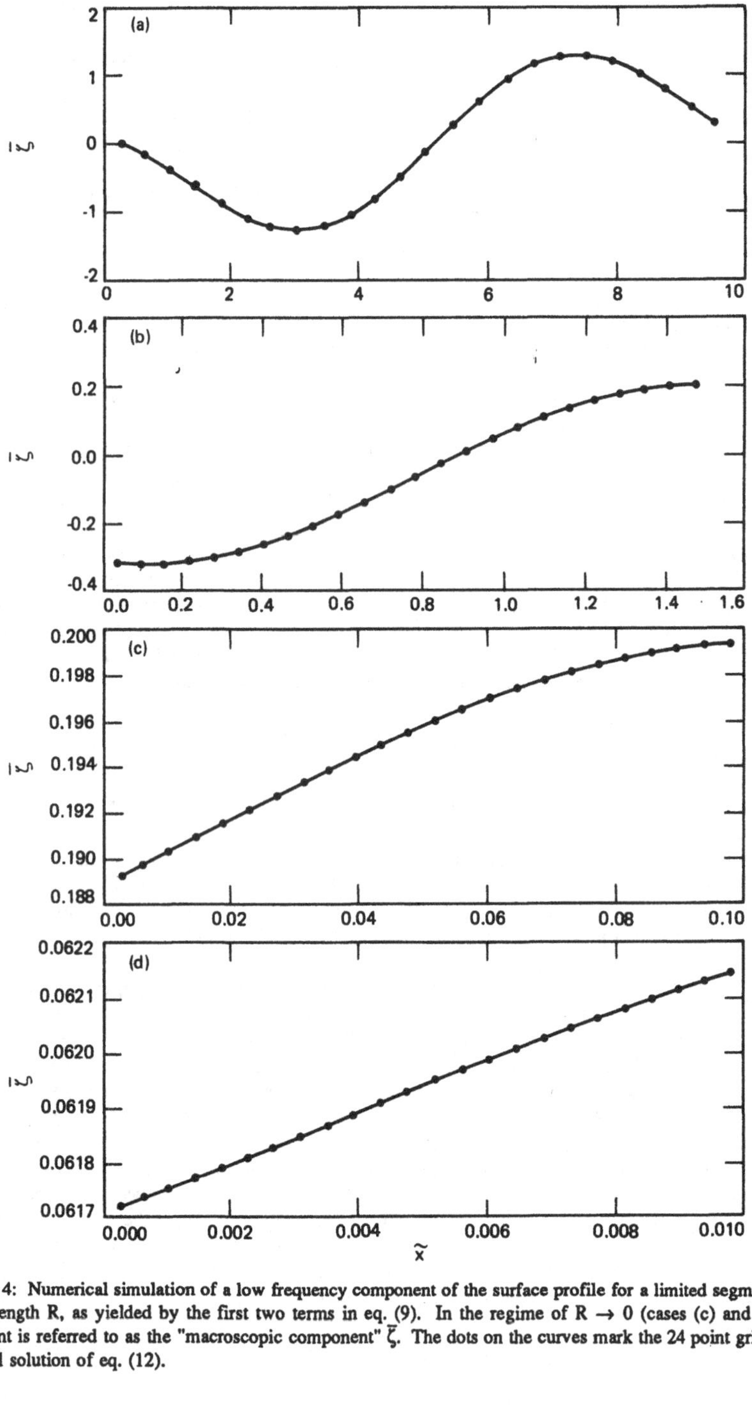

Fig. 4: Numerical simulation of a low frequency component of the surface profile for a limited segment of the relative length R, as yielded by the first two terms in eq. (9). In the regime of R → 0 (cases (c) and (d)), this component is referred to as the "macroscopic component" $\bar{\zeta}$. The dots on the curves mark the 24 point grid used in numerical solution of eq. (12).

$$\zeta' = \sum_{i \geq 2} \lambda_i^{1/2} z_i(x) \Theta_i \tag{15}$$

as a definition of the microscopic, fractal, component. The first two terms in eq. (9), that pertain to the macroscopic field $\bar{\zeta}$, are excluded from this representation.

5. ACKNOWLEDGEMENTS

This work was performed at the Jet Propulsion Laboratory, California Institute of Technology, under contract with the National Aeronautic and Space Administration. The author thanks Dr. Peter Weichman of Caltech's Physics Department for fruitful discussions of the project.

6. REFERENCES

Adler, R. J., 1981: The Geometry of Random Fields. Wiley, New York, 279 pp.

Barenblatt, G.I. and I.A. Leykin, 1981: On the self-similar spectra of wind waves in the high-frequency range. Izvestiya, Atmosphere and Ocean Physics. (English translation), 17(1), 35-41.

Berry. M. V., Diffractals, J. Phys. A, 12(6), 781-797, 1979.

Glazman, R. E., 1985: Mathematical modeling of breaking wave statistics in The Ocean Surface: Wave Breaking, Turbulent Mixing and Radio Probing. pp.145-150. Edited by Y. Toba and H. Mitsuyasu, D. Reidel Publishing Co., Boston.

Glazman, R. E., 1986: Statistical characterization of sea surface geometry for a wave slope field discontinuous in the mean square. J. Geophys. Res., 91(C5), 6629-6641.

Glazman, R. E., G. Pihos and J. Ip, 1988: Scatterometer wind speed bias induced by the large-scale component of the wave field. J. Geophys. Res., 93(C2), 1317-1328.

Glazman, R., and P. B. Weichman, 1989: Statistical geometry of a small patch in a developed sea, J. Geophys. Res., 94(C4), 4998-5010.

Hasselmann, K., et al. (16 more names), 1973: Measurements of wind-wave growth and swell decay during the Joint North Sea Wave Project (JONSWAP), Deutsch. Hydrogr. Z., suppl. A., 8(12).

Holloway, G., 1986: Eddies, waves, circulation and mixing: Statistical geofluid mechanics. pp. 91-148, in Ann. Rev. Fluid Mech., Eds: M. Van Dyke et al., Annual Reviews Inc., Palo Alto, Calif.

Kwoh, D. S. and B. M. Lake, 1984: A deterministic, coherent and dual-polarized laboratory study of microwave backscattering from water waves, Part 1: Short gravity waves without wind. IEEE J. Oceanic Eng. OE-9, 291-308.

Kwoh, D. S. and B. M. Lake, 1985: The nature of microwave backscattering from water waves. In: The Ocean Surface: Wave Breaking, Turbulent Mixing and Radio Probing. Y. Toba and H. Mitsuyasu (eds.), 249-256, D. Reidel Publishing Co., Dordrecht.

Lyzenga, D. R., A. L. Maffett, and R. A. Schuchman, 1983: The contribution of wedge scattering to the radar cross section of the ocean surface, IEEE Trans. Geosci. and Remote Sensing, GE-21(4), 502-505.

Meecham, W. C. and W. C. Clever, 1977: Use of C-M-W representations for nonlinear random process applications, Lecture Notes in Physics: Statistical Models and Turbulence, p. 205-229, Springer, New York.

Phillips, O. M., 1977: The Dynamics of the Upper Ocean, 2nd Ed., Cambridge University Press, New York, 336 pp.

Zakharov, V. E. and N. N. Filonenko, 1966: The energy spectrum for stochastic oscillation of a fluid's surface. Doklady Academii Nauk S.S.S.R., 170(6), 1292-1295, (in Russian).

Zakharov, V. E. and V. S. L'vov, 1975: Statistical description of non-linear wave fields. Radiophysics and Quantum Electronics, 18, 1984-1097, (Translated from: Izvestiya VUZ, Radiofizika, 18(10), 1470-1487, 1974).

Zakharov, V. E. and M. M. Zaslavskii, 1982a: The kinetic equation and Kolmogorov spectra in the weak turbulence theory of wind waves. Izvestiya, Atmospheric and Oceanic Physics, (English translation), 18(9), 747-753.

Zakharov, V. E. and M. M. Zaslavskii, 1982b: Ranges for generation and dissipation in the kinetic equation for a low-turbulence theory of wind waves. Izvestiya, Atmospheric and Oceanic Physics, (English translation), 18(10), 821-827.

Zakharov, V. E. and M. M. Zaslavskii, 1983: Shape of the spectrum of energy carrying components of a water surface in the weak-turbulence theory of wind waves. Izvestiya, Atmospheric and Oceanic Physics, (English translation), 19(3), 207-212.

FRACTAL LINEAR MODELS OF GEOPHYSICAL PROCESSES

O.G. Jensen, J.P. Todoeschuck*, D.J. Crossley and M. Gregotski
Geophysics Laboratory
Department of Geological Sciences
McGill University
3450 University Street
Montreal, Canada H3A 2A7

ABSTRACT. Linearized models for the analysis and inversion of geophysical data are extended to allow for a stochastic component which is spectrally self-affine or fractal. The fractal noises, characterized by spectral intensity densities of the form $1/|f|^K$, are very common in nature. Classical white Gaussian noise, Johnson thermodynamic noise, Brownian motion and random walks and flicker noise belong to this class. Of these, flicker noise for which K=1 is especially important: it is that fractal noise which is most heavily weighted in long periods while being strictly stationary in its statistical measures. Mandelbrot noted the ubiquity of flicker noise in nature. Here, we describe Gauss' classical least-squares method for the inversion of linear data models extended to allow for a flicker-noise excitation or corruption. The methods described for flicker noise are easily modified to account for any other fractal noise.

1. INTRODUCTION

Measurements made on many natural phenomena can be regarded as a combination of a system response with a random or stochastic excitation or error. In description of such measurements we form a model having two elements: one involving the stochastic component and the other describing the purely deterministic component. In some common problems in geophysics, we may regard the stochastic element as an additive noise or error in observing the response. In this case, the deterministic component, representing the uncorrupted response, is usually described as a generally non-linear equation parameterized by the essential physical measures relevant to the phenomenon. We form the model of the observed responses as a sum of the uncorrupted response and the corrupting noise:

observations = uncorrupted response + noise (model 1)

In another class of problems in geophysics, the stochastic component is the unknown but natural excitation of the phenomenon while the deterministic component describes its natural mechanism of response to this excitation. Generally, the deterministic part can be described in terms of a complex system of non-linear differential equations. As physicists, we attempt to describe the essence of the system as parsimoniously as possible. Usually, this leads to a linearization of the differential system which allows us to adequately model the phenomenon for small excitations as the superposition or convolution (here indicated by the symbol "*") of the response and excitation:

observations = system response * excitation (model 2)

---

* Now at Defence Research Establishement Pacific, FMO, Victoria, B.C., Canada V0S 1B0.

D. Schertzer and S. Lovejoy (eds.), Non-Linear Variability in Geophysics, 227–239.

The stochastic excitation component in this model and the noise component in the additive linear model, above, are determined by a sufficient set of statistical measures which describe the essence of the process. Since any particular observation of a phenomenon involves only a single realization or sample of the process of excitation or additive noise, we usually must infer the process statistics from sample-based estimates.

Classically, in analysis of model (1), we seek to obtain those physical parameters which determine the theoretical or uncorrupted response most closely corresponding to the observations. As a criterion for solution of this problem, we often require minimum noise power, or alternately, the smallest sum of squared errors. Similarly, in analysing model (2), we seek to establish yet-undetermined parameters of the deterministic physical system or particular scales for the statistical measures of the process of excitation. The parameters of the deterministic system arise as coefficients of our differential system. The scales of the statistical measures may, for example, describe the level, power or distribution of the excitation.

Clearly, we must use statistical methods of analysis. At this point, we may follow either of two strategies: we may attempt to optimize our method for the particular realization of the phenomenon at hand or we may attempt to optimize our method for all possible realizations, the process as a whole. Geophysical analysts divide into two camps according to their choice of model and strategy: they attempt, by model (1) above, to describe the details of a particular phenomenon (e.g., anomaly modelling) or they attempt, using either of the two models above, to develop global methods of analysis (e.g., data processing). In industrial geophysics, the former strategy dominates the field of mineral prospecting while the latter dominates in petroleum prospecting largely because of the overwhelming amount of data involved in seismic reflection surveying.

Because we often desire methods applicable to any particular example or realization of the geophysical phenomenon, we reasonably optimize our methods of analysis to deal with the entire process of the phenomenon rather than that part of it which is the realization immediately at hand. Many geophysicists, who have derived their methods from the practical theories of electrical systems, control and communications engineering and the statistical theory of time-series analysis, use classical stochastic linear systems or linear data models to describe the measurements they seek to analyse. Essentially, they employ model (2) above to describe their measurements and invert the data in accordance with this model in order to separate or deconvolve its determinism and randomness.

2. THE FRACTAL NOISES

A geophysical experiment or survey often has, as goal, the establishment of some basic properties of the determinism of the process. Taking an example from elementary physics, we may observe a swinging pendulum in order to estimate its period which theoretically provides a measure of the local gravitational acceleration. We enforce near-linearity on the deterministic component of this system by allowing only very small amplitudes of swing. We may attempt to reduce the continuous viscous and frictional forces by placing the pendulum in vacuum and through use of low-friction bearings. An elementary pendulum, or a similar mechanical oscillator, is the heart of some of the most sophisticated electromechanical geophysical instruments. In observatory gravimeters, for example, mechanical oscillators are now being used to monitor the gravitational acceleration at certain sites on the earth's surface with a view to detecting fluid-motions in the earth's outer liquid core. The best of these instruments are able to reproduce acceleration measurements to one part in 10^{11}, while making measurements several times per minute. At this contemporary level of precision we can easily maintain a sufficient degree of linearity in the system. But, we find our observations of the temporal variations of gravitational acceleration are dominated by "noise". This noise, which provides a stochastic excitation to the system, derives from several sources.

Thermodynamic noise (equivalently Johnson noise resulting from inherent mechanical resistance or friction in the oscillator) is spectrally white and with variance or, more properly, power scaled linearly by temperature and frequency bandwidth of observation. Typically, such noise provides for the dominant component of continuing excitation of gravimeters (and other geophysical instruments based on pendulums and mechanical oscillators) at periods shorter than a few minutes. At longer periods, observations made by these instruments, in common with all other physical devices which are capable of D.C. measurements, become dominated by flicker noise. This noise is most directly characterized by its 1/f power spectral density. It arises as a conspiracy of innumerable sources including internal crystal dislocations in the mechanical elements of the instrument and in solid-state components and resistors of the electronic instrumentation, geological crustal motions, variations in the overlying air-mass column, variations in the rotational dynamics of the earth, etc. Geological tectonic processes may be involved in slowly and episodically translating or uplifting the observatory site. These processes may appear as a random walk over extremely long time scales.

As the gravimeter is removed from or brought closer to the earth's center of mass by the tectonism, the acceleration of its mass element varies. Apart from the body tides, the Brownian tectonic noise and more particularly the flicker noise dominate the excitation of gravimeters at periods beyond about 1 hour. Earthquakes and large man-made explosions can also produce very large amplitude seismic wave transients but these are necessarily composed of periods less than the earth's gravest 54-minute normal mode. Superposed on all of these incidental excitations is that time varying force provided by the motion of density variations in the liquid core, by resonant elastic waves throughout the body of the earth and by possible dynamics of the earth's solid inner core. It is these very long period excitations which the geodynamicists are now seeking to distinguish.

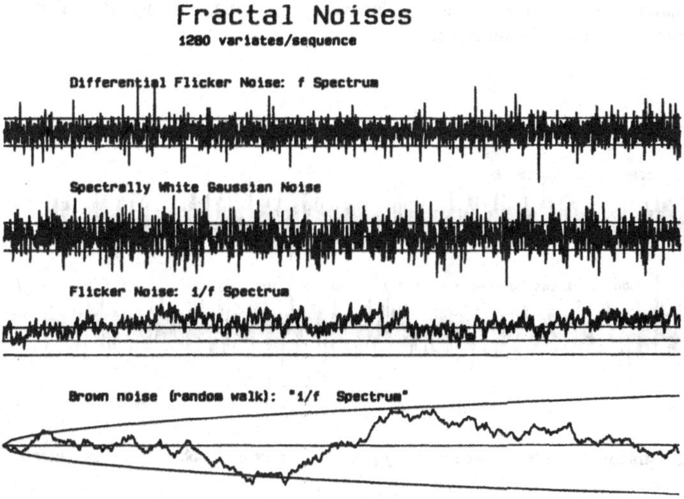

Fig. 1: Various Fractal noises and the illustration of their self-affinity.

Through the example of the gravimeter presented just above, we have introduced three particular noise processes: white Gaussian noise, flicker noise and Brownian noise. Brownian noise, named in reference to its common expression as Brownian particle motion, is very heavily weighted in low frequencies; it is often called brown noise in analogy to a long-wavelength dominated spectrum of light. These are members of a class of stochastic processes which we shall here call the fractal noises. Stationary random processes possessing a power density spectrum of the form $1/|f|^K$, with $K \leq 1$ and with random Fourier phase components uniformly distributed between 0 and 2π are spectrally self-affine (Mandelbrot, 1983). Transformations on such processes which simultaneously adjust their time and amplitude scales can be found which maintain all orders of their statistical measures. Extending this class to non-stationary random processes possessing squared Fourier amplitude spectra of the form $1/|f|^K$, an extended-sense spectral self-affinity can be shown to hold for all K. K=1 determines flicker noise. K=2 determines a

Wiener-Bachelier process which we here call Brownian or brown noise. This is the classical random walk in which the probability density function of step amplitudes is Gaussian. We discover that these three, perhaps the most common natural noise phenomena, are members of a class related through their spectral self-affinity. We are philosophically drawn to the expectation that it is this spectral self-affinity, that is, their fractal nature, which is their most fundamental quality. It may, in fact, be the most basic property of natural randomness. That these most common noises (i.e., Johnson noise, Brownian noise and flicker-noise) are of the class of spectrally self-affine noises offers the hope that analytical methods could be developed which are optimized in such a way as to employ their similar qualities in an objective criterion for the separation of the deterministic and stochastic components of the processes.

Figure 1 shows four fractal noises with K = -1, 0, 1 and 2 which are appropriate to the description of some common geophysical phenomena. That all these are self-affine is easily demonstrated. Given a sequence of one of these noise series, we may obtain an average of N successive variates and scale this average by $N^{(1-K)/2}$ to obtain a new variate. The sequence of such new variates is statistically indistinguishable from the original sequence. Figures 2, 3 and 4 show that this property holds for white Gaussian noise (K = 0), flicker noise (K = 1) and Brown noise (K = 2) respectively with N = 100. It necessarily also holds for ideal differential flicker noise (K = -1) but a sequence of this noise is very difficult to generate with sufficient precision to show the self-affinity under averaging by 100's. One remarkable property of flicker noise which was noted by Mandelbrot and McCamy (1970) is that no amplitude scaling is required to maintain its statistical measures under averaging. A much stronger condition of self-affinity, i.e., self-congruency, holds for flicker noise.

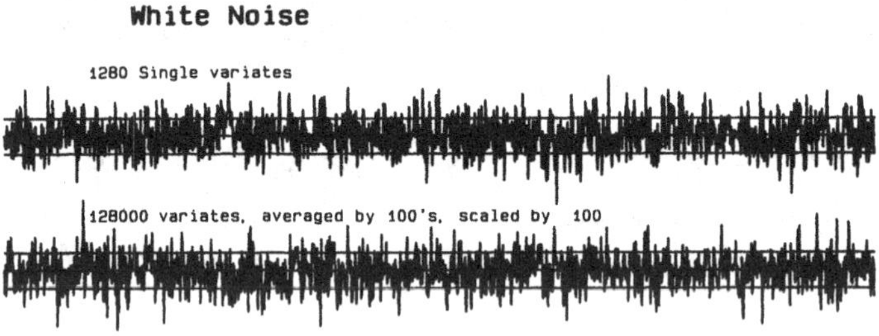

Fig. 2: Illustration of white noise dominating generally errors in geophysical measurements.

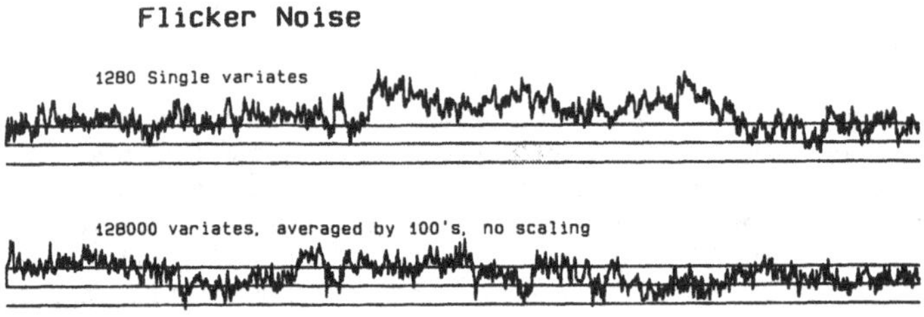

Fig. 3: Illustration of flicker noise which is a good candidate of stochastic process for long period geophysical variations.

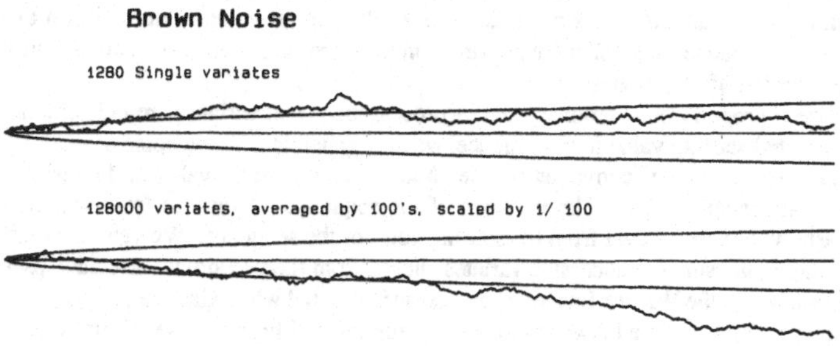

Fig. 4: Illustration of brown noise describing local topography especially in hilly or mountainous regions.

Each of the noise processes shown in fig. 1 and figs. 2-4 characterize essential properties of some geophysical process. Common spectrally white Gaussian noise, purely random noise, generally dominates errors in geophysical measurements except at long periods. Its quality of being most uncertain of all noises has justified its inclusion in modelling geophysical data since Gauss (1839) first published a solution of a geophysical inverse problem in describing the worldwide observations of the geomagnetic field as a linear combination of a finite series of associated Legendre functions. He implicity assumed uncorrelated, zero-mean errors which were measured by a single statistic the variance: what we now call spectrally white (i.e., uncorrelated) Gaussian random noise.

Flicker noise very commonly dominates long period geophysical variations. Mandelbrot and McCamy (1970) and Jensen and Mansinha (1984, 1987) have argued that the earth's equilibrium pole of rotation measured in geographical coordinates follows a 2-vector flicker-noise path. The variation of the earth's rotation rate is an almost flicker noise process (see Lambeck, 1980). The rate variations of the atomic clocks used to measure the earth's rotation rate are themselves dominated by flicker noise at periods beyond about 1 second (Vessot, 1974). The ubiquity of flicker noise in astronomy, geophysics and elsewhere has been noted by Press (1978). One especially interesting phenomenon characterized by flicker noise is the acoustic impedance function of depth in sedimentary basins (Hosken, 1980; Walden and Hosken, 1985; Todoeschuck, Jensen and Labonté, 1988). Because the seismic reflectivity sequence is approximately a differencing of the acoustic impedance function, the reflectivity sequence has a differential flicker noise form (Todoeschuck and Jensen, 1988a, 1988b, 1988c) with a nearly-f^1 spectrum (see fig. 1).

Random walk processes or brown noise with a $1/f^2$ spectrum abound in geophysics. Local topography, especially in hilly or mountainous regions, is well described by brown noise (Mandelbrot, 1983). Tectonic motions, fault slipping and episodic geological uplifting is commonly brown noise-like. Drift of certain geophysical instruments such as exploration gravity meters tends towards a brown noise form. This class of fractal noises describes common stochastic phenomena; we believe that we will find advantages in analysing geophysical data and interpreting geophysical measurements by recognizing that the stochastic elements involved are closely related geometrical phenomena.

2.1. Modelling the fractal noises

In order to create synthetic data models, that is, to simulate geophysical data, we require algorithms for the generation of the fractal noises. Using digital computers, recursive, periodic (but with extremely long period) pseudo-random processes can be easily generated. Many common algorithms exist which will produce long (often 2^{32}-1 points) almost uncorrelated sequences of random variates, uniformly distributed on the interval [0,1) or [-1/2,+1/2). We shall assume that an adequate generator of such random variates is available. The most common algorithm for generating a pseudo-Gaussian, pseudo-uncorrelated random sequence from the uncorrelated uniform distribution is to simply sum 12 successive variates selected from the distribution. Since the variance of the uniform distribution with range of 1 unit is 1/12, the variance of the pseudo-Gaussian variates is 1. Its mean is 12 times the central value of the uniform distribution and can be subtracted to produce an excellent approximation to a zero-mean, uncorrelated random sequence with a Gaussian distribution of amplitudes.

Brown noise is properly non-stationary and its process mean and variance are indeterminate. However, given a prior-selected starting value for a sequence, we may generate a brown noise sequence which evolves from this value. Typically, for convenience, we choose 0 as a starting value and consider this to be an equivalent process "mean" value. The variance of the sequence of variates from this equivalent mean increases linearly with their interval from the starting point of the sequence. We generate such a process by simply accumulating the sum of successive variates chosen from the pseudo-uncorrelated, pseudo-Gaussian generator. Brown noise, the Wiener-Bachelier process is integrated white Gaussian noise.

Flicker noise is intermediate between the white Gaussian and brown noises (Barnes and Allen, 1966). One crude method for its generation is to accomplish a half-order integration of a white Gaussian sequence by multiplication of its Fourier transform coefficients of frequency, f, by $1/|f|^{1/2}$. The major problem in this method arises through the approximation of the strict Fourier transform by periodic discrete Fourier transforms usually realized by the classical fast-Fourier transform algorithms. Voss (Gardner, 1978) has described a better algorithm which allows for recursive generation of a sequence of flicker variates. He shows, that given N dice, we may generate a sequence of flicker variates as follows. We first throw all N dice and sum their values for a zeroth variate. We then save N-1 dice from the previous step, throw the remaining die and sum the N values for the first variate. To obtain the next (i.e., 2nd) variate, we save N-2 of the original dice, throw the remaining 2 dice and sum the N values. We next (i.e., 3rd variate) save the N-2 remaining original dice, one of the two thrown in the previous step, throw the remaining one and sum the value. We then (i.e., 4th variate) save N-3 of the original dice, throw the remaining three dice and sum. We next (i.e., 5th variate) save the N-3 original dice, 2 of those thrown in the previous step, throw the remaining one and sum the N values. For the 6th variate, we save the original N-3 dice and one of the pair saved in the previous step, throw the remaining two and sum the N values. The number of dice to be thrown at any step is determined by a sequence N, 1, 2, 1, 3, 1, 2, 1, 4, 1, 2, 1, 3, 1, 2, 1, 5, 1, ... Given N dice, this sequence has a period of 2^N which is the longest interval of correlation between variates. Voss described this method with 6 dice, each having six sides.

Jensen and Mansinha (1984) generalized this algorithm to allow for up to 31 dice, each in principle having as many sides as there are separate variates provided by a generator of uniformly distributed random numbers. As a consequence of the central limit theorem, the probability density function of the flicker variates approaches a Gaussian form as N becomes large. The Gaussianity of the sequence is well approximated for N > 8 or so. It is clear that by this method, the sum value becomes progressively less correlated with the original, zeroth value. Subtracting N times the mean value of the uniform distribution used in generation and scaling each variate by $(12/N)^{1/2}$, we obtain a zero-mean, unit-variance, pseudo-Gaussian flicker-noise sequence. The $1/|f|$ form of the spectrum is extraordinarily well preserved over a range of periods between 1 interval and $2^N/4$ or so. For large N, this range may represent as many as 10 decades (more properly N-2 octaves) in frequency.

Now from flicker noise, we may generate an f-noise or differential flicker-noise process almost directly. Approximating strict differentiation by a running difference between the present and previous value of a unit-variance flicker-noise process and rescaling amplitudes to obtain a useful approximation to a zero-mean, unit-variance, differential-flicker-noise process.

2.2. Geophysical data models and their inversion

We may often describe a linear (or linearized) model of our geophysical data or measurements:

$$\mathbf{m} = \mathbf{A}\mathbf{p} + \mathbf{e}, \tag{1}$$

where the vector **m**: $(m_0, m_1, ..., m_N)$ are N+1 measurements distributed in time or place, **p**: $(p_1, p_2, ..., p_M)$ are M parameters which sufficiently describe the geophysical system, **A**, a matrix with elements a_{ij}, i = 1, 2, ..., M, and j = 0, 1, ..., N, which approximately relate each parameter to each measurement and **e**: $(e_0, e_1, ..., e_N)$, the vector of errors between the actual measurements and those predicted by the parameterized model **p** through the design matrix **A**. Assuming the error components to be unpredictable in detail, we seek to determine some set of parameters (i.e., **p**) which "best" in some sense describe the measurements. Our choice of criterion for deciding which is the best solution usually involves establishing some special quality of the errors, **e**; many criteria on **e** are being used in current geophysical analysis. We may, for example, require the average magnitude of the components of **e** to be minimum (the linear programming solution), the average squared magnitude to be minimum (the least-squares solution), the sample skewness to be extreme (extreme-skewness solution) or the sample kurtosis to be maximum (maximum-kurtosis or varimax solution). Implications of these criteria in the analysis of geophysical seismic and other data have been discussed in detail by Vafidis and Jensen (1987, 1988) and Jensen and Vafidis (1987). Of these criteria, the least-squares solution is most commonly used for both practical (its solution involves only linear algebraic operations) and philosophical (it describes the least power stochastic composition) reasons. The well known solution for **p**,

$$\hat{\mathbf{p}} = (\mathbf{A}^t \mathbf{V}_e^{-1} \mathbf{A})^{-1} (\mathbf{A}^t \mathbf{V}_e^{-1}) \mathbf{m}, \tag{2}$$

minimizes the expected power of the error component if we know, *a priori*, the variance-covariance matrix, $\mathbf{V}_e$ which describes the paired-product relationship between the various error components. Note that in this solution for **p**, any scaling of the matrix $\mathbf{V}_e$ is immaterial; we therefore are required only to know its normalized form. If, as we often require, the components of **e** are stationary (i.e., their statistical measures do not change with component index) we may describe a normalized Toeplitz variance-covariance matrix having a unit diagonal. The variance-covariance matrix of the components of the actual error vector **e** are then obtained by scaling with the sample variance obtained as

$$s^2 = \frac{\hat{\mathbf{e}}^t \mathbf{V}_e^{-1} \hat{\mathbf{e}}}{(N-2M+1)} \tag{3}$$

where $\hat{\mathbf{e}}$ are estimates via the solution for **e** as

$$\hat{\mathbf{e}} = \mathbf{m} - \mathbf{A}\,\hat{\mathbf{p}}\,. \tag{4}$$

It is by our *a priori* selection of the appropriate, normalized variance-covariance matrix for the components of the error vector **e** that we may optimize our solution for a particular fractal noise. Gauss (1839) chose uncorrelated or white noise as appropriate to his modelling and we have often blindly followed this lead ever since irrespective of our knowledge of the stochastic elements involved in the generation of the measurements. Uncorrelated or white Gaussian noise is simply introduced by assuming the normalized variance-covariance matrix of error components is an identity matrix, i.e.,

$$\mathbf{V}_e = \mathbf{I}. \tag{5}$$

The unnecessary multiplications by this matrix in the solution equation (2) above are avoided by simply removing the $\mathbf{V}_e$ from the equation. In many circumstances, this prior assumption of the fractal properties of the involved stochastic component of the model can be justified on geophysical, physical (often thermodynamic arguments apply) or philosophical grounds. Unfortunately, and more commonly, the conscious prior choice of the appropriate fractal noise, which is the most important assumption to be justified at this stage in the process of solution, is avoided or ignored. We now recognize that other fractal noise processes are common contributors to the stochastic quality of our geophysical measurements. Fortunately, the inclusion of the appropriate, if known and justifiable, fractal noise is almost trivial. Providing we may expect the noise or error to be stationary and Gaussian, we need only describe, *a priori*, its autocorrelation function. If we have reason to expect any one of the fractal noise processes described above to be representative of the stochastic element in the grophysics, we would simply describe a classical Toeplitz variance-covariance matrix whose first row is the prior assumed autocorrelation function of the particular fractal noise. Approximately, this can be most directly accomplished by forming an idealized spectrum of the desired fractal noise and then, via a good Fourier transformation approximation, compute the equivalent autocorrelation function. Jensen and Mansinha (1984, 1987), Todoeschuck and Jensen (1988a, 1988b, 1988c) have used more elaborate (more properly, less crude) methods in forming or implying the presence of the variance-covariance matrix appropriate to their particular problems.

2.3. An example, decomposition of the rotation pole-path record

The rotation axis of the earth is not fixed to the body of the earth but rather moves relative to the geographical and temporal reference frames with both periodic and apparently secular motions (see Jensen and Mansinha, 1987). These motions, manifest as variations in the length of day and variations in latitude of any particular place on the earth's surface, constitute a wobbling of the earth (see Munk and MacDonald, 1960; Lambeck, 1980). The polar motion relative to the geographical reference frame follows a path which shows seasonal, annual and 14-month periodic components with significant amplitude along with a rich spectrum of other minor components. The 14-month period is called the Chandler wobble; it is essentially the free Eulerian wobble mode of the body of the earth. This wobble is excited whenever the polar axis of the earth, which is its axis of maximum moment of inertia, is displaced from the rotation axis. Because energy is removed from the wobble by a variety of dissipation mechanisms involving the anelasticity of the earth, its internal fluid core and effects due to the ocean and atmosphere, any transient excitation of the pole produces a decaying spiral towards the equilibrium rotation axis. The excitation mechanism, however, is an essentially continuous random process and therefore we cannot observe the decaying spiral of the Chandler wobble in isolation. Precise knowledge of its period and decay rate would provide an extremely valuable fiduciary for our understanding of the condition of the earth.

We describe the Chandler wobble path as a linear superposition of the form

$$z(t) = c(t) * p(t) \tag{6}$$

where z(t) represents the pole-path position, p(t) is the apparently stationary, stochastic displacement function of time of the equilibrium centre of the axis of figure from the rotation pole and c(t) is damped Chandler resonance spiral. We use a right-handed coordinate system describing the pole path as a complex-valued time function with

$$z(t) = x(t) + iy(t), \tag{7}$$

where x(t) is the displacement of the equilibrium pole along the Greenwich (0°) meridian and y(t) is the displacement along the 90°E meridian. The Chandler transient, c(t), is causal and stable, possessing a minimum-phase or minimum-delay characteristic and purely resonant. It therefore has a purely anti-resonant causal, stable, minimum-phase inverse, $\gamma(t)$, such that

$$\gamma(t) * c(t) = \delta(t) \tag{8}$$

where $\delta(t)$ is the Dirac delta function and we may deconvolve z(t) to obtain the excitation pole function (Smylie et al. 1970). If we normalize b(t) such that

$$b(t) = \delta(t) - \gamma(t) \tag{9a}$$

with

$$b(0) = 0 \tag{9b}$$

we may describe the pole-path function by a continuous analog of an autoregressive linear data model (see Jensen and Mansinha, 1984, 1987):

$$z(t) = b(t) * z(t) + p(t) \tag{10}$$

Sampled without aliasing with interval T, this continuous analog reduces to the classical discrete, infinite-order autoregressive data model:

$$z_n = b_n * z_n + p_n , \tag{11}$$

or alternately,

$$\gamma_n * z_n = p_n \tag{12}$$

b_m , m = 1, 2, ... is the infinite-order, autoregressive, one-step forecasting operator; z_n , n = 0, 1, 2, ... is the pole-path function, now sampled at regular time interval, T. Practically, we will have only a finite number of pole-path observations, say n = 0, 1, ...N. Properly, the Chandler resonance is described by a forecasting operator having a single coefficient. That is,

$$b_0 = 0, \tag{13a}$$

$$b_1 = e^{i\sigma T} \tag{13b}$$

$$b_m = 0; m > 1 \tag{13c}$$

where

$$\sigma = \omega_0 + i / \tau , \tag{13d}$$

where, further,

$$P = 2\pi / \omega_0 \tag{13e}$$

is the period of the Chandler resonance and τ its dampling time constant. Geophysical theory provides that the actual excitation of the wobble will be that which possesses the minimum power whatever its self-correlation characteristics may be. The reduced autoregression equation for the pole path, now, reduces as

$$\mathbf{z} = \mathbf{Z}\,\mathbf{b} + \mathbf{p}, \tag{14}$$

where **z** is the n+1 element data vector, **Z** is an n+1 x 1 matrix of data elements (Note that the **Z**-matrix is similar to the **z**-vector except that its 1-row element indices are all decremented by 1.), **b** is the 1-element forecasting operator vector and **p** is the unknown n+1-element excitation vector. This equation (14) has the form of equation (1), above. Solution for **b** minimizing the excitation power in **p** is obtained according to equation (2) above providing we are able to prior-assign the appropriate variance-covariance matrix governing the presumed-stationary excitation pole displacement function. That is, we obtain an estimate for **b** which provides for the minimum power of excitation by a fractal pole displacement function:

$$\hat{\mathbf{b}} = (\mathbf{Z}^t\, \mathbf{V_p}^{-1}\, \mathbf{Z})^{-1}\, (\mathbf{Z}^t\, \mathbf{V_p}^{-1})\, \mathbf{z} \tag{15}$$

Mandelbrot and McCamy (1970) showed and Jensen and Mansinha (1984; 1987) argued that the excitation pole path is essentially a complex-valued, flicker noise process having a $1/|f|$ spectral form over 3 or 4 decades of frequency. In solving equation (15), Jensen and Mansinha assigned an appropriate normalized variance-covariance matrix, $\mathbf{V_p}$ for such a prior-assumed noise with the additional constraint that the real and imaginary components were uncorrelated with each other. The variance-scale, s_p^2, of the variance-covariance matrix which is equivalently a measure of the excitation power of the pole displacement function is obtained by deconvolving the pole-path record to obtain an estimate on **p** as

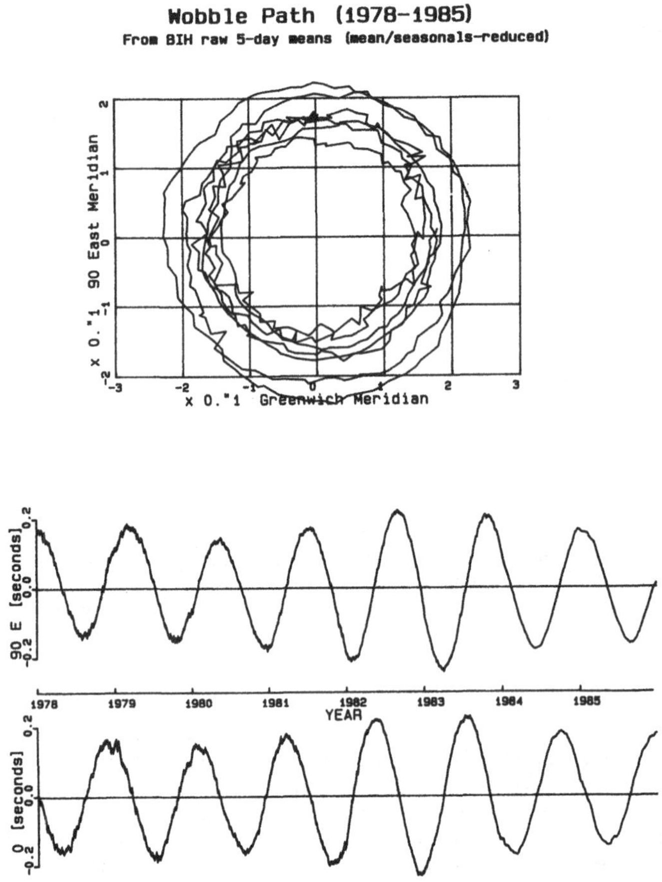

Fig. 5: Plot of the record of raw 5-day means from the years 1978-85 (source: Bureau International de l'Heure, Paris) after removal of the data mean and seasonal components to the 6th annual harmonic.

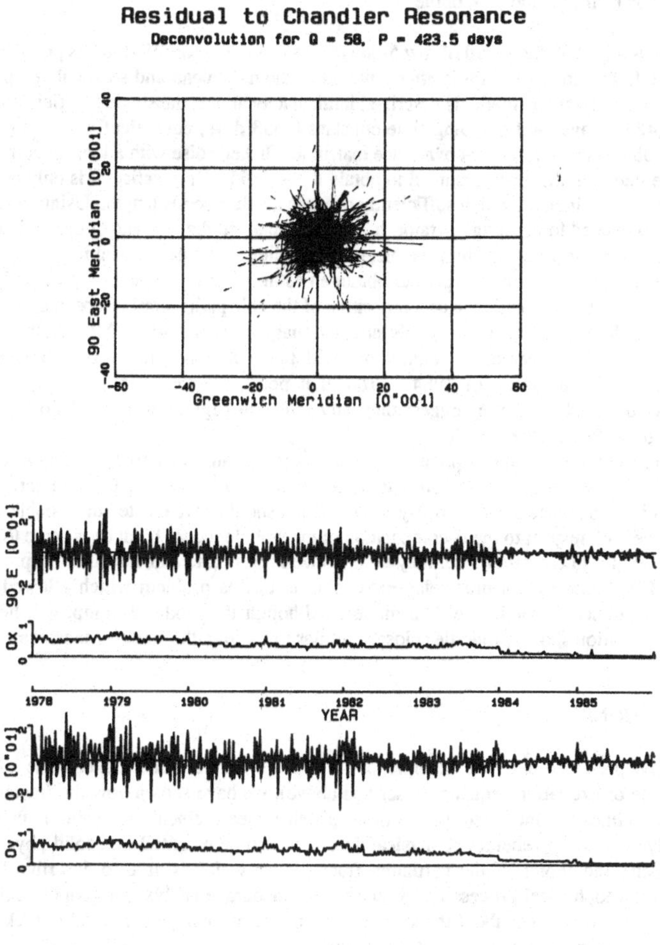

Fig. 6: Estimation of the excitation pole path via equation (16) where we used the determined value of the complex-valued forecasting operator, b.

$$\hat{\mathbf{p}} = \mathbf{z} - \mathbf{Z}\,\hat{\mathbf{b}} \tag{16}$$

and forming

$$s_p^2 = \frac{\hat{\mathbf{p}}^t\,\mathbf{V}_p^{-1}\,\mathbf{p}}{(n-1)}\,. \tag{17}$$

2.4. Resolution of the Chandler Wobble

Figure 5 shows a plot of the record of raw 5-day means from the years 1978-85 as published by the Bureau International de l'Heure (BIH), Paris after removal of the data mean and seasonal components to the 6th annual harmonic. Using the methods described immediately above, the single coefficient which determines the period (423.5 days) and damping time constant (1553 days, equivalently, Q = 58) of the Chandler wobble was obtained. A variance-covariance matrix for flicker-noise with a longest correlation interval of 5000 sample intervals was prior assumed to obtain this solution. In practice, it is only necessary to assign a longest correlation interval which sufficiently exceeds the data series length. Using the determined value of the complex-valued forecasting operator, **b** , we then obtained the estimate of the excitation pole path via equation (16) shown in figure 6. In principle, and according to the data model implied in the development of the solution above, we would expect to obtain an estimate of the excitation pole path showing a clear flicker-noise-like quality. While the decomposition of the pole path record has been optimized to seek such an excitation pole, that found is obviously not stationary in its statistics. Measurement noise, probably essentially uncorrelated, decreasing in amplitude with the evolution of the record dominates the estimated pole path until the beginning of 1984. From that point, the excitation pole-path record shows its flicker-noise-like qualities. We suggest that only after the beginning of 1984 do we begin to see the excitation process forcing the wobble.

We can say little more at this point about the process because this clear excitation record is short and uneventful so that we recognize no correlations between its variations and those major, transient physical processes which took place during this 2-year period and should have excited the wobble. This experiment is inconclusive with respect to the decomposition of the wobble record but it does serve to show how a data decomposition of a linear data model may be optimized for a particular fractal noise process. Jensen and Mansinha (1987) described a more elaborate solution to this problem which allowed for a stationary additive measurement error in the data model. Although the model decomposed, here, is much less complete, its solution shows properties closely similar to those in the Jensen-Mansinha solution.

3. CONCLUSIONS

While we have presented only one example, one which is optimized for a flicker-noise innovation (i.e., the error sequence or excitation sequence described above), we have shown how the fractal noises might be introduced into linear or linearized models of geophysical measurements and data. In using these methods for the analysis of more general geophysical data, it would be essential that solid physical or geological evidence justify the choice of the particular fractal noise which is used to describe the indeterminism inherent in the geophysical process under study. We encourage analysts of geophysical data to question their traditional assumptions about the stochastic elements of their problem and to seek to discover their possibly true fractal properties through direct study.

4. ACKNOWLEDGEMENTS

This work was supported by the Natural Sciences and Engineering Research Council of Canada through operating research grants to two of the authors (O.G.J. and D.J.C). We would like to thank Profs. S. Lovejoy, G. Austin and D. Schertzer, Visiting Professor, of the Department of Physics of McGill University for their interest and help and for convening the exciting Workshop on Scaling, Fractals and

Nonlinear Variability in Geophysics held during August 25-29, 1986 at McGill University, Montreal, which led to the organization of the ideas we have presented in this article.

5. REFERENCES

Barnes, J. A., and D. W. Allan, 1966: A statistical model of flicker noise. Proc. I.E.E.E. 54, pp. 176-178.

Gardner, M., 1978: Mathematical games: white and brown music, fractal curves and one-over f fluctuations, Scientific Amer., 238(4), pp. 16-32.

Gauss, C. F., 1839: Allgemeine Theorie des Erdmagnetismus, Leipzig (Republished 1877: Gauss Werke 5, Gottingen).

Hosken, J. W. J., 1980: A stochastic model of seismic reflections, presented at the 50th Annual Meeting of the Society of Exploration Geophysists, Houston. (Abstract G-69, Geophysics, 46, 419).

Jensen, O. G., and L. Mansinha, 1987: Excitation of geophysical systems with fractal flicker noise. Time Series and Econmetric Modelling, pp. 165-188, Reidel, Dordrecht.

Jensen, O. G., and L. Mansinha, 1984: Deconvolution of the pole path for a fractal flicker-noise residual, in Proceedings of the International Association of Geodesy (IAG) Symposia 2, pp. 76-99, Ohio State University Press, Columbus.

Jensen, O., G. and A. Vafidis, 1987: Inversion of seismic data using extreme skewness and kurtosis, submitted to Geophysical Prospecting, 1987-02-18.

Lambeck, K., 1980: The Earth's Variable Rotation: Geophysical Causes and Consequences, Cambridge University Press, Cambridge.

Mandelbrot and McCamy, 1970: On the secular pole motion and the Chandler wobble. Geophys. J. R. Astr. Soc., 21, pp. 217-232.

Mandelbrot, B. B., 1983: The Fractal Geometry of Nature, Freeman, San Francisco.

Munk, W. H., and G. J. F. MacDonald, 1960: Rotation of the Earth, a Geophysical Discussion, Cambridge.

Press, W. H., 1978: Flicker noises in astronomy and elsewhere, Comments Astrophys., 7, pp. 103-119.

Smylie, D. E., G. K. C.Clarke, and L. Mansinha. 1970: Deconvolution of the pole path, in Earthquake Displacement Fields and Rotation of the Earth, Astrophysics and Space Science Library Series, Reidel, Dordrecht.

Todoeschuck, J. P., and O. G. Jensen: 1988a: Joseph geology and seismic deconvolution, Geophysics, (in press).

Todoeschuck, J. P., and O. G. Jensen, 1988b: Scaling Geology and seismic deconvolution, Pure and Applied Geophysics, (in press).

Todoeschuck, J. P., and O. G. Jensen, 1988c: 1/f geology and seismic deconvolution, Pure and Applied Geophysics, (in press).

Todoeschuck, J. P., O. G. Jensen and S. Labonté, 1988: Scaling geology: evidence from well logs (in preparation).

Vafadis, A., and O. G. Jensen, 1987: Non-Gaussian seismic data models: direct inversion for extreme skewness and kurtosis of the reflectivity sequence, submitted to Geophysics, 1987-02-25.

Vafidis, A., and O. G. Jensen: Non-Gaussian seismic data models: inversion using prior statistical estimates, submitted to I.E.E.E. Transactions in Remote Sensing, 1988-01-15.

Vessot, R. F. C., 1974: Lectures on frequency stability and clocks and on the gravitational red-shift experiment, in Experimental Gravitation, Proceedings of the International School of Physics 'Enrico Fermi', Course LVI, p. 111-1162, Academic Press, New York.

Walden, A. T., and J. W. J. Hosken, 1985: An investigation of the spectral properties of primary reflection coefficients, Geophysical Prospecting, 33, pp. 400-435.

EXTREME VARIABILITY OF CLIMATOLOGICAL DATA: SCALING AND INTERMITTENCY

Ph. Ladoy, S. Lovejoy*, D. Schertzer**
SCEM/CLIM, Météorologie Nationale,
2 Av. Rapp, 75007, Paris

ABSTRACT. The atmosphere displays linked spatial/temporal variability over wide ranges in scale. In this paper we study the scaling behaviour of the fluctuations. These properties are characterized by spectral exponents, and the extreme fluctuations by exponents of the corresponding probability distributions. Using both temperature and rain rate data from climatologically representative locations, we find very similar behaviour including relatively small dispersions in the estimated exponents. We also show that the effect of spatial averaging is primarily to reduce the amplitudes of both the fluctuations and their spectra by constant factors, a behaviour which is consistent with global space/time scaling. Finally we argue that these non-standard statistical procedures may be indispensable in taming extremely variable climatic and meteorological data.

1. INTRODUCTION

We are primarily interested in studying the variability of climate over "regional" spatial scales (e.g. up to the scale of France) and temporal scales up to the order of several decades of years. However, atmospheric fields are extremely variable over a wider range[1]. The phenomenology of atmospheric phenomena includes existence of sharp gradients, sudden transitions, erratic fluctuations, many with structures spanning a wide range of scales. This extreme variability cannot be tamed with standard statistical techniques involving assumptions about the existence of characteristic decorrelation times and distances (hence implicitly of exponential decays of correlations) as well as the existence of characteristic amplitudes of fluctuations (hence of exponential decays of the probability distributions). This extreme variability (intermittency) of climatological data (temperatures and precipitations) discussed here has two basic features:

1- The fluctuations (ΔX) of the fields X span a wide range of scales. The energy spectra displays regimes over which the energy varies in a simple power law manner: over this range, there is no characteristic time scale; the field is scale invariant (scaling).

2- At a given scale the fluctuations ΔX may span a wide range of intensities. The probability distributions are fat-tailed (intermittent) and large fluctuations will occur; the sample spectrum (e.g. in one dimension the modulus squared of the corresponding fourier component) will have large random peaks.

Scaling and intermittency have until recently been associated isotropic (self-similar) processes defined by a single fractal dimension. Such models have been useful in modelling various geophysical fields such as landscape(topography) or clouds. Recent developments have shown that such mono-scaling behaviour is the exception rather the rule. Instead, multifractal fields with a hierarchy of fractal dimensions generically arise as a result of cascade processes of broadly the same type as those that concentrate energy, moisture and other conserved fluxes into smaller and smaller regions of the atmosphere. These multifractal cascades have many interesting properties including very strong intermittency.

---

* Physics dept., McGill University, Montréal, Canada.
** EERM/CRMD, Météorologie Nationale,Paris, France.

[1] 10 orders of magnitude in spatial scale (1mm to 10,000km), coupled with 7 orders of magnitude in temporal scale (1 second to 1 month if we consider climatological variations, the range would be much larger)

D. Schertzer and S. Lovejoy (eds.), Non-Linear Variability in Geophysics, 241–250.

The object of this paper is to characterize the scale invariant regimes of the climate with a few exponents. As long as we are interested in the behaviour of single moments (such as the exponent of the second order moments that characterize the scaling of the spectra), a single exponent is sufficient. More refined (multifractal) analyses will be performed in future. The data used here are three hourly or daily average temperatures and daily accumulations of precipitation from recording stations in France, Saint-Pierre and Miquelon, New-Caledonia, and Antilles. The work reported here is an elaboration of results previously reported in Ladoy (1986) and in Ladoy et al. (1986).

2. METHODS FOR STUDYING SCALE DEPENDANCE

2.1. Simple scaling

A specific type of scaling called "simple scaling" or "scaling of the increments" may be defined as follows. For a function of a single variable X(t) (T(t) for temperature, R(t) for the rain rate), a particular type of scale invariance arises when fluctuations (ΔX) at small scales ($\Delta t/\lambda$, $\lambda>1$) are related to those at large scale (Δt) via the following relation:

$$\Delta X(\Delta t/\lambda) \overset{d}{=} \Delta X(\Delta t) / \lambda^H \tag{2.1}$$

where:

$$\begin{aligned} \Delta X(\Delta t) &= X(t_1) - X(t_0) & \Delta t &= t_1 - t_0 \\ \Delta X(\Delta t/\lambda) &= X(t_2) - X(t_0) & t_2 &= t_0 + (t_1 - t_0)/\lambda \end{aligned} \tag{2.2}$$

and the sign "$\overset{d}{=}$" indicates equality in probability distributions. The random variables are equal in this sense when $Pr(u>q) = Pr(v>q)$ for any threshold q. "Pr" means probability. The parameter H is a constant[2] called the - unique - scaling parameter, and is $0 \leq H \leq 1$.

2.2. Energy spectra

The energy spectrum $E(\omega)$ of fluctuations ΔX where ω is a frequence is useful. The spectrum is scaling when it varies in a power law manner, i.e. it is of the form:

$$E(\omega) \propto \omega^{-\beta} \tag{2.3}$$

When $E(\omega)$ is of this form over a given frequency range, there is no characteristic time and hence within the range, the process is scale invariant ("scaling"). In simple scaling (and when the variance $<\Delta X^2>$ is finite), the exponents H and β are related by the following formula[3]:

$$\beta = 2H+1 \tag{2.4}$$

For example, the familiar Kolmogorov spectrum $E(\omega) \approx \omega^{-5/3}$ for the spectrum of turbulent wind fluctuations implies H=1/3.

2.3. Probability distributions

The direct analysis of probability distributions is best accomplished by using log-log plots such as those shown in fig. 1.

---

[2] Either in the framework of monofractality or for the extreme fluctuations.

[3] The formula is a simple consequence of 2.1 and the fact that the energy spectrum is the Fourier transform of the autocorrelation funcyion .

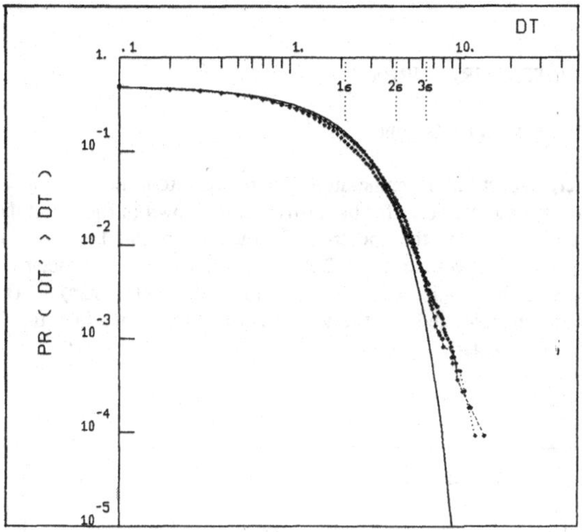

Fig. 1: The probability distributions of daily temperature differences in daily mean temperatures from Macon France for the period 1949-1979 (10,957 days). Daily means are computed by averaging consecutive three hourly data). Positive and negative differences are shown as separate curves. A best fit gaussian is shown for reference indicating that the extreme fluctuations correspond more than 7 standard deviations; which for a gaussian, would have a probability level of $\approx 10^{-20}$. The distribution is far from gaussian, since it rather has hyperbolic tails as displayed by the extremes which are nearly straight on this log-log plot.

This method enables[4] us to estimate α, the hyperbolic intermittency parameter (the negative slope for large ΔX in fig. 1). Hyperbolic intermittency means that large fluctuations in ΔX are distributed as follows:

$$\Pr(\Delta X' > \Delta X) \propto \Delta X^{-\alpha} \qquad (2.5)$$

$\Pr(\Delta X' > \Delta X)$ is the probability of a random fluctuation $\Delta X'$ exceeding a fixed ΔX. When the fluctuations are of this type, the phenomena is so intermittent that high order moments $\langle \Delta X^h \rangle$ diverge (tend to ∞ as the sample size is increased, see Feller (1971)) for all $\alpha \geq h$ (this gives rise to the "pseudo-scaling" (Schertzer and Lovejoy 1987, Lavallée et al., this volume). Fluctuations of this type with scaling spectra and hyperbolic intermittency are expected to occur due to the action of cascade processes concentrating energy fluxes, temperature variance fluxes (more generally of the dynamically relevant conserved flux), from large to small scales. Ever since Richardson, the idea of cascade processes transfering energy from large to small scales has been central to theories of fully developed turbulence (leading notably to Kolmogorov's 1941 $\omega^{-5/3}$ energy spectrum). Extreme variability results because as the number of cascade steps tends to infinity, the energy becomes distributed over a (mathematically) singular measure which may be characterized by a hierarchy of fractal dimensions. In these multiplicative (and multifractal) processes, the hierarchy of exponents specifies the variation with scale of the statistical moments (and hence probability distributions). In such processes, equation 2.1 (involving a single parameter H related to a single fractal dimension) applies only to the extreme tails of the probability distributions. In spite these

4 it also allows us to evaluate H, as well as the limits to scaling.

approximations, this paper will use the simple scaling of eq. 2.1 to study climatological data over a range of time and space scales.

3. CLIMATOLOGICAL TEMPERATURES

3.1. Scale invariance of the local temperature

Figure 2 shows the energy spectrum of temperature fluctuations from the climatological recording station in Macon France over the same period as for the distributions shown in fig. 1. At the high frequency end (6 hours)$^{-1}$ to about (2 weeks)$^{-1}$, the spectrum follows a straight line on a log-log plot (fig. 2) corresponding to $E(\omega) \propto \omega^{-1.7}$. Over the range (2 days)$^{-1}$ to (6 hours)$^{-1}$, the spectrum is dominated by the diurnal peak and various sub harmonics, especially (12 hours)$^{-1}$, and (8 hours)$^{-1}$. This "red noise" is very close to the $\omega^{-5/3}$ spectrum predicted for the temperature fluctuations in a turbulent fluid when the temperature acts as a "passive scalar".

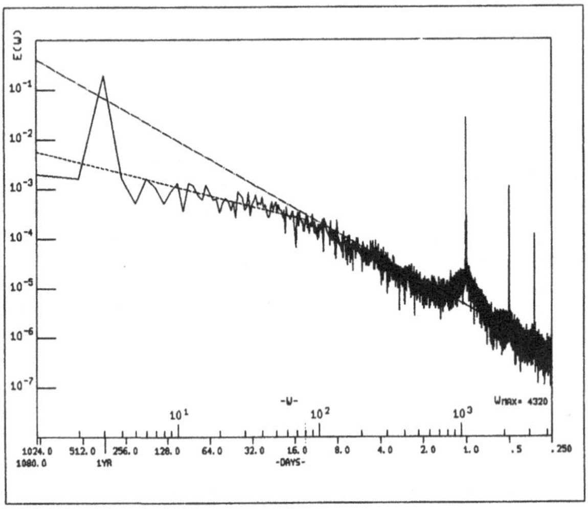

Fig. 2: The average of ten consecutive 3 years spectra for the three hourly temperature data used in fig. 1. Note the annual, diurnal peaks as well as the sub-harmonics of the latter. The two straight lines are fit to the data corresponding to periods greater and less than 14 days, and correspond to $\omega^{-0.72}$ and $\omega^{-1.78}$ respectively

The high frequency scaling regime clearly breaks down for $\omega \leq$(2 - 3 weeks)$^{-1}$ and the spectrum becomes nearly flat. This "spectral plateau" - quasi "white noise" spectral region (Lovejoy and Schertzer 1986b) -is very roughly constant spectrum at these time scales. The smooth transition at $\omega \approx$(3 weeks)$^{-1}$ is the "synoptic maximum" - see Kolesnikova and Monin (1965) - these authors estimated the period of the maximum to be between 1 and 3 weeks. It is generally agreed that this is the minimum time scale for the planetary sized fluctuations.

Figures 3, 4 and 5 display spectra of temperature variance for the three very climatologically different recording stations those of Noumea in New Caledonia, St. Pierre and Miquelon and the historical Paris-Montsouris station. We can see the same general shape of the spectra, β varying from $\approx$1.45 to$\approx$1.8.

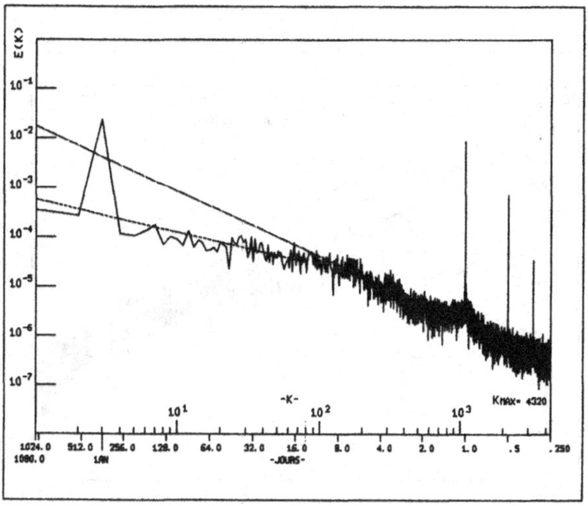

Fig.3: The average of ten consecutive 3 years spectra for the three hourly temperature at Noumea, New Caledonia. Note the annual, diurnal peaks as well as the sub-harmonics of the latter. The two straight lines are fit to the data corresponding to periods greater and less than 14 days, and correspond to $\omega^{-0.66}$ and $\omega^{-1.45}$ respectively.

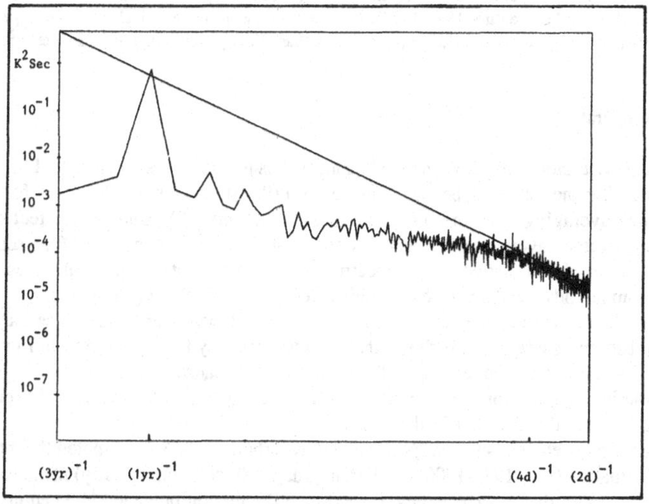

Fig.4: The average of ten consecutive 3years spectra for the daily temperature at Saint Pierre and Miquelon. Note the annual, diurnal peaks as well as the sub-harmonics of the latter. The straight line is fit to the data corresponding to periods less than 14 days and corresponds to $\omega^{-1.80}$.

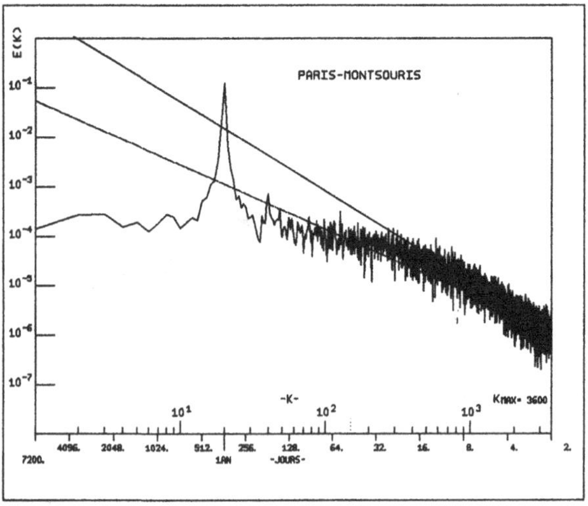

Fig 5: The average of 5 consecutive of 20 years (≈7,200 days) spectra for the daily temperature at Paris-Montsouris, (for the last century:1873-1972). Note the annual peak (and lack of 11-year peak). The straight line is fit to the data corresponding to smaller periods less than 20 days and corresponds to $\omega^{-1.8}$.

3.2 Regional temperatures

Temperature is of necessity a spatially averaged quantity. As temperatures are averaged over larger regions (eventually covering the entire globe, see Jones et al. (1982)), the amplitude of ΔT for a fixed Δt will decrease because averaging smooths out large local variations. To study the effect of such spatial smoothing, we performed averages of 53 daily recording stations over France (fig. 6), a scale refered to as regional. The general level of the energy spectrum is less than that of the local series. Note that the synoptic maximum has advanced somewhat to higher frequencies (to about $(10\text{days})^{-1}$).

This hyperbolic fall-off is quite different than that of a gaussian distribution. The fairly low value of the empirical (absolute) slope ($\alpha \approx 4.5$) shows that the intermittency is very strong. In figure 7 we plotted the probability distribution of the regional ΔT. We note that because of scaling, the only difference with the local distribution is a constant factor a linear shift on a log-log plot. The spatial averaging smooths out the fluctuation hence the amplitude is decreased.

To illustrate the difference between gaussian and hyperbolic tails, we computed the return probability of a extreme fluctuation.Over 1951-1980, the extreme (daily) fluctuation in (daily) mean temperatures was found to be 14°C. With the hyperbolic law, the extreme fluctuation predicted in a century long record is 17.8°C, whereas using the conventional Gumbel law for return probabilities, the return period for such a large fluctuation is 237 years!

The critical meteorological situation on January the 12th in 1987 in France is also very illustrative. Indeed, a seventeen days long cold wave had been severe for 6 days with dayly average temperature less than -10°C. The Loire river began to froze and threaten to prevent the incoming of fresh water indispensible to cool down the nuclear core of the power plant Saint Laurent des Eaux. It turns out that Electricité de France, the state owned agency who runs the power plant, didn't forecast such a drastic event in underestimating the occurences of extrem fluctuationes.

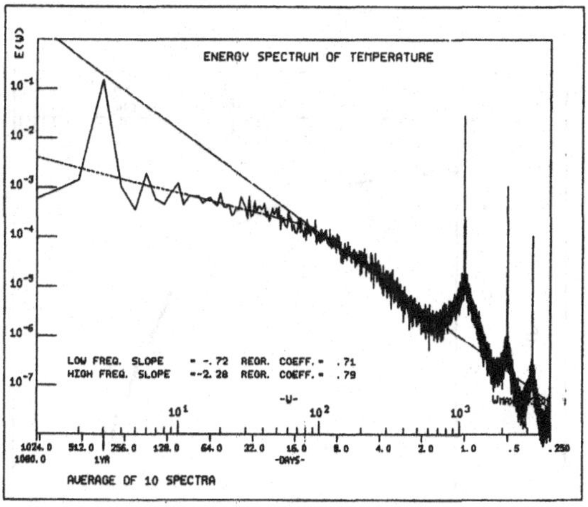

Fig.6: The average of ten consecutive 3 years spectra for the three hourly regional temperature obtained by averaging over 53 meteorological measuring stations. Note that the spectrum maintains its shape, but is displaced towards lower energies. In particular, the slope is unaffected, the straight line fit to the data corresponding to periods less than 22 days which corresponds to $\omega^{-1.8}$. As expected,because time and space are statiscally linked larger frequencies ($\leq 4(\text{day})^{-1}$) are damped by the spatial averaging

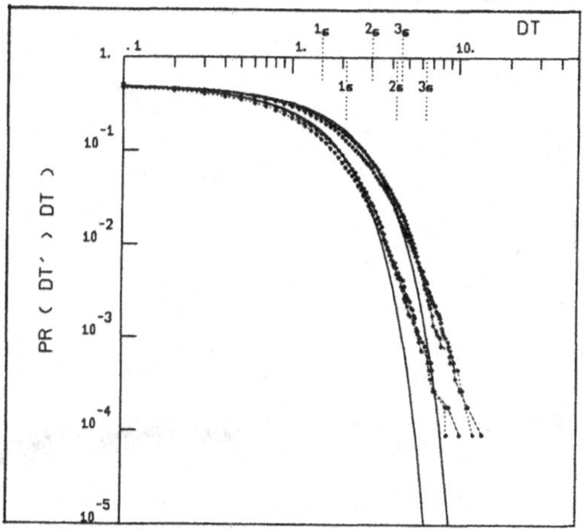

Fig. 7: The probability distributions of daily temperature differences (in daily mean temperatures) from Macon, France (rightmost curves), and regionally averaged in space over 53 stations (leftmost curves) for the period 1949-1979 (10,957 days). Daily means are computed by averaging consecutive three hourly data. Positive and negative differences are shown as separate curves. A best fit gaussian is shown for reference indicating that the extreme fluctuations correspond to more than 7 standard deviations; which for a gaussian, would have a probability level of $\approx 10^{-20}$. Both distributions are far from gaussian, as shown by their nearly straight extremes on this log-log plot indicating that they are hyperbolically distributed. Notice that the two sets of curves are virtually identical except for a left/right shift corresponding to a constant factor. This is exactly what we would expect if the temperature is scaling in timt and space.

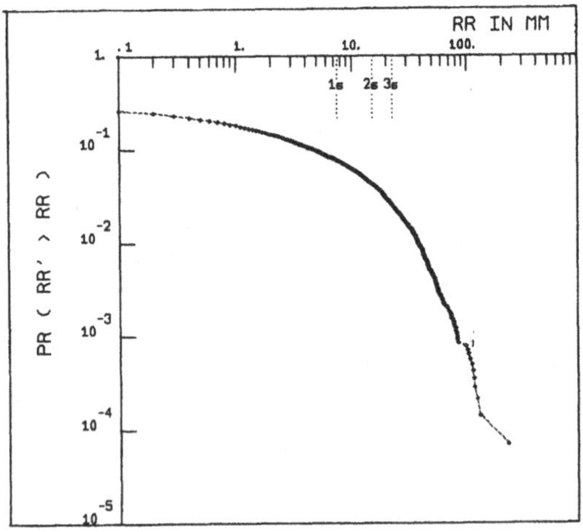

Fig. 8: The probability distributions of the fluctuations of the daily rain accumulation at Nimes-Courbessac for 40 years (1949-1988 hence 14,245 days) in a log-log plot. This distribution displays an asymptotic hyperbolic behaviour, with exponent $\alpha \approx 2.6$.

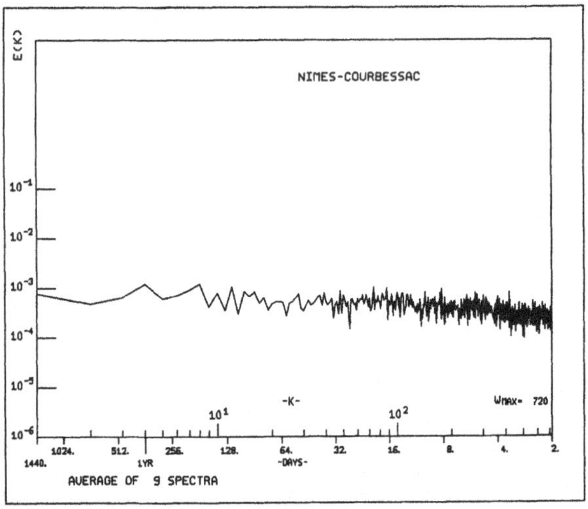

Fig.9: The average of six consecutive 4 years spectra of the daily rain accumulation at Nimes-Courbessac. Note there is no obvious annual peak, neither synoptic maximum. The scaling regime seems to span the whole range, but with a small slope $\beta \approx .3$.

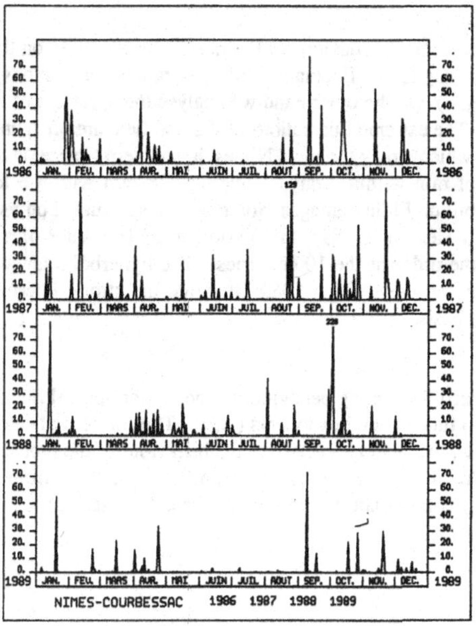

Fig.10a: Total amount of daily precipitation (in mm) over a period of four years.

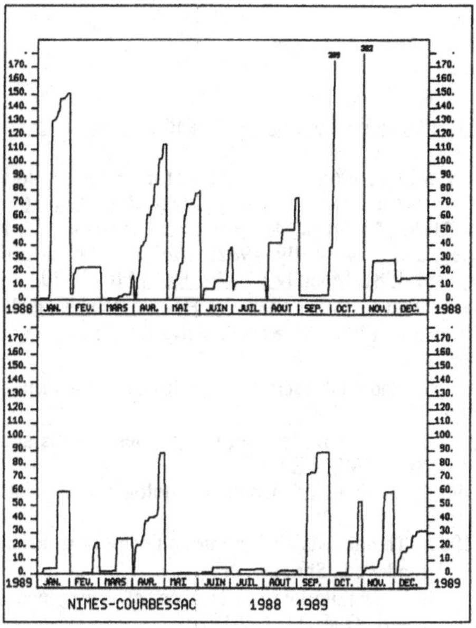

Fig.10b: Monthly accumulations of precititation (in mm) for two years

4. PRECIPITATION

In order to study the climatological behaviour of the rain field, we study on the one hand the probability distribution Pr(ΔR'>ΔR) for day to day fluctuations of daily rain accumulations, which shows a hyperbolic fall-off with $\alpha \approx 3.5$ (see fig. 8). On the other hand we analyse the spectrum of this fluctuations (see fig. 9) which is rather flat ($\beta \approx .3$). The extreme fluctuations of the rain field are often underestimated. However, the rather recent sudden flood which struck the town Nimes, in southern France, is one of the many vivid Natur manifestations: 420 mm of rain within a day, including 228 mm within 6 hours ... and consequently several casualties and 1.5 billion FF in damages. Some people speculated on return period for such a large fluctuation. According to Hémain (1989), this period is of the order of a century, in opposition to optimistic estimations of the order of the 10 centuries! The hyperbolic law estimated from our data set confirms the pessimistic point view. Figs. 10a,b illustrate the strong intermittancy of precipitation on both the daily and monthly basis.

5. CONCLUSIONS

In this paper we characterized the scaling behaviour of the climatological fluctuations of temperature and rain rates with the help two fundamental exponents (H,α). The former indicates how the amplitude of the fluctuations depend on scale, the latter characterizes the behaviour of the probability distribution, hence the extreme fluctuations. We find that quantitatively similar behaviour from different climatologically representative locations. It is important to realize that these findings could not be obtained by means of standard statistical procedures.

6. ACKNOWLEDGMENTS

We thank our colleagues of the Météorologie Nationale R. Astoul, N. Bourdette, A. Gorse, G. Payen, G. Mayencon, P. David, N. Besleaga and S. Janicot for stimulating discussions and help preparing data. We also acknowledge discussions with K. Laval, D. Lavallée, R. Viswanathan, J. Wilson.

7. REFERENCES

Feller, W., 1971: An introduction to probability theory and its applications. vol.2, Wiley, New-York, 567pp.

Kolesnikova, V. N. ,and A. S. Monin, 1965: The spectra of micrometeorological synoptic and climatic oscillations of meteorological fields. Meteorological Research, 16, 30-56. Soviet. Geophys. Committee, Acad. Sci. USSR, Canadian Meteorological Translation 17, 1971.

Jones P. D., T. M. L. Wigley, and P. M. Kelly, 1982: Variations in surface air temperatures: Part I, Northern hemisphere, 1881-1980. Monthly Weather Rev., 110, 59-70.

Hémain, J. C., 1989: L'évenement du 3 octobre 1988. La Houille Blanche, 6, 423-428

Ladoy, Ph., D. Schertzer, S. Lovejoy, 1986a: Une étude d'invariance locale-régionale des températures, La Météorologie, 7, 23-34.

Ladoy, Ph., 1986: Approche non-standard des séries climatologiques, invariance d'échelle et intermittence. Thesis, Paris IV, 194pp.

Lovejoy, S., D. Schertzer, 1986a: Scale invariance, symmetries fractals and stochastic simulation of atmospheric phenomena. Bull AMS, 67, 21-32.

Lovejoy, S., D. Schertzer, 1986b: Scale invariance in climatological temperatures and the local spectral plateau. Annalea Geophysicae, B-4 , 401-410.

Schertzer, D., S. Lovejoy, 1985: The dimension and intermittency of atmospheric dynamics. Turbulent Shear Flow, 4, 7-33, B. Launder ed., Springer, NY.

Schertzer, D., S. Lovejoy, 1987: Singularités anisotropes, et divergence de moments en cascades multiplicatifs. Annales Math. du Qué., 11, 139-181.

ON THE EXISTENCE OF LOW DIMENSIONAL CLIMATIC ATTRACTORS

M. A. H. Nerenberg, Turab Lookman and Christopher Essex
Department of Applied Mathematics
University of Western Ontario
London, Ont., Canada N6A 5B9

ABSTRACT. Recent work has highlighted the possibility of using certain ideas from dynamical systems theory to study climate. We review the evidence for chaos in climatic changes based on estimates of the dimension of the reconstructed attractor from time series data.

1. INTRODUCTION

Interest has recently focused on whether a complicated dynamical system, such as that governing climate, may be deterministic. If records of climatic change show the signs typical of a simple deterministic system, then it should be describable by a simple model with a few degrees of freedom. The time evolution might be represented by a low dimensional attractor, a trajectory which stays on a low dimensional submanifold of the total available phase space.

Analyses of hydrodynamic experiments indicate that a relatively small number of degrees of freedom (6-10) are sometimes enough to characterize observed turbulent flows. Guckenheimer and Buyza (1983) analyzed the transition to geostrophic turbulence in a rotating annulus to find that the dimensionality could increase to as high as 11. Studies by Sreenivasan and Strykowski (1984) on turbulence in open flow systems suggest that, at least at Reynolds numbers not too far above the transition to fully developed turbulence (~6625), the attractor is relatively low-dimensional (around 6). Experiments on weakly turbulent Couette flow also indicate values between 2 and 5 depending on the Reynolds number. Thus, evidence for the existence of a low-dimensional attractor has gradually accumulated, albeit in relatively simple laboratory systems. The flows encountered in the atmosphere are typical of a hierarchy of structures that range from laminar to turbulent motions. If the basic structures of the observed flows depend only on a finite, possibly small number of degrees of freedom, then it suggests that, in spite of its complexity, the system of the atmospheres and oceans that determines climate, may be described by a nonlinear model depending on a few variables.

Noisy signals have traditionally been analyzed using Fourier spectra, correlation functions, and other methods developed for linear systems. However, these techniques are inadequate for characterizing chaos in deterministic nonlinear systems. More appropriate properties for characterizing chaos are the Lyapunov exponents, entropies and dimensions of the attractors. Methods have been developed for computing these properties from time series data obtained in experiments.

Evidence for chaos in climatic changes has so far been based on estimates of the dimensions of the reconstructed attractor from time series data. Using relatively few data points (184), Nicolis and Nicolis (1985) analyzed the isotope record of deep sea cores to estimate the number of degrees of freedom (≥4) which govern the observed long term climate evolution during the past million years, Grassberger (1986) has since claimed that the conclusions of Nicolis and Nicolis are based on too few, too finely sampled and too much smoothed data and that there is no sign of a low dimensional attractor. However, using a far more extensive data set of the daily 500mb geopotential over a 40-year span and two variants of the basic method, we find evidence for an attractor with as few as 7 degrees of freedom. A review of these studies and the current status of the topic forms the subject matter of this paper.

D. Schertzer and S. Lovejoy (eds.), Non-Linear Variability in Geophysics, 251–255.

2. DYNAMICAL INVARIANTS

The most striking attribute of chaos is sensitivity to initial conditions. One way of quantifying this is by evaluating the Lyapunov numbers. It must be shown that nearby trajectories separate exponentially. The largest Lyapunov number is zero for periodic or multiperiodic behaviour but positive for chaotic behaviour. Methods have been proposed for estimating the non-negative Lyapunov exponents from measurements of a single dynamical variable in the form of a time series. In a method developed by Wolf et al. (1984) the largest Lyapunov number is determined by monitoring the separation between initially nearby points $P_1(t_0)$ and $P_2(t_0)$ in phase space. Before the separation becomes too large, a new point $P'_2(t_1)$ is chosen so that the separation is again small and with the constraint that it must lie as nearly as possible in the same direction as the point it replaces (fig. 1). The procedure is repeated until the trajectory is followed to the end of the time series. The largest Lyapunov number λ_1 is then given by

$$\lambda_1 = 1/T \sum_{i=1}^{N} \log\left(\frac{L_i}{L'_{i-1}} \right) \tag{1}$$

where $T = t_N - t_0$ is the experiment duration, $L_k = | P_2(t_k) - P_1(t_k) |$ is the length evolved from a length $L'_{i-1} = | P'_2(t_{k-1}) - P_1(t_{k-1}) |$, and the summation is over the number of replacement steps. Exponents computed by this method for several model systems show good agreement with those obtained by simultaneously integrating both the equations of motion and the linearized equations for each model (Shimada and Nagashima (1979)).

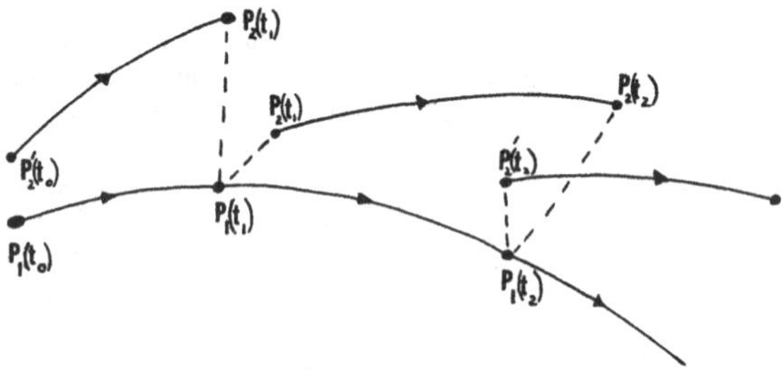

Fig. 1: Evolution of nearby points in phase space. A new point P'_2 is chosen when $| P_2 - P_1 |$ becomes large.

For systems with several positive Lyapunov exponents, the entropy is perhaps of greater interest than the largest Lyapunov exponent. The information entropy provides a lower bound to the Kolmogorov entropy which is conjectured to be equal to the sum of the positive Lyapunov exponents. The Kolmogorov entropy is the long-time average rate at which information about the system is lost due to the system dynamics. If the actual state X(t) of the system in phase space is measured with accuracy l and given that all the properties of the system except the initial condition X(0) are known, then the information entropy can be written as

$$S(l) = -\sum_{i=1}^{M(l)} p_i \ln p_i \qquad (2)$$

where p_i is the probability for X(t) to fall into the ith cell and $M(l)$ is the number of cells of length l needed to cover the attractor. It can be shown that

$$S(l) \sim -\sigma \ln l\,, \qquad (3)$$

where σ is the entropy dimension. Two related measures proposed for determining the geometrical properties of attractors are the Fractal or Hausdorff dimension, δ and the correlation dimension ν. The quantities ν, σ and δ are related by $\nu \leq \sigma \leq \delta$. Several methods have been proposed for computing ν, σ and δ. In principle, δ can be computed straightforwardly from the definition: the phase space is divided into cubes of side r and the number of cubes N(r) that contain a part of the attractor is determined; δ is given by $N(r) \sim r^{-\delta}$ for small r. In practice, however, the number of cubes required increases exponentially with δ and the computation of δ greater than 3 is very difficult (Greenside et al., 1982).

3. CORRELATION DIMENSION FROM TIME SERIES

The measure that we shall focus on is the correlation dimension, ν, introduced by Grassberger and Procacia (1983). Even though we are currently undertaking calculations of Lyapunov exponents, the characterization of chaos in climate has up to now relied largely on estimates of ν. The correlation function is defined by

$$C(r) = 1/N^2 \sum_{\substack{i=1 \\ i \neq j}}^{N} \sum_{j=1}^{N} \Theta(r - |X_i - X_j|), \qquad (4)$$

where $\Theta(u) = 1$ for $u > 0$, $\Theta(u) = 0$ for $u \leq 0$, and $\{X_i\}$ is the set of N points on the attractor constructed from a time series of $M \geq N$ observations. The correlation function counts the number of pairs with distance $|X_i - X_j|$ smaller than r. If, for sufficiently small r,

$$C(r) \sim r^{\nu} \qquad (5)$$

then ν is referred to as the correlation dimension of the attractor. This is a natural definition, since if points are uniformly distributed on a d-dimensional object, embedded in a D-dimensional space, $(D \geq d)$, then the number of points in any hypersphere of radius r, centered at a point on it behaves as r^d.

A time series consists of the measurements of a single dynamical variable of a multivariable system. A method is thus required to estimate ν from such a series. It has been proposed (Grassberger and Proccacia, 1983) that points X_i on the attractor can be constructed by using elements of the time series as components of X_i. For example, one way of choosing X_i in a D-dimensional space is to choose D consecutive measurements as components of X_i. For each embedding dimension, D, C(r) is obtained as a function of r. The region of interest is that for which

$$C(r) \sim r^{p}. \qquad (6)$$

If p approaches a limiting value ν as D is increased, then ν is identified as the correlation dimension.

Nicolis and Nicolis (1985) used 500 values from the isotope record of the deep sea core V28–238 to conclude that ν is 3.1. If this were true, one might be able to create a model predicting climatic changes of the last million years with only four independent variables. It is to be noted that only 184 actual measurements were used, the rest of the data was interpolated. Grassberger (1986), using exactly the same data, has found that he is unable to reproduce the results of Nicolis and Nicolis. He concludes that the

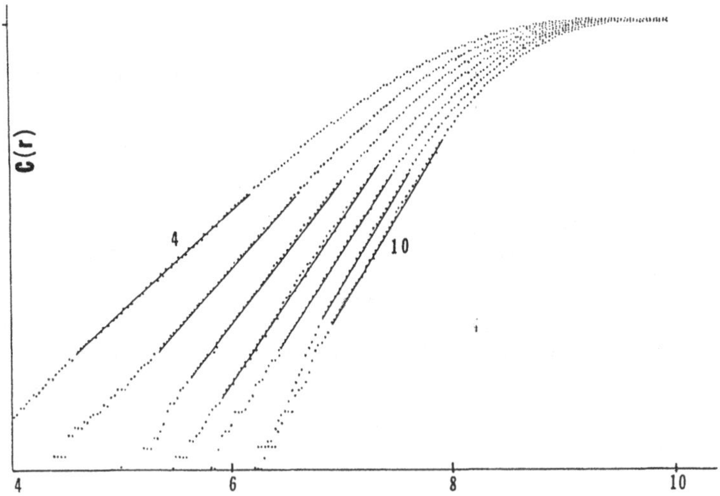

Fig. 2: Plot of ln C(r) versus ln r for embedding dimension, D, from 4 to 10 as obtained by Method 1. Note the convergence of slopes as D increases.

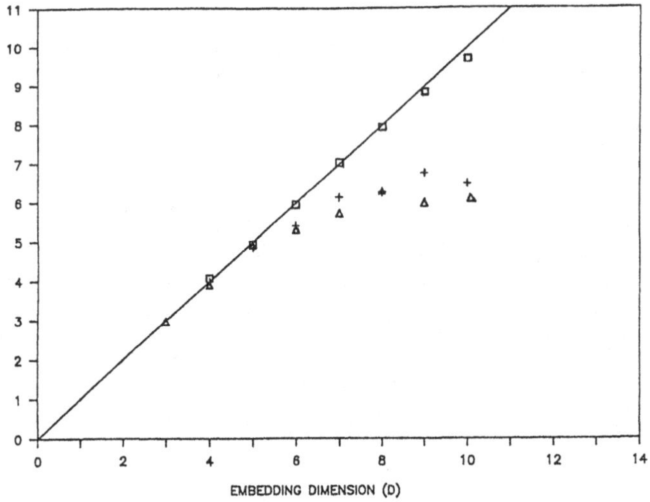

Fig. 3: Summary of results of scaling exponent, p, as a function of embedding dimension, D, obtained by two methods (shown as triangles and pluses). A random number set of the same size as the data set was used as a control (shown as squares). Note the saturation of the exponent arising from the data, for both methods, while there is not saturation for the random number set. This limiting value is identified as ν.

results suffer from too much smoothing and that it is difficult to distinguish the results from those of a random signal treated in the same way as the data.

Any algorithm developed to determine dimension, entropy, Lyapunov exponents will yield a number, but in practice it is often difficult to establish whether the number is reliably obtained and whether it is not an artifice of the method of calculation. We have recently concluded a study that strongly suggests the existence of an attractor with an embedding dimension as low as 7. It is a first step towards a study of the requirements on the quantity of data needed to determine ν (see companion paper of Essex in this volume). The data we analyzed were the daily geopotential values at 500mB taken at 12UT extending from 1946 to 1982. It is thus over a time scale much shorter than the one million years of Nicolis and Nicolis. Twelve sets of 12084 measurements were used. Each set represented the observations at one specific site and the cluster of sites fell within eastern North America.

Two distinct methods were used to construct the points X_i on the attractor. In Method 1 the twelve sets of measurements were appended to produce one long time series. The D consecutive daily measurements then formed the components of X_i in D-dimensional space. In Method 2 the measurement at a given time from each site formed a component of X_i. Thus, embedding in three dimensions only required three sets of measurements. In this way the number of points X_i was always maintained at 12084, regardless of the embedding dimension. We found that the dimensionality is not sensitive to the manner in which X_i is formed.

To increase the extent of the scaling region, we incorporated a 'boundary correction'. Expansion points which are close to the boundary of the data set can be surrounded by others only for small values of r. Such points contribute to a false saturation of C(r) when r is still small, and in general mask the scaling region. However, they can be discarded in the outer sum to produce a boundary corrected C(r). This has the effect of increasing the range in r for which $C(r) \sim r^p$.

4. RESULTS AND SUMMARY

The results are summarized in figs. 2 and 3. Figure 2 shows that the straight line fit to ln C(r) versus ln r, using Method 1, is good over a reasonable range of r.

Figure 3 indicates that, despite the rather different methods employed in selecting points X_i, both yield similar results. The value of ν saturates between 6 and 7. We therefore conclude that on short time scales (i.e., decades) climate might be represented as a system with as few as 7 degrees of freedom. Since the results are based on localized measurements on the earth, the estimate of ν may be considered to be only locally valid. It would be interesting to examine the variation in ν over the whole globe, to see if even approximately it is a global constant.

We also note that our estimate of ν is the same as that obtained at Reynolds number 6625 in a study on the transition to turbulence in coiled tubes (Sreenivasan and Strykowski, 1984). The study finds that around Reynolds 6625 there exists a threshold below which $\nu < 6$ and above which ν strongly increases with Reynolds number. However, one must be careful in extrapolating such a result to the atmosphere.

5. REFERENCES

Grassberger, P., 1986: Nature, 322, 609–612.

Grassberger P. ,and I. Proccacia, 1983: Phys. Rev. Lett., 50, 346-349.

Greenside, H., et al., 1982: Phys. Rev., A25, 3453.

Guckenheimer, J., and G. Buzyna, 1983: Phys. Rev. Lett., 51, 1438-1441.

Nicolis, C., and G. Nicolis, 1984: Nature, 311, 529-532.

Shimada, I. , and T. Nagashima, 1984: Prog. Theo. Phys., 61, 1605-1616.

Sreenivasan, K. R., and P. J. Strykowski, 1984: in Turbulence and Chaotic Phenomena in Fluids. Ed. Tatsumi, N. Holland.

THE THIRD GENERATION WAM MODELS FOR WIND-GENERATED OCEAN WAVES.

Will Perrie
Department of Fisheries & Oceans
Bedford Institute of Oceanography
P.O. Box 1006
Dartmouth, N.S., Canada. B2Y 4A2

ABSTRACT. A simplified efficient procedure has recently been set up to integrate the Boltzmann integrals for nonlinear transfer due to wave-wave interactions in wind-generated surface waves. The integration process uses a geometrically spaced polar grid over the spectral area such that loci and coefficients inside the integrated scale by multiples of the geometric scaling factor. This is presented with a view to implementing it in the latest generation of models for surface waves.

1. INTRODUCTION

As described in Hasselmann and Hasselmann (1985), WAM models give explicit representation to source functions for wind input (S_{IN})nonlinear transfer due to wave-wave interactions S_{NL}, and dissipation due to wave breaking/whitecapping processes S_{DS}. The energy conserving wave transport equation may be written as:

$$\frac{\partial F}{\partial t} + \underset{\sim}{C}_g \cdot \nabla_{\underline{x}} F = S_{IN} + S_{NL} + S_{DS} \tag{1}$$

where $F(\underline{k}, \underline{x}, t)$ is the two-dimensional ocean wave spectrum, as a function of wavenumber $\underline{k}$, position $\underline{x}$, and time t and $\underset{\sim}{C}_g$ is the group velocity.

Wind input follows Snyder et al (1981),

$$S_{IN}(\underline{k}) = \max\left\{0;\ 0.25\ \frac{\rho_2}{\rho_w}\ (\frac{U_5}{c}\cos\theta - 1)\ \omega F(\underline{k})\right\} \tag{2}$$

where ρ_2 and ρ_w are densities of air and water, $\omega = 2\pi f$, c is phase velocity, U_5 is the velocity at 5m, and θ is the angle between the wind vector and the wave propagation direction.

Dissipation follows the general white-capping functional of Hasselmann (1974),

$$S_{DS}(\underline{k}) = -\, c\bar{\omega}\left[\frac{\omega}{\bar{\omega}}\right]^2 \cdot \left[\frac{\alpha}{\alpha_{PM}}\right]^2 \cdot F(\underline{k}) \tag{3}$$

where

$$\bar{\omega} = \frac{\iint F(f, \theta)\ \omega\ df d\theta}{\iint F(f, \theta)\ df d\theta} \tag{4}$$

D. Schertzer and S. Lovejoy (eds.), Non-Linear Variability in Geophysics, 257–259.

α_{PM} is a constant based on Pierson-Moskowitz spectrum and is empirically determined as in Komen, Hasselmann and Hasselmann (1984), through numerical integrations of the wave transport equation which effectively tune S_{DS}.

The nonlinear transfer S_{NL}, is parameterized according to Hasselmann and Hasselmann (1985). The basic closure is brought about by assuming that a random linear wave field can be Gaussian, and that weak nonlinear interactions do not counter this lowest order Gaussianity, which follows Hasselmann (1967).

2. DISCUSSION

Shallow water models are constructed by taking into account shoaling (the effect due to depth-dependent propagation velocity), modification of deep water source terms, as in Herterich and Hasselmann (1980), addition of bottom dissipation, and adoption of various high frequency parameterizations.

Several times faster computation seem possible when a geometrically progressive polar grid is used in wavenumber space, such that the locus for interacting wavenumbers also scales geometrically, as in Tracy and Resio (1982).

In terms of action density, the nonlinear transfer due to 4-wave interactions is given by:

$$\frac{\partial n_4}{\partial t} = \iiint d\underline{k}_1\, d\underline{k}_2\, d\underline{k}_3\, C\, \delta(\underline{k}_1 + \underline{k}_2 - \underline{k}_3 - \underline{k}_4)\, \delta(\omega_1 + \omega_2 - \omega_3 - \omega_4) \times \times [n_1 n_3(n_4 - n_2) + n_2 n_4(n_3 - n_1)] \tag{5}$$

where n_i is the action density at wave number k_i, and frequency ω, δ is the Dirac delta function, and C is a complicated coupling coefficient:

This may be represented in terms of a transfer function $T(\underline{k}_1, \underline{k}_3)$.

$$\frac{\partial n_1}{\partial t} = \int d\underline{k}_3\, T(\underline{k}_1, \underline{k}_3). \tag{6}$$

For given $\underline{k}_1$, and $\underline{k}_3$, the latter may be calculated for $\underline{k}_2$ on the locus of interaction,

$$0 = \omega_1(\underline{k}_1) + \omega_2(\underline{k}_2) - \omega_3(\underline{k}_3) - \omega_4(\underline{k}_1 + \underline{k}_2 - \underline{k}_3) = W(\underline{k}_2) \tag{7}$$

and in terms of orthogonal coordinates relative to the interaction locus, (ξ, η)

$$T(\underline{k}_1, \underline{k}_3) = \iint d\xi d\eta \, \frac{\delta(\eta)}{\left|\frac{\partial W(\xi, \eta)}{\partial \eta}\right|} \cdot C \cdot [n_1 n_3(n_4 - n_2) + n_2 n_4(n_3 - n_1)] \tag{8}$$

which may be represented as the contour integral,

$$T(\underline{k}_1, \underline{k}_3) = \oint \frac{d(\xi)}{\left|\frac{\partial W(\xi, \eta)}{\partial \eta}\right|} \cdot C \cdot [n_1 n_3(n_4 - n_2) + n_2 n_4(n_3 - n_1)] \tag{9}$$

The action density term $n_1 n_3(n_4 - n_2) + n_2 n_4(n_3 - n_1)$ depends on the spectra considered and is well represented by Hasselmann et al. (1973) parameterizations.

With some manipulation, it may be shown that integration on a geometrically progressive grid, as in fig.1, causes the locus equation, phase space factor, and coupling coefficient to scale by powers of λ_o.

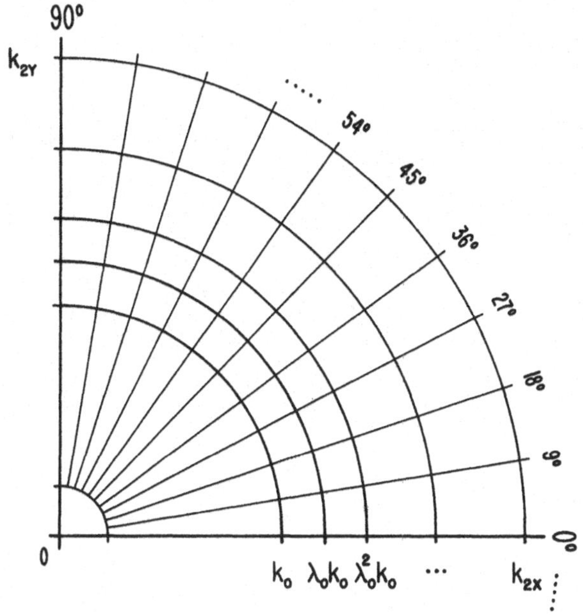

Fig. 1: Geometrically spaced progressive grid.

REFERENCES.

Hasselmann, K., 1967: Nonlinear interactions by methods of theoretical physics. Proc. R. Soc. Land. A299, 77-100.

Hasselmann, K. . 1974: On the spectral dissipation of ocean waves due to whitecapping. Boundary-Layer Met. 6, 107-127.

Hasselmann, K., T. P. Barnett, E. Bouws, H. Carlson, D. E. Cartwright, K. Enke, J. A. Ewing, H. Gienapp, D. E. Hasselmann, P. Kruseman, A. Meerburg, P. Muller, D. J. Olbers, K. Richter, W. Sell, and H. Walden, 1973: Measurements of wind-wave growth and swell during the Joint North Sea Wave Project (JONSWAP). Deutsche Hydrogr. Z. A 8, No. 12.

Hasselmann, S. and K. Hasselmann, 1985: Computations and parameterizations of the nonlinear energy transfer in a gravity-wave spectrum. Part I: A new method for efficient computations of the exact nonlinear transfer integral. J. Phys. Oceanogr.· 15, 1369-1377.

Herterich, K. and K. Hasselmann, 1980: A similarily relation for the nonlinear energy transfer in a finite-depth gravity-wave spectrum J. Fluid Mech. 97, 215-224.

Komen, G. J., S. Hasselmann and K. Hasselmann, 1984: On the existence of a fully developed wind-sea spectrum. J. Phys. Oceanogr. 14, 1271-1285.

Snyder, R. L., F. W. Dobson, J. A. Elliott and R. B. Long, 1981: Array measurements of atmospheric pressure fluctuations above surface gravity waves. J. Fluid Mech. 102, 1-59.

Tracy, B. A. and D. T. Resio, 1982: Theory and calculation of the nonlinear energy transfer between sea waves in deep water. U.S. Army Engineer Waterways Experiment Station WIS Report 11, 47 pp.

l/f GEOLOGY AND SEISMIC DECONVOLUTION

J. P. Todoeschuck* and O.G. Jensen
Geophysical Laboratory
Department of Geological Sciences
McGill University
3450 University Street
Montreal, P.Q., Canada H3A 2A7.

ABSTRACT. The reflection seismic problem in geophysics involves the generation near the surface of a seismic wavelet which propagates downwards to reflect from layers in the ground. The returning reflections carry information about depth and velocities but they also generate multiple reflections, making the resulting trace at the surface very hard to interpret. The signal must be deconvolved with an appropriate filter. The common practice in seismic deconvolution is to assume that the reflection sequence is uncorrelated; that is, it has a white power spectrum and a delta function auto-correlation. This means that the acoustic impedance function must have a spectrum proportional to l/f^2 (where f is frequency) which is characteristic of a non-stationary Brownian process. We propose that the power spectrum of the acoustic impedance function is l/f, which is stationary. This corresponds to a reflection sequence having a power spectrum proportional to f and a negative auto-correlation at small lags. This behaviour in reflection sequences has been seen before and we show it again in a well off Newfoundland.

If the power spectrum is proportional to f then the first term of the discretized auto-correlation function is -.405 of the zero lag term while higher-lag terms are negligible. We construct a filter analogous to the prediction error filter (PEF) commonly used in geophysics using this extra term. The method requires one additional term in the system of normal equations which are solved interatively. When used to deconvolve an artificial seismogram based on the well log, it recovered the reflection sequence with approximately half the error of the PEF.

1. INTRODUCTION

A common assumption in the deconvolution of seismic reflection sequences is that they are essentially uncorrelated; that is, their power density spectra are white (Robinson, 1957). This means that the acoustic impedance functions have power spectra proportional to $1/f^2$. Under the realistic case that the reflection coefficients are small, if the power spectrum of the acoustic impedance behaves as l/f^{β}, then the power spectrum of the reflection sequence behaves as $l/f^{\beta-2}$.

If r(t) are the (small) reflection coefficients and V(t) is the acoustic impedance as function of time t then we may write

$$r(t) = 1/2\, d\,(\ln V(t))/dt \qquad (1)$$

(Peterson et al. 1955) which may be integrated to give

* Now at Defence Research Establishement Pacific, FMO, Victoria, B.C., Canada V0S 1B0.

D. Schertzer and S. Lovejoy (eds.), Non-Linear Variability in Geophysics, 261–268.

$$\ln (V(t)/V(0)) = 2\int_0^t r(t')dt' \tag{2}$$

If the reflection coefficients are small then V(t) is never much different from V(0) and we may write

$$\frac{V(t)}{V(0)} = 1 + \frac{U(t)}{V(0)} \tag{3}$$

where U(t)/V(0) is a small quantity so that we may use the relation $\ln(1+x) \approx x$ to rewrite (2) as

$$\frac{U(t)}{V(0)} \approx 2\int_0^t r(t')dt' \tag{4}$$

which we Fourier transform to yield

$$\tilde{U}(f) \approx \frac{V(0)}{\pi i f}\,\tilde{r}(f) \tag{5}$$

where $\tilde{U}(f)$ is the Fourier transform of U(t) and so on, from which we obtain

$$|\tilde{U}(f)|^2 \approx \left(\frac{V(0)}{\pi f}\right)^2 |\tilde{r}(f)|^2 \tag{6}$$

that is, the power spectral density of the acoustic impedance function is approximately that of the reflection sequence divided by the square of the frequency.

A $1/f^2$ power spectral density is characteristic of one dimensional Brownian or Wiener-Bachelier processes (Mandelbrot, 1983, p. 351). The important thing about this from the geological point of view is that Brownian processes are not stationary. To put it another way, in an infinite layered half-space with a Brownian acoustic impedance, we expect the velocities to wander far from the initial values. The boundary between stationary and non-stationary for processes of the $1/f^{\beta}$ type is around $\beta = 1$. Therefore, for stationary geology, the power spectral density of the impedance function must have β less than or equal to unity. On the other hand, as β approaches zero, we approach the case of uncorrelated noise. This is not geologically reasonable. For any small depth step, we expect the geology to be correlated; geological strata have finite thicknesses. In a sedimentary basin, rock types tend to be repetitive: limestones, sandstones and shales, all with similar acoustic impedances. (This is not the case in the crystalline basement, but then you are out of the basin.) Therefore, we would not be surprised to find cases where β was approximately unity.

With this in mind, fig. 1 is striking. A reflection sequence was calculated from a velocity log from well Amoco IOE A-1 Puffin B-90 on the Grand Banks of Newfoundland (44° 39' 09" N, 53° 42' 28" W). Its power density is roughly proportional to spatial frequency (in travel time co-ordinates). Similar power spectra were obtained by Hosken (1980)). The combination of these observations with at least some theoretical justification leads naturally to a consideration of the effects of an 1/f geology upon the deconvolution of seismic signals.

2. A GENERALIZATION OF THE PREDICTION ERROR FILTER

The oldest and most widely used deconvolution technique is predictive decomposition (Robinson, 1957). Let us briefly review this method so as to establish the notation. We regard a set of seismic observations $\{x_k\}$ as a result of a convolution of wavelet $\{w_k\}$ with a reflection sequence $\{r_k\}$ representing the geology. We seek the prediction error filter (PEF) $\{a_k\}$ which, when applied to the observations, produces an output $\{e_k\}$ which we desire to be the same as $\{r_k\}$.

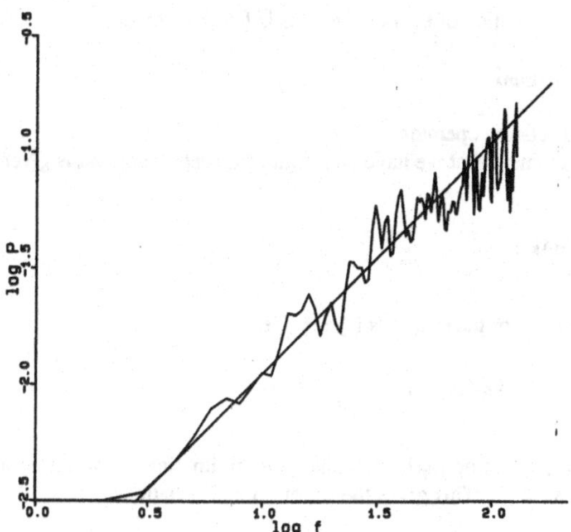

Fig. 1: Power spectrum of reflection sequence calculated from a velocity log of the Puffin well. Fourier transform of the Hanning-windowed (N=256) auto-correlation function. Power is proportional to frequency along the straight line.

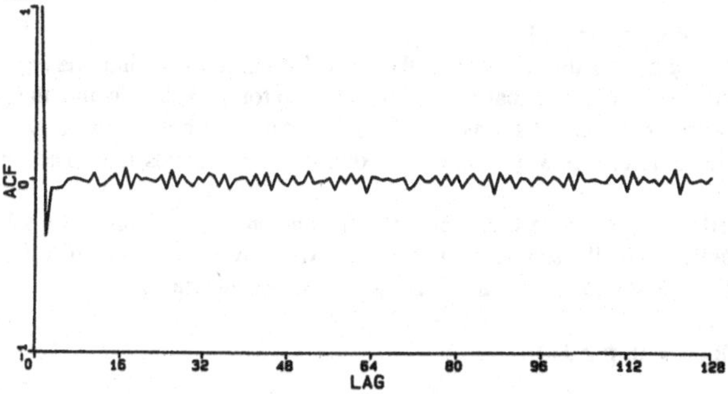

Fig. 2: Part of the auto-correlation function of the Puffin well. The negative trough reaches -.31 of the initial peak.

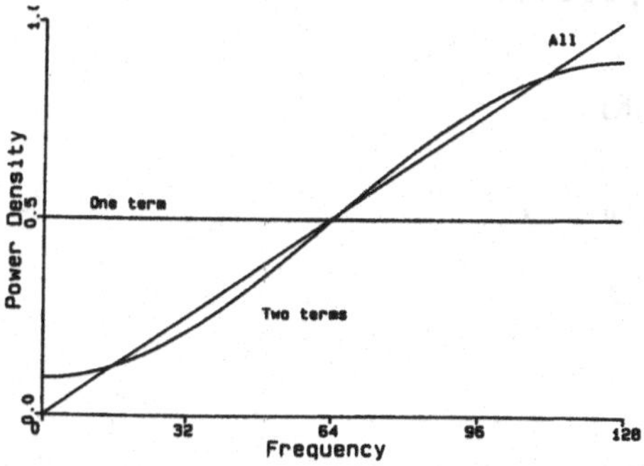

Fig. 3: Power spectra obtained by Fourier transforming the auto-correlation function of an f power spectrum truncated at one and two terms compared with the exact spectrum.

The auto-correlation function of the observations is $\{R_m\}$ given by

$$R_m = E[x_k x_{k-m}] \tag{7}$$

where E denotes the expectation operator.

Assuming for the moment that we have an infinite data set, the output is given by

$$e_k = \sum_n a_n x_{k-n} \tag{8}$$

The auto-correlation Y_j of the output is $E[e_k e_{k-j}]$ so

$$Y_j = \sum_m a_m \sum_n a_n R_{m+j-n} \tag{9}$$

The prediction error operator $\{a_n\}$ is obtained by minimizing Y_o with respect to each parameter a_n, $n \geq 1$. (We normalize $a_o = 1$). This gives the set of normal equations

$$\sum_n a_n R_{p-n} = 0,\ p \geq 1 \tag{10}$$

which are solved to yield the filter $\{a_n\}$.

We see immediately on substituting each of the normal equations back into the expression for the Y_j's that the prediction error filter requires that all the Y_j's are zero for $j \geq 1$, that is, that the power spectrum of the reflection sequence is white. As was noted above, this may not be the case. Since predictive decomposition is widely used, it is worthwhile to modify it as little as possible to accommodate non-zero Y_j's.

Suppose the reflection sequence has a number of non-zero Y_j's,say for $0 \leq j \leq J$. (The reason for proposing this formulation will become clear below.) We may still minimize Y_o with respect to the parameters a_n, $n \geq J+1$. Doing so leads to a reduced set of normal equations

$$\sum_n a_n R_{p-n} = 0,\ p \geq J+1 \tag{11}$$

which only implies that the Y_j's with $j > J$ are zero. These equations are identical with those of the PEF. The normal equations with $p \leq J$ can be obtained directly from the equations for the Y_j's.

From (9)

$$Y_J = a_o \sum_n a_n R_{J-n} \tag{12}$$

so, since by our normalization $a_o = 1$

$$\sum_n a_n R_{J-n} = Y_J. \tag{13}$$

Likewise

$$Y_{J-1} = a_o \sum_n a_n R_{J-1-n} + a_1 \sum_n a_n R_{J-n} \tag{14}$$

so

$$\sum_n a_n R_{J-1-n} = Y_{J-1} - a_1 Y_J \tag{15}$$

while

$$\sum_n a_n R_{J-2-n} = Y_{J-2} - a_1 Y_{J-1} - (a_2 - a_1^2) Y_J \tag{16}$$

and so on. We thus recover another full set of equations corresponding to the normal equations for the PEF.

3. A FILTER FOR l/f GEOLOGY

If the power spectrum P(f) of the reflection coefficients is proportional to frequency up to a frequency f_o, that is, if

$$P(f) = P_o \frac{f}{f_o} \, , \; f < f_o$$

$$= 0 \, , \; f > f_o \tag{17}$$

then the auto-correlation Y(t) of this reflection sequence is given by

$$Y(t) = \frac{P_o}{2\pi f_o t} \{\cos(2\pi f_o t) + 2\pi f_o t \sin(2\pi f_o t) - 1\} \tag{18}$$

By L'Hopital's rule

$$Y(0) = P_o f_o \tag{19}$$

In the discretized case f_o is the Nyquist frequency $f_o = 1/2\,\Delta t$, where Δt is the sampling interval of the reflection sequence. The sine term is zero for non-zero lags and the auto-correlation function is determined by the remaining terms so that

$$Y_j = -4 \frac{Y_o}{j^2 \pi^2} \, , \; j \text{ odd}$$

$$= 0, \; j \text{ even} \tag{20}$$

The $1/j^2$ means that the auto-correlation dies off rapidly with lag. The first few values are 1, -.405, 0, -.045, 0, -.016. It has been known for some time that negative values at small lags are characteristic of the auto-correlation function of reflection sequences generated from well logs (O'Doherty and Anstey, 1971). Figure 2 shows the auto-correlation function corresponding to the power spectrum in fig. 1, which are in reasonable agreement with the exact case.

The form of the auto-correlation function for l/f geology suggests an interesting approach to the problem. Figure 3 shows the power spectra corresponding to truncation of the auto-correlation at various values of j. The standard PEF results from truncating at the first term, represented in the figure by the horizontal line. This is one explanation of why the PEF is so successful; regardless of the true form of the power spectrum, a straight line through the average will be good enough to begin with. We see that the fit to our theoretical spectrum is greatly improved by the addition of one extra term and hardly at all by additional terms. We will therefore treat the case where the only non-zero values of the auto-correlation function are Y_o and Y_1, with $Y_1 = -.405\, Y_o$. It may also be noted that this approximation is not overly sensitive to the exact value of ß. Figure 4 shows the variation of the early terms in the auto-correlation function with ß. The approximation will clearly hold good for quite a wide range of slopes.

From (11), (13) and (15) the normal equations become

$$\sum_n a_n R_n = Y_o - a_1 Y_1 \,,$$
$$\sum_n a_n R_{1-n} = Y_1 \,, \tag{21}$$
and
$$\sum_n a_n R_{p-n} = 0 \;,\; p \geq 2$$

The first of these equations provides the means of establishing the scale of the Y's (since we know the ratio between Y_o and Y_1 but not their size). The presence of a filter coefficient on the right-hand side of the equation can be dealt with by an iterative method. We will generally be interested in obtaining a filter of finite length, say of length L + 1, so we can write the remaining equations in matrix form as

$$\begin{bmatrix} R_o & R_1 & \cdots & R_{L-1} \\ R_1 & R_o & \cdots & R_{L-2} \\ \cdot & \cdot & \cdot & \cdot \\ \cdot & \cdot & \cdot & \cdot \\ R_{L-1} & R_{L-2} & \cdots & R_o \end{bmatrix} \begin{bmatrix} a_1 \\ a_2 \\ \cdot \\ \cdot \\ a_L \end{bmatrix} = \begin{bmatrix} Y_1 - R_o \\ -R_1 \\ \cdot \\ \cdot \\ -R_{L-1} \end{bmatrix} \tag{22}$$

The matrix is of Toeplitz form which of course offers advantages in the solution of the system of equations. Our technique has been to assume initially that Y_1 is zero, calculate the filter coefficients using the FORTRAN subroutine EUREKA (Robinson, 1978, pp. 44-45), use these to evaluate Y_o and therefore Y_1 and, finally, to substitute the new value of Y_1 into (22). The solution converges satisfactorily after a small number (typically less than 10) of iterations.

4. EXAMPLE

We generated an artificial seismogram from the reflection sequence calculated for the Puffin well. Figure 5 shows the reflection sequence and the seismogram. We then deconvolved the seismogram with both the PEF and the new filter, as shown in fig. 6. A filter length of 3 proved to be optimal for both filters. The error between the recovered reflection sequence and the true one is shown for both filters in fig. 7. The new filter is clearly superior. The root mean square (RMS) of the error divided by the RMS of the true reflection sequence, is 0.20 with the prediction error filter and 0.09 with the new filter.

5. CONCLUSION

Do reflection sequences world-wide have a typical power spectrum? If so, what is it? The few examples given clearly cannot settle the matter. We think that an argument based on the overall stationarity of geology is sound and that therefore, in the absence of evidence to the contrary, reflection sequences are better modeled as having a power spectrum proportional to frequency rather than as being white. The well logs, of course, provide only a limited band-width of evidence, but on the other hand, the band is the region of seismic interest.

The filter presented here is only one of the ways of introducing frequency behaviour into the deconvolution problem. There are others, such as spectral factorization, but the filter has the advantage of being only a slight modification of a well-developed technique. Further, it seems to work well and is computationally fast.

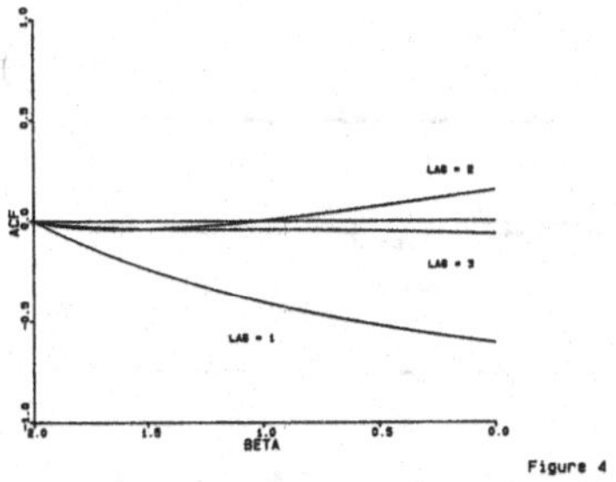

Fig. 4: Variation with ß of the lag = 1, 2, 3 terms of the auto-correlation function of an f^{β} process.

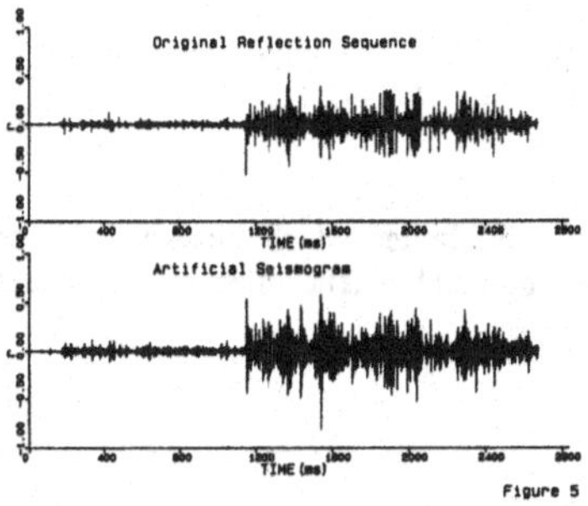

Fig. 5: (Top) Reflection sequence from the Puffin Well. First 160 ms are in the casing of the well. (Bottom) Artificial seismogram.

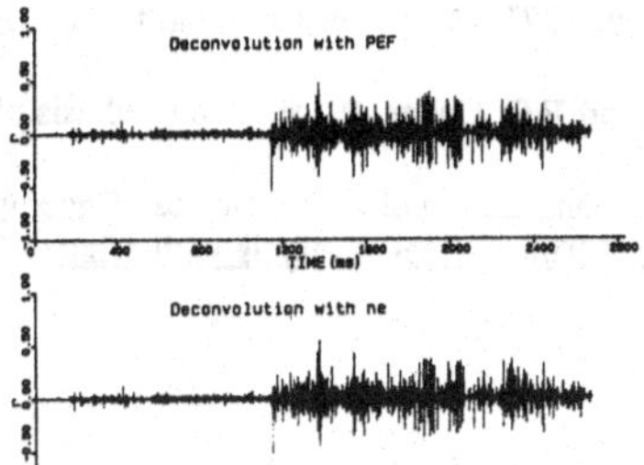

Fig. 6: (Top) Reflection sequence recovered from the seismogram by deconvolution with a prediction error filter (PEF). (Bottom) Reflection sequence recovered with the new filter.

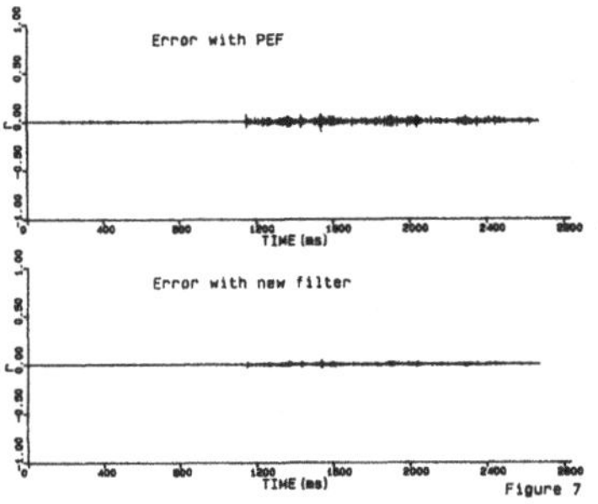

Fig. 7: (Top) Error in the reflection sequence for the prediction error filter. The RMS error is 20%. (Bottom) Error in the reflection sequence for the new filter. The RMS error is 9%.

6. ACKNOWLEDGEMENTS

The authors wish to thank Imperial Oil Limited which supported this project through its University Research Program and which provided the well log. The reflection coefficients were derived from the log by Stephane Labonté.

7. REFERENCES

Hosken, J. W. J., 1980: A stochastic model of seismic reflections: 50th Annual Meeting Soc. Explor. Geophys., Houston. (Abstract G-69, Geophysics, 46, 419.)

Mandelbrot, B. B., 1983: The Fractal Geometry of Nature, W.H. Freeman & Co.

O'Doherty, R. F., and N. A. Anstey, 1971: Reflections on amplitudes, Geophysical Prospecting, 19, 430-458.

Peterson, R. A., W. R. Fillippone, and F.B. Coker, 1955: The synthesis of seismograms from well log data, Geophysics, 20, 516-538.

Robinson, E. A., 1957: Predictive decomposition of seismic traces, Geophysics, 22, 767-778.

Robinson, E. A., 1978: Multichannel Time Series Analysis with Digital Computer Programs. Revised edition, Holden-Day.

THE STOCHASTIC COHERENCE AND THE DYNAMICS OF GLOBAL CLIMATE MODELS AND DATA

R. Viswanathan[1], C. Weber[2], P. Gibart[3]
EERM/CRMD, Météorologie Nationale,
2 Ave. Rapp,
Paris 75007, France.

ABSTRACT. We investigate the scaling behaviour of climatological fields. On the one hand, we empirically discuss the dimensionality of the climate attractor with the help of rather long (N=22,000) time series of temperatures in France. On the other hand, we analyze the intermittent behaviour of the fluctuations of the temperature field. We compare them with those produced by a Global Climate Model, which we show to be much more smooth.

1. INTRODUCTION

Geophysical systems such as the atmosphere require a large number of degrees of freedom ($\approx 10^{27}$) for their description, which is far beyond existing computational capacities, since the largest core memory is only (!) of the order of 2^{29}. However, the extreme variability of climatological fields over a wide range of scales, involves only a very small fraction of the available degrees of freedom, leading to the phenomena of intermittency, i.e. sudden jumps from quiescence to activity, order to chaos. This property may lead to the extreme point of view that this variability is concentrated on a low dimensional attractor, i.e. the knowledge of a very limited number (i.e. of the same order of the dimension of the attractor) of parameters would permit one to obtain a fairly accurate description of the evolution of chaotic atmospheric phenomena. In particular, Nicolis and Nicolis (1984) reported the empirical findings of a low value (3.1) obtained with several hundred points, thanks to a simple technique for estimating the dimension of such attractors (Grassberger and Procaccia, 1983). However the small number of independent points used (just a bit more than one hundred) brings the relevance of the result into question (Grassberger, 1986; Essex, this volume). This is the reason why we study the climate attractor on much more important sets of points. Our inconclusive result leads us to explore the alternative that climate variability and especially intermittency could result rather from an infinite dimensional attractor, although only a small fraction of the available degrees of freedom are effectively exited. We explore this alternative, using the same data set but looking for the extreme fluctuations which might be then characterized by hyperbolic tails of the probability distributions, and consequently divergence of the higher order statistical moments.

---

[1]Now at MFIL, London.
[2]Now at SNCF, Rennes, France.
[3]Now at CCF, Paris.

D. Schertzer and S. Lovejoy (eds.), Non-Linear Variability in Geophysics, 269–278.

2. ANALYSIS OF THE DIMENSION OF THE CLIMATE ATTRACTORS

2.1. Method

To calculate the dimension of the climate strange attractor, we made use of the three hourly temperature series spanning the 30 year period 1950-1980 at the French climatological station Marignane. With these data, we implemented the Grassberger-Procaccia (1983) algorithm of lagged coordinates to build up phase spaces of higher and higher dimension. The method uses a (discretised) signal X(t), with a lag τ to create "embeddings" of the signal in phase spaces with higher and higher dimensions d (d is an integer). The sequence X(t), X(t-τ), X(t-2τ), ... X(t-dτ), is used to model the successive derivatives of the signal. The resulting plot can be regarded as a Poincaré section of the full phase portrait in a d+1 dimensional phase space. For each embedding dimension d, we estimated the fractal dimension of the set of points D(d) using the correlation dimension as an estimate of the fractal dimension. The latter is obtained by determining the average number of point pairs within circles of increasing radii from points on the set of interest. When d is sufficiently large, D(d) approaches a constant equal to the (finite) dimension D_∞ of the strange attractor. As already pointed out, many authors have interpreted the asymptotic value as being of the order of the number of independent variables needed to describe the system.

We worked mainly with 8,400-point data sets, although occasionally 22,000-point sets were used to improve our confidence in our slope estimates (D). Unfortunately, the computer time made systematic use of the larger data set prohibitive. We studied the influence of the lag τ which we varied from 3 hours to 3 days, first while gradually increasing the embedding dimension (d), and then, after obtaining globally homogeneous results, we considered a case with d>>D. We also sought to observe the evolution of D with respect to d for small values of the latter.

2.2. Results of the dimension calculations

This study seems to indicate convergence of the algorithm to a value $D \approx 9$ (for τ=2 days), a value which is fairly close to that obtained by Nerenberg et al. (this volume) who obtained $D \approx 6$ for daily temperatures over 37 years. We note that for the various values of τ used, the algorithm tends to a limit, but with different values of D which are not compatible. In fact, there is no way that we know of to estimate the optimal value of τ. Recall that the lagging operation is supposed to simulate the derivative of the series, and should be of no fundamental importance in a scaling regime.

The limit D_∞ (=D(d) with $d \rightarrow \infty$) is attained for fairly small values of d ($\approx$25). To study the robustness of the pair of values (9, 25), we recalculated everything with the longer (22,000-)point series (the longest of any empirical series used for such a purpose of which we are aware). The result was in conformity with the value found in the 8,400-point series (which in turn agreed with 16,400-point series). Another estimate of robustness was performed with d=40. Here also, there was a rather good agreement with the previous results, independent of the value τ.

Finally, we sought to evaluate D(d) for small d. To do this, we plotted log D against log d (fig. 1). Note that at least for d <20, we obtain $D(d) \approx d^{0.6}$, after which, we obtain a plateau corresponding to stable values of D.

2.3. Comparison of Global Climate Model with climate data

One of the objects of the preceding section was to provide a framework in which to quantify the comparison of Global Climate Models (GCM's) with climate data so as to provide a statistical evaluation of the former. The model that we studied was the Climatological version of Emeraude the NWP model at the Météorologie Nationale which furnished us with daily temperature values over nearly 9 simulated years (3,285 days).

To make the comparison, we calculated the dimensions with two sets of equivalent series (each with 3,285 points), one for the model, and one for the data. We found that the model attractor had a smaller dimension than the real attractor, since it remains lower than 3.5. This is partly due to the parametrisation used in the model. In effect, the model cannot represent phenomena spanning wide ranges in scale without parametrising (smoothing and truncation) of the sub-grid structures.

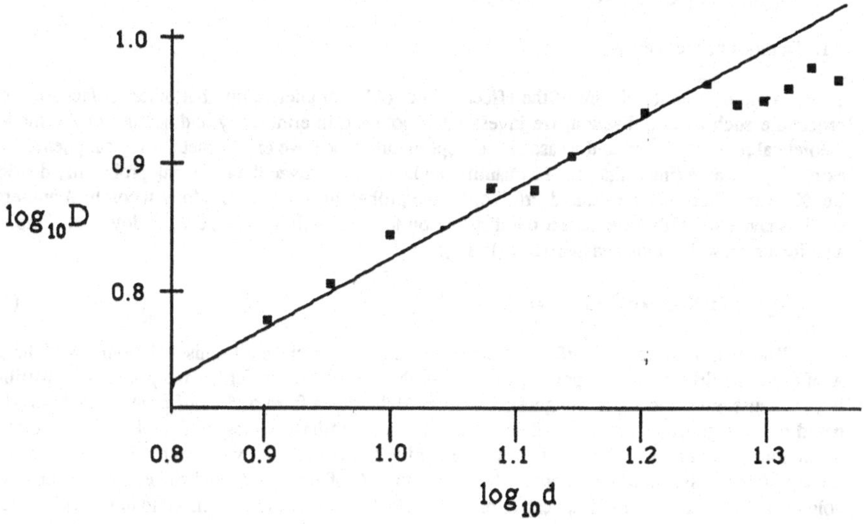

Fig. 1: Showing the log embedding dimension (d) versus the D, showing the line $D=d^{0.6}$ for reference.

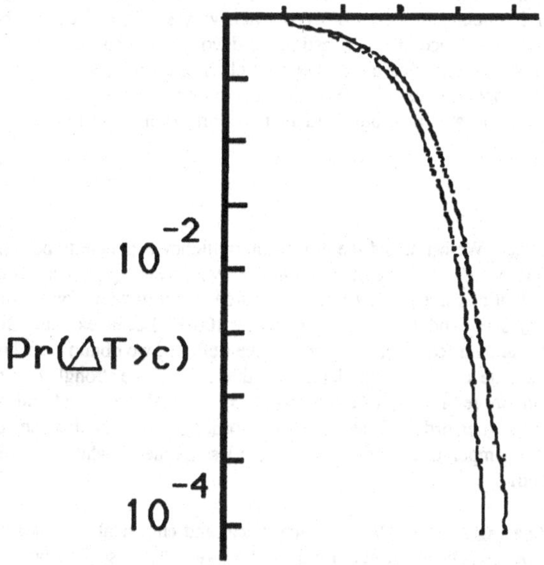

Fig. 2: Shows the model temperature fluctuation distribution (left curve) as well as data (right) curve. The horizontal scale is logarithmic and the same as the vertical. Note the value of $\alpha \approx 5$ (Ladoy et al., 1987) is roughly followed. In the following, we reproduce this analysis and build upon it by studying the moments and the asymmetry for both positive and negative large threshold c on both the model and real data sets.

3. INTERMITTENCY/SCALE INVARIANCE

3.1. Outlines of the theory

To get a more accurate picture of the effects of not only parametrisation but other artifacts of the model procedure such as initialisation, we investigated both the intermittency and scaling of the model. For theoretical reasons, linked to the cascade concept in turbulence, we expect meteorological phenomena to be extremely variable (intermittent). To quantify the latter, we analyze the following probability distributions. Let X be the field of interest, and Pr(X>c) be the probability that X>c. Many theoretical and empirical studies have shown that we expect the distribution to be hyperbolic (see, e.g., Ladoy et al. (this volume) and, for a review, Lovejoy and Schertzer (1986)):

$$Pr(X>c) \approx c^{-\alpha} \text{ for } c \rightarrow \infty \quad (1)$$

This behaviour is very different from that of a gaussian, since all statistical moments of the variable X of order $\geq\alpha$ diverge. More precisely the lower the value of α the further the probability distribution is from a gaussian. Contrary to the gaussians, Lévy stable laws form a class of probability laws which are based on divergence of moments. These variables seem to have widespread applications in many fields (economics, astronomy, turbulence). They are the limit laws defined by the central limit theorem when the finite variance assumption is dropped (c.f. appendix A of this paper and Schertzer and Lovejoy (this volume)). In the theory of multiplicative processes stable laws are closely related to universal subgenerators (Schertzer and Lovejoy, 1987; Schertzer et al., 1988). The extreme (tail) behaviour of stable laws probability distribution function is linked to the divergence of moments. Therefore parameter estimates for stable laws are crucial for application in meteorological fields. Unfortunately statistical estimates are complicated by the infinite variance syndrome (i.e. considering infinite variance as unnatural). Contrary to this syndrome, outliers occur frequently in samples, even when very small, and cannot be rejected as experimental noise. Due to the importance of this question we devote an appendix to the analysis of the two classes of statistical estimators of the Lévy laws. The first class of estimators considered is based on density or distribution function properties (mainly asymptotic behaviour, scaling, ...). The second class is based on the characteristic function and is subdivided in two parts (norm minimization and explicit estimators).

3.2. Results

3.2.1. Evaluation of α_m=α model. We estimated the model intermittency exponent to be $\alpha_m \approx 10$. In fact, as we shall discuss below, the model data, unlike the real data, are very nearly gaussian. This implies that the strongest temperature fluctuations are greatly underestimated in the model; for example, ten year extremes are underestimated by 50%, and 100 year extremes by 100%. Let us examine the criterion of gaussianity for the model and real data ($\alpha_r \approx 5$). Concerning the kurtosis, we obtain respectively k_m=3.2 for the model and k_r=4.7 for the time series. This indicates that the real data are strongly non-gaussian (k=3 for a gaussian variable). Furthermore, a study of the moments shows that the model underestimates the moments, particularly those of higher order. The moment of order 3 indicates that the real data have pronounced asymmetry (drops in temperature are more brutal, but less frequent), whereas the corresponding model results are nearly symmetric.

3.2.2. Evaluation of the scaling exponent H. Various theoretical and empirical evidence shows that, in general, small and large scale structures of the fields are related by a scale invariant exponent. Mathematically, the simplest relationship of this type which is of interest is the following simple scaling:

$$\Delta X(\Delta t/\lambda) \stackrel{d}{=} \Delta X(\Delta t)/\lambda^H \quad (1)$$

where $=^d$ means equality in probability distributions and $\Delta X(\Delta t) = X(t+\Delta t) - X(t)$. This equality means that contracting scales by factor λ decreases the fluctuations by the factor λ^H. Exponents of this type have already been reported in the literature (e.g., Lovejoy and Schertzer, 1986; Ladoy et al., 1987, this volume). We calculated them using fourier energy spectra. If the spectral exponent at high frequencies is β, then $\beta=2H+1$. For the real data, we have $\beta_r=1.8$ for midlatitudes for frequencies >(20 days)$^{-1}$, whereas for the model, at high frequencies we obtain $\beta_m=5.4$ whereas at low frequency, we find $\beta\approx 0$. Once again, we find the extremely strong effect of the sub-grid scales which corresponds to the high frequencies.

4. CONCLUSIONS

From either the dynamical or probabilistic point of view, the temperature series are complex and chaotic. This raises serious problems for deterministic numerical modelling such as in GCM's because the models strongly smooth small (high wavenumber) structures as well as truncate extreme fluctuations (such as meteorological "fronts"). We quantify both of these effects by measuring respectively spectral and probability exponents (β, α) of termperature fluctuations. For the model, we find spectral exponents ≈ 0 for low wavenumbers (corresponding to nearly white noise), and ≈ 5.4 for high wavenumbers (which is a very sharp cut-off at small scales). In comparison the data shows a continuous scaling spectrum with exponent ≈ 1.8. For the probability distributions we find exponents of ≈ 10 and 5 for the model and the data respectively reflecting a sharp truncation of the extremes.

Finally, we used a standard technique for estimating the dimension of strange attractors to study the variation of the dimension of the occupied space (D) as the number of degrees of freedom (i.e. the embedding dimension, d) is increased. For $d\leq 20$, we obtained the interesting relation $D\approx d^{0.6}$, i.e. the space is never uniformly filled - even for very small d. This would be quite difficult to explain if the dynamics was governed by a small number of coupled ordinary differential equations as is sometimes supposed. For d>20, D(d) flattens off as expected since there is not nearly enough data for meaningful estimates.

APPENDIX A: METHODS FOR ESTIMATING LEVY DISTRIBUTION PARAMETERS

A.1. Stable laws: general properties

Stable laws are defined by their characteristic functions of the form:

$$L(t) = \exp[\, i\partial t - \gamma|t|^{\alpha}\{1 + i\beta\, |t|/t \tan(\pi\alpha/2)\}] \quad \text{(A1)}$$

where α is the Lévy exponent[1] ($2\geq \alpha > 0$, when $\alpha = 2$ the law is gaussian), β is the skewness factor ($-1 \leq \beta \leq 1$, $\beta = 0$ for symmetric distributions), γ is the scale factor ($\gamma > 0$, $\gamma = \sigma^2/2$ for a gaussian law), ∂ is the location parameter (when the mean exists (i.e $\alpha > 1$), ∂ is the mean). The gaussian law is the only stable distribution with finite variance, for $\alpha < 2$ the variance is infinite (and moments of order $p \geq \alpha$ diverge) since the Lévy stable laws have hyperbolic tails of order α. Indeed, let Pr{A} denote the probability of the event A then:

$$\lim_{x\to -\infty} \Pr\{X< x\}.|x|^{\alpha} = p \text{ and } \lim_{x\to +\infty} \Pr\{X>x\}.x^{\alpha} = q$$
$$\beta = (q-p)/(q+p). \quad \text{(A2)}$$

These relations show that if we represent the probability distribution of the variable X being over a threshold x versus x in a log-log plot its tail becomes a line of slope $-\alpha$. If $\alpha < 1$ and $|\beta| = 1$, the probability density is concentrated on either $]0, \infty[$ or $]-\infty, 0[$, and its Laplace transform is rather easy to obtain. The Lévy variables satisfy the following generalized central limit theorem:

---

[1]The case $\alpha=1$ is slightly different and we will not treat it for the sake of simplicity.

If X1, ...,Xn is a sequence of n independent identically distributed (i.i.d.) stable variables then $n^{-1/\alpha} \sum Xi$ has the same distribution as X1.

(A3)

A.2. The Fama-Roll estimators

In this section, estimation techniques developed by Fama and Roll (1972) will be presented. They limited their estimation to the symmetric case ß=0 with characteristic exponent $\alpha \geq 1$.

A.2.1. Estimation of the mean. When $\alpha > 1$, ∂ is the mean of the sample. It can be readily estimated as the empirical mean. Since the tail of the distribution has great influence in the Lévy case there will always be sampling problems to picture correctly these tails. Fama and Roll introduced the g-truncated mean which is the average of the middle g % of the ordered observations in the sample, the rest being discarded. Their results indicate that for values of α close to 1 the .25 truncated mean is the best estimator and as α increases up to 2 the .5 and .75 estimators are more efficient. Finally, the sample mean is the best estimator when the law is gaussian.

A.2.2. Estimation of the scale parameter. The authors justify their approach considering two different points. On the one hand, the sample fractiles are asymptotically unbiaised and consistent estimates of population fractiles. On the other hand, the interquartile ranges are approximately constant for different values of α. They selected the interquartile range $(X_{.72} - X_{.28})$ since the theoretical values range from 0.824 to 0.830 for $1 \leq \alpha \leq 2$ for symmetric laws:

$$c = \gamma^{1/\alpha} = 1/0.827 \quad (X_{.72} - X_{.28})/2. \tag{A4}$$

This estimator has an asymptotic bias of less than 0.4%. Since it is a linear combination of order statistics it is asymptotically normally distributed with normally distributed variance:

$$\sigma^2 (c) = (0.28)(0.72-0.28) / (N\, f(\alpha,0.72)^2\, 1.654^2) \tag{A5}$$

N is the sample size, f the density of probability at 0.72 (which in the symmetric case is equal to $f(\alpha,0.28)$).

A.2.3. Estimator of the characteristic exponent α. This estimator is also based on order statistics:

$$Z_f = (X_f - X_{1-f}) / 2c \tag{A5}$$

Z_f is an estimator of the f-fractile of the standardized symmetric stable distribution. A table with theoretical values for some different values of α must be then computed. Fama and Roll proved with Monte Carlo simulations that high values of f work best (.93-.97). They also tried the mean of the results obtained for different values of f but with poor success. The stability property provides a way to define an exponent estimator. Since Lévy laws are stable under addition of random variables the sum of N i.i.d. variables is a Lévy stable law with the same exponent, the same skewness factor and a different scale $\gamma_n = n\gamma$. It can be easily proven that this property, written for interquartile ranges reads, $R_{f,i}$ being the f interquartile of the sum of i random variables:

$$R_{f,n} = n^{1/\alpha} R_{f,1}, \quad \alpha = \log n / \log R_{f,n} - \log R_{f,1} \tag{A6}$$

Experimentally, f = 0.75, 0.8, 0.95 and n = 2, 4, 7, 11 have been tried and yield good results.

A.3. An estimation of ß

The estimator is based on the following theorem. X is a stable Lévy random variable with $\partial = 0$ and α is known, then:

$$\beta = \tan(\pi\alpha(1/2 - \rho)) / \tan(\pi\alpha/2), \text{ where } \rho = \Pr\{X>0\} \quad (A7)$$

It is easy to see that ρ does not depend on γ. The proof consists in evaluating ρ in its integral form and using the Parseval relation for the Fourier transform, $Tf(t) = Ff(2\pi t)$:

$$\int_{-\infty}^{+\infty} g(y)\, f(y)\, dy = \int_{-\infty}^{+\infty} Ff(x)\, Fg(x) dx; \quad Ff(x) = \frac{1}{\sqrt{2\pi}} \int_{-\infty}^{+\infty} \exp(ixy)\, f(y)\, dy \quad (A8)$$

The Fourier transform of the Heavyside function H(x) (= 0 if $x<0$ and = 1 if $x \geq 0$) is in the sense of distributions (pp is the principal part and δ is the Dirac mass in 0):

$$TH(t) = -1/2\pi i \text{ pp } 1/t + 1/2\, \delta \quad (A9)$$

Let f be the density of a stable law and L its characteristic function:

$$\rho = \int_{-\infty}^{+\infty} H(x).f(x)\, dx = \int_{-\infty}^{+\infty} L(t).TH(t)\, dt \quad (A10)$$

After a couple of changes of variables and noting that $\int_{]0,\infty[} \exp(-xu) \sin u\, du/u = \pi/2 - \arctan x$ for $x>0$ the theorem (Eq.A7) can be proved. One of its interesting spin off is that if $\alpha<1$ and $|ß| = 1$ then $\rho = 0$ or $\rho = 1$. But if $\alpha>1$ and $|ß| = 1$ then $\rho = 1/\alpha$ or $\rho = 1-1/\alpha$ .This relation defines an estimator of ß by estimating ρ. It is simple to find a confidence interval for ρ (thus ß) noticing that this problem is similar to the classical Bernouilli scheme of estimation on the random variable signX.

A.4. McCulloch 's estimators

A quite similar approach to Fama and Roll has been developed by McCulloch (1986). The particularity of his method, also based on ordered statistics, is that estimates can be found even in the asymmetric case. To avoid any sampling bias McCulloch suggested an easy interpolation to derive the sample fractile value X_p. Let F be the empirical cumulative distribution function and $\{X_i\}$ be the ordered observations, then

$$X_p = X_i + (p - \text{int}(n.p)) * (X_{i+1} - X_i) \quad (A11)$$

We can proceed to the estimation of α and β, we define:

$$\mu\alpha = (X_{95} - X_{05}) / (X_{75} - X_{25});$$
$$\mu\beta = (X_{95} + X_{05} - 2*X_{50}) / (X_{95} - X_{05}) \quad (A12a)$$

Both $\mu\alpha$ and $\mu ß$ are independent of γ and ∂. As mentioned above:

$$\hat{\mu}\alpha = (\hat{X}_{95} - \hat{X}_{05}) / (\hat{X}_{75} - \hat{X}_{25});$$
$$\hat{\mu}\beta = (\hat{X}_{95} + \hat{X}_{05} - 2*\hat{X}_{50}) / (\hat{X}_{95} - \hat{X}_{05}) \quad (A12b)$$

are consistent estimators of $\mu\alpha$ and $\mu\beta$. Tables giving $\mu\alpha$ and $\mu\alpha$ for different values of α and β can be computed. Since $\mu\alpha$ (resp. $\mu\beta$) is a strictly decreasing function of α for each given value of β (resp. increasing of β), the estimates $\mu\alpha$ and $\mu\beta$ give us a good fix on α and β. The relations:

$$\mu\alpha = f1(\alpha,\beta) \; ; \; \mu\beta = f2(\alpha,\beta) \tag{A13}$$

can be inverted to yield:

$$\alpha = g1(\mu\alpha,\mu\beta) \; ; \; \beta = g2(\mu\alpha,\mu\beta) \tag{A14}$$

New tables can then be derived where α and β are given as functions of $\mu\alpha$ and $\mu\beta$. Thus simply by calculating a couple of sample quantiles, once convenient tables are available, estimators of α and β can be obtained just by reading the tables and interpolating. The results proved to be quite interesting although the tables presented by the author seemed unsatisfactory. Computation of more accurate tables will help improving the results obtained with this method.

The same principles are used to derive γ and ∂. They will only briefly be presented here. A table for $\mu c = (X_{75}-X_{25}) / c$ is computed for different values of α and β. If $\mu c = f3(\alpha,\beta)$ then:

$$\hat{c} = (\hat{X}_{75}-\hat{X}_{25}) / f3(\hat{\alpha},\hat{\beta}) \tag{A14}$$

is a consistent estimator for c. Similarly a table for $\mu\partial = (\partial - X_{50}) / c$ is constructed so that the location parameter ∂ can be estimated as $\hat{c}+f4(\hat{\alpha},\hat{\beta})+\hat{X}_{50}$. McCulloch also computed tables for estimation of the asymptotic covariance matrix of the parameters.

A.6. Norm minimization and parameter estimates

The basic idea in this section is to compare the empirical and the theoretical characteristic functions. Boldface typing shall denote theoretical values. The estimate of the four parameters of a stable law minimize the (adequately chosen) distance between those two functions:

$$d(\mathbf{L},L) = \int_{-\infty}^{+\infty} | \mathbf{L}(t)-L(t) |^2 \exp(-|t|^p) \, dt \tag{A15}$$

This norm d is judicious. The exponential term forces convergence of the integral and it gives more weight to the regions near $t = 0$ which correspond to the tail of the distributions[1]. Since tails are of primary interest this distance seems suitable. For $p = 1, 2$, it leads to simple calculus since the integral can be evaluated using polynomial approximations. For example, if $p = 2$, using the hermitian polynomials we have:

$$\mu(t) = | L(t) - \mathbf{L}(t) |^2 \; ; \quad \int_{-\infty}^{+\infty} \mu(t) \exp(-t^2) \, dt = \Sigma_{1,m} \, w_i \, \mu(u_i) \tag{A16a}$$

u_i are the zeros of the polynomials and w_i the associated weights. The value 20 was chosen for m:

$$Z(a,b,c,d) = \Sigma_{1,20} \, w_i \, \mu_{a,b,c,d}(u_i) \tag{A16b}$$

When the scale and mean parameters are too different from (1, 0), because of the flatness of the surface (A15), slight changes on this surface result in large variations of the estimates. To circumvent this drawback, Leicht and Paulson (1972) introduced a standardization procedure with the gradient search for the minimum, if $\{X_{i0}\}$ are the observations, successively standardized arrays will be:

[1]This argument has a drawback because $t = 0$ corresponds to the tails then we need a large sample to have enough extremes for a correct representation of the tails.

$$X_{i1} = (X_{i0} - \partial_1) / \gamma_1^{1/\alpha_1} ;$$
$$X_{ik} = (X_{i0} - (\partial_1 + \Sigma_{2,k}(\partial_j \Pi_{1,j-1} \gamma_i^{1/\alpha_i})) / \Pi_{1,k} \gamma_i^{1/\alpha_i} \quad \text{(A17)}$$

Since the parameters of the stable laws belong to a specific domain it would have been appealing to use a constrained gradient method. Nevertheless it is more satisfactory to use an unconstrained method and to verify *a posteriori* that the estimates fall within the prescribed range. The gradient was calculated by means of small variations of the parameters in (A16b) and the Armijo method was used to derive the next point in the gradient direction. This procedure ceases when two successive points are close enough relative to *a priori* criteria. This method assumes to converge rapidly and correctly that the starting point is good otherwise the minimum attained by this technique might be local only. The efficiency highly depends on the initial values of the parameters deduced by other estimators.

A.7. Explicit estimators for stable laws

Let $\{X^i\}$ be a sample of n independent realisations of the Lévy stable law with parameters α, β, γ, ∂ and:

$$\hat{}L(t) = 1/n \, \Sigma \, e^{iX_j t} \quad \text{(A18)}$$

Since four parameters are unkown it is natural to estimate the empirical characteristic function for two values of t which yields four relations. t^1 (resp. t^2) is a positive number:

$$| L(t) | = \exp(-\gamma t^{\alpha}) \;\; ; \;\; \log[\log| L(t) |] = \alpha \log t + \log \gamma \quad \text{(A19)}$$

hence:

$$\alpha = \log(t^1/t^2)^{-1} (\log(\log|L(t^1)|) - \log(\log|L(t^2)|))$$
$$\gamma = \exp [\log(t^1/t^2)^{-1} [\log(t^1) \log(\log|L(t^2)|) - \log(t^2) \log(\log|L(t^1)|]] \quad \text{(A20)}$$

By applying these formulas to $\hat{}L(t^1)$, $\hat{}L(t^2)$ and $\hat{}L(-t^1)$, $\hat{}L(-t^2)$ we have explicit unbiased estimators $\hat{}\alpha$ and $\hat{}\gamma$. The natural idea to estimate β and ∂ is to evaluate the logarithm of the phase of the characteristic function. But removing the indetermination of $2k\pi$ seems untractable. To overcome this difficulty we will use derivatives of the characteristic function. Suppose $t>0$, then:

$$dL/dt = [- \gamma\alpha \, t^{\alpha-1} \, (1 + i\beta \tan(\pi\alpha/2)) + i\partial t] \, L(t)$$
$$G(t) = (t \, L(t))^{-1} \, dL(t)/dt \, ; \;\; \mathrm{Im} \, G(t) = - \beta\gamma\alpha \tan(\pi\alpha/2) \, t^{\alpha-2} + \partial \quad \text{(A20)}$$

Evaluating G at t^1 and t^2 yields estimators of β and ∂. The relations (A20) define linear regression type estimators of all four parameters when different values of t are chosen. These estimators are extensions of those already found previously. For example:

$$\hat{}\alpha = \mathrm{Cov} (\log\text{-}\log|\hat{}L(t^i)|, \log t^i) / \mathrm{Var} (\log t^i) \quad \text{(A21)}$$

A.8. Confidence interval for estimators

$n\hat{}L(t)$ is a sum of bounded (thus of finite variance) i.i.d. random variables ($E(n\hat{}L(t)) = nL(t)$). The estimators ($\hat{}\alpha$, $\hat{}\beta$, $\hat{}\gamma$, $\hat{}\partial$) are functions of the empirical characteristic function evaluated at t^1, t^2 and their opposite values. Since:

$$E(\hat{}L(u)\hat{}L(v)) = 1/n^2 \, \Sigma\Sigma \exp(iX^k u)\exp(iX^l v) = 1/n \, [L(u+v) - L(u)L(v)] \quad \text{(A22)}$$

we can compute the covariance matrix C of the vector estimator $\wedge Z = (\wedge L(t^1), \wedge L(t^2), \wedge L(-t^1), \wedge L(-t^2))$. $¥(A)$ denotes the law of the random variable A. The central limit theorem implies that $¥(n(\wedge Z(t)-Z(t))/\sqrt{n})$ converges to the 4-variate normal distribution with covariance matrix nC. We have $\wedge\alpha = m(\wedge Z)$ with $\alpha = m(Z)$ (the estimator is unbiased). So by a classical result in statistics $¥(\sqrt{n}(\wedge\alpha-\alpha))$ converges to the normal distribution $N(0,\varpi)$:

$$\varpi^2 = \sum\sum n\, c^{ij}\, \partial m/\partial z^i\, \partial m/\partial z^j \tag{A23}$$

Similar results can be derived for the other estimators. The search of optimal values of t^1 and t^2 is difficult. Nevertheless one should note that these explicit estimators can be compared to the Fischer information matrix. When n points t^i are used, C becomes a 2n x 2n matrix, estimators are functions of a 2n vector $\wedge Z$ but the formal expression of confidence intervals is the same.

A.8. A brief review of experimental results

A.8.1. Estimation of α. The log-log plot method of estimating requires a huge amount of data (at least several thousands) and is of dubious practical interest. The Fama-Roll type techniques are costly too in terms of data. α is estimated decently (less than 10% error) with 2,000 points when it is in the range 1.4-2. The norm minimisation approach works extremely well for the same range of α with less a thousand points provided the algorithm is initialized with good starting values. This remark holds for all parameters in this method. Results of explicit estimators depend on the choice of the t^i but for reasonable values of t^i (in the sense of Shannon or Nyquist theorems) it is the method which requires the lower amount of data (less than 500 in the range studied).

A.8.2. Estimation of β. It is the most difficult parameter to estimate. When α is near 2, β has almost no incidence on the probability distribution. The estimation based on the relation we proved is fairly accurate. For α around 1.5, β is estimated with less than 10% error with 500 points. Similar results are obtained with minimisation techniques. Results with explicit estimators are highly sensitive on the choice of t^i.

A.8.3. Estimation of γ. All estimators provide good results for the scale parameter but the last 2 need a lower amount of data.

A.8.4. Estimation of ∂. For $\alpha > 1$, the estimation of the mean is consistent with all methods. But when $\alpha < 1$, since the mean is infinite only the gradient algorithm finds sensible values of ∂ while the performance of explicit estimators rely heavily on a good choice of t^i.

REFERENCES

Essex, C., 1990: Correlation dimension and data sample size. (this volume).
Fama, E., R. Roll, 1972: Parameter estimates for symmetric stable distributions. JASA, 66, 331-338.
Grassberger, P., I. Procaccia, 1983: On the characterization of strange attractor. Phys. Rev. Lett., 50, 346
Grassberger, P., 1983: Are there really climate attractors? Nature, 322, 609.
Ladoy, P., D. Schertzer, S. Lovejoy, 1986: Une étude d'invariance locale-régionale des températures. La Météorologie, 7, 23-34.
Ladoy, P., S. Lovejoy, D. Schertzer, 1990: Extreme variability of climatological data, scaling and intermittency. (this volume).
Leitch, R., M. Paulson, 1975: Estimation of stable law parameters. JASA, 70, 670-677.
Lovejoy, S., D. Schertzer, 1986a: Scale invariance, symmetries fractals and stochastic simulation of atmospheric phenomena. Bull. AMS, 67, 21-32.
McCulloch, J., 1986: Simple consistent estimators of symetric stable distribution parameters. Working paper. Ohio State University.
Nicolis, C., G. Nicolis, 1984: Is there a climate attractor? Nature, 311, 529.
Schertzer, D., S. Lovejoy, 1987b: Physically based rain and cloud modeling by anisotropic, multiplicative turbulent cascades. J. Geophys. Res., 92, 9693-9714.

5 - REMOTE SENSING OF THE CLOUDY ATMOSPHERE: THEORY AND OBSERVATIONS

LANDSAT OBSERVATIONS OF FRACTAL CLOUD STRUCTURE

Robert F. Cahalan
Laboratory for Atmospheres,
Goddard Space Flight Center,
Greenbelt, MD 20771

ABSTRACT. LANDSAT Thematic Mapper (TM) data, with 30m spatial resolution, has been employed to study the spatial structure of boundary-layer and ITCZ clouds. Statistical parameters characterizing the cloud size, shape, and spacing have been determined. The probability distributions of cloud areas and cloud perimeters are found to be approximately power-law, with the power related to the fractal dimension. Stratocumulus clouds conform to a single power more closely than do fair weather cumulus, which exhibit a clear change in the fractal dimension at a diameter of about 0.5 km. The fractal dimension also changes with the reflectivity threshold. As the threshold is raised from cloud base to cloud top, the perimeter fractal dimension increases, perhaps indicative of the increased turbulence at cloud top. The distribution of nearest-neighbor spacings between cloud centers is shown to deviate from that expected for independently distributed centers due to cloud clustering. The cloud spacing distribution is better approximated by a Levy flight. Results of this study are being used to develop statistical simulations of broken cloud fields. Properties of the associated radiation fields will then be computed by Monte Carlo methods.

1. INTRODUCTION

None of the three major aspects of clouds – their optical properties, their spatial and temporal extent, and their precipitation – are well understood. Indeed, the treatment of cloudiness remains one of the greatest sources of uncertainty in models of the atmosphere and its climate. Difficulties arise for two reasons: First, cloudiness is produced by a poorly understood interaction between small-scale turbulent convective processes and large-scale dynamics and thermodynamics. Second, cloudiness produces large changes in absorbed solar and outgoing infrared radiation, resulting in changes in net radiation. Both problems affect estimates of climate sensitivity, since one must know how cloudiness changes under a given perturbation, as well as whether the change produces a positive or negative radiative feedback.

This paper focuses on cloud spatial structure – how it varies with a change in scale or resolution for various cloud types. A number of theoretical studies have shown the sensitivity of cloud radiative properties to their spatial structure. (See for example Aida, 1977; Davies, 1978; Harshvardhan and Weinman, 1982; McKee and Cox, 1974, 1976; McKee and Klehr, 1978; Reynolds et al., 1978). To produce a tractable problem, such studies typically restrict themselves to a lattice of bar clouds or cubic clouds, having a small number of adjustable length scales. By contrast, actual cloud fields are typically observed to be extremely irregular and/or fragmented at all scales.

A useful notion for discussing such cloud structure is that of a "scaling fractal" (Mandelbrot, 1983 and references therein), a geometrical pattern (other than Euclidean lines, planes and surfaces) with no intrinsic scale, so that no matter how closely you inspect it, it always looks the same, at least in a statistical sense. If one measures the perimeter of a scaling fractal at successively higher resolution, the result increases at a rate determined by the "fractal dimension".

S. Lovejoy (1982) has presented empirical evidence that clouds behave as scaling fractals from 1000 km down to 1 km, with a fractal dimension of approximately 4/3. If this were strictly true, than a satellite image of a cloud field, taken over a uniform dark ocean background with no evident geographical features,

D. Schertzer and S. Lovejoy (eds.), Non-Linear Variability in Geophysics, 281–295.

would be statistically indistinguishable from an image of any small portion of the scene if it had the same number of pixels as the original. It is important to consider the consequences and limitations of this idealization.

A number of questions arise, of which we mention three here. First, scaling must clearly break down at some scale, since microscopic cloud structure differs from the macroscopic structure observed by eye or by satellite. At what scales and in what manner does this breakdown occur, and what physical processes control the scales?

A second question involves the universality of the fractal dimension. Various climatological regimes exhibit cloud types having distinct spatial and temporal characteristics. For example, in the North Pacific Ocean one finds a band of storm track cloudiness extending from the coast of Japan, a region of stratocumulus off the California coast, fair weather cumulus in the subtropics, and a band of deep convective clouds centered at 5N. (See for example Cahalan, 1981, 1984 and Hahn et al., 1982.) It would seem natural to expect such distinct structures to exhibit distinct fractal characteristics. Do different cloud types have different fractal dimensions?

Finally, clouds are three-dimensional objects, and cloud structure often varies considerably from cloud base to cloud top. In the case of fair weather cumulus this vertical variation is evident to the eye. They are flat at the base where parcels reach the lifting condensation level, but have considerable small-scale variations at cloud top. Vertical variations in structure are less obvious for nocturnal stratocumulus, but show up for example in the temperature variance, which has a sharp peak at cloud top, where the radiative loss drives the turbulence (Caughey et al., 1982). How are such vertical variations in cloud structure reflected in fractal dimensions or other scaling parameters?

While this paper does not fully answer these questions, we do report some progress on each of them. The availability of high resolution LANDSAT data has been essential in achieving this progress. As we shall see, the size distributions in typical cumulus cloud fields increase rapidly down to the smallest detectable size. (See also for example Blackmer and Serebreny, 1962; Plank, 1969; Shenk and Salomonson, 1971; Lopez, 1976, 1977a and 1977b; Kunkle et al., 1977 ; Gifford, 1978; and Hozumi et al., 1982). In addition, the nearest-neighbor distributions (e.g. Cahalan, 1986) show that better than 500 meter resolution is needed to "see between" subtropical fair weather cumulus clouds, and thus directly observe the associated sea surface temperature without contamination by sub-resolution clouds. LANDSAT's 30 m resolution thus provides a unique tool for study of the cloudy marine boundary layer.

In the next section we describe the LANDSAT data and summarize our analysis techniques. Section 3 describes results for the size distributions and fractal dimensions, including dependence on cloud type and vertical level. Section 4 gives results for the nearest neighbor spacings and contrasts them with spacings of random points. Finally, section 5 summarizes our results and conclusions.

2. DATA AND ANALYSIS TECHNIQUE

The major source of data for this investigation is the LANDSAT multispectral scanner (MSS) and thematic mapper (TM) instruments. Considerable attention has been devoted to LANDSAT images with minimum cloud cover over land. Meanwhile, cloud studies have relied upon coarser-resolution meteorological satellite data or ground and aircraft observations with limited areal coverage. We find LANDSAT useful as an ultra-high-resolution probe for the study of cloud spatial structure, particularly over oceanic surfaces where the relatively uniform background and high contrast simplify the cloud observations. The resolution approaches that of aircraft, while the field of view is large enough to show meteorological variations.

In fields of marine cumulus one often requires a resolution on the order of 0.1 km in order to resolve a significant fraction of the clouds, and one requires a field of view on the order of 100 km in order to observe many mesoscale clusters. LANDSAT TM data has a resolution of 0.12 km in the thermal emitted band (10.4 to 12.5 microns) and 0.03 in the visible and near IR reflected bands (0.45 to 0.52, 0.52 to 0.59, 0.63 to 0.67, 0.76 to 0.90, 1.55 to 1.75, and 2.08 to 2.35 microns). The field of view is 185 km, somewhat smaller than a GCM gridbox, with about 6100 pixels on a side for the reflected bands, and 1600 for the thermal band. The MSS lacks a thermal band and has somewhat less spatial and radiometric resolution, but has a data base beginning in 1972, ten years before TM. Analysis of the MSS data is reported in Joseph and Cahalan (1987). The present paper will be restricted to TM results.

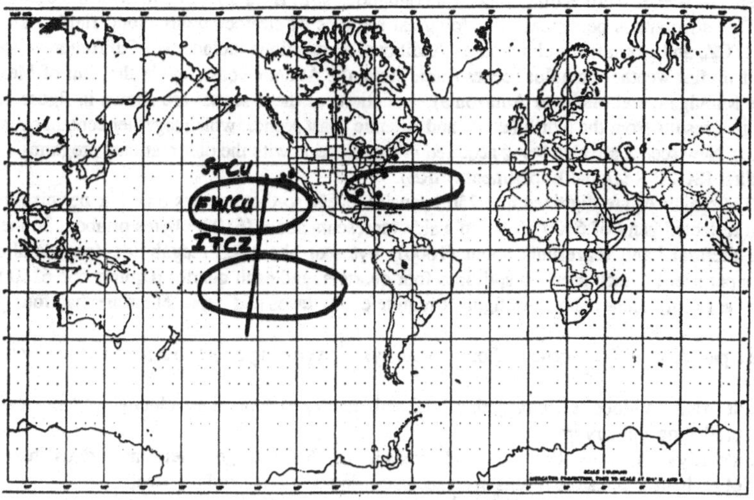

Fig. 1: Location of LANDSAT Thematic Mapper scenes: Gulf Stream off Cape Cod; Gulf Stream off South Carolina; Gulf Stream off Florida Keys; Gulf of Mexico; southern California stratocumulus; central Pacific strip centered on 140W and beginning in California stratus and continuing southward through fair weather cumulus and ITCZ into Southern Hemisphere storm track.

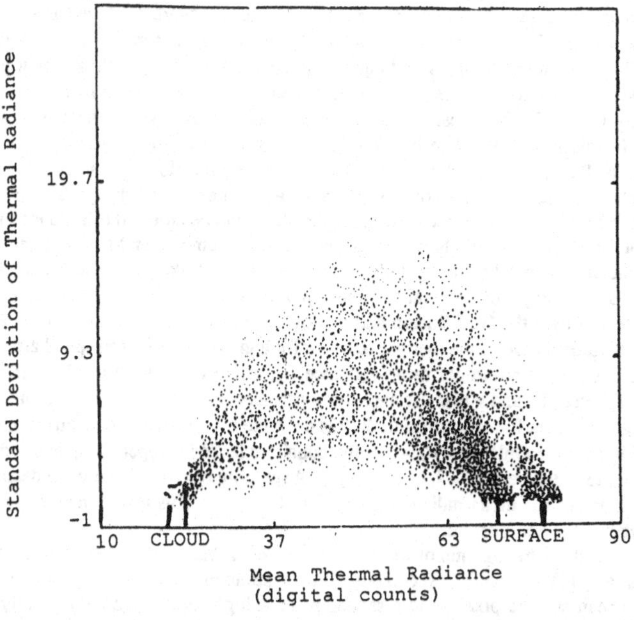

Fig. 2: Spatial coherence plot for two regions in the Cape Cod scene – one over the Gulf Stream and one over the Labrador current. Points with minimum standard deviation fall into two groups : those with minimum mean radiance Ic corresponding to completely cloud-covered pixel arrays, and those with maximum mean radiance Is corresponding to completely cloud-free pixel arrays. The colder cloud top temperatures occur over the Gulf Stream, where the boundary layer is deeper. For each region of single-layer clouds, the cloud fraction is given by (I-Ic)/(Is-Ic), where I is the region's mean radiance.

Eleven TM scenes were acquired along a contiguous strip from 25N to 27S over the eastern Pacific Ocean. This strip spans persistent climatological regimes of stratocumulus cloud fields, fair weather cumulus, ITCZ, and storm track cloudiness. Additional acquisitions include two California stratocumulus scenes and six fair weather cumulus scenes (three along the Atlantic coast, one in the Gulf of Mexico, one over Lake Michigan, and one over Amazonia). Locations of these scenes are shown in figure 1. In this paper we give results from the North Pacific and Atlantic coast scenes, which together contain over 100,000 cloud areas. New scenes are also being acquired in coordination with the observational programs associated with the First ISCCP Regional Experiment (FIRE).

The TM radiances are digitized into 256 intensity levels, so that one has one byte per pixel per band. Thus each reflected band of 6100 pixels on a side represents about $(6100)^2$ bytes or about 36 megabytes, while the thermal band represents about $(1600)^2$ bytes or about 2.5 megabytes, for a total of 218.5 megabytes per scene. Each scene is split into four quarters, and each quarter requires one 6250 bpi tape. Therefore the twenty scenes shown in figure 1 require eighty tapes and represent more than four gigabytes of data.

Once a particular TM subscene is chosen for analysis, processing of the data proceeds in several steps:

1. determine a cloud reflectivity threshold and identify each pixel as "cloudy" or "not cloudy", thus producing a binary image;
2. identify contiguous clear and cloudy areas in the binary image, determine various characteristics of each area, and store results in an ancillary file with one record per area;
3. determine statistical distributions from each ancillary file and fit the distributions to determine cloud parameters;
4. compute statistics of cloud parameters for various cloud types, thresholds, etc.

We briefly explain the first two steps here, and leave the last two steps for Sections 3 and 4.

Imagine an image of scattered cumulus over a dark ocean background, displayed as a binary image with black below some cloud reflectivity threshold and white above. If the threshold is steadily reduced, the cloud areas steadily grow until the threshold equals the ocean reflectivity, below which point the whole scene suddenly becomes white. On the other hand, if the threshold is steadily increased, the cloud areas steadily shrink until only the brightest cloud tops are visible, above which the whole scene suddenly becomes black. In this study, we choose a "cloud base" threshold just above the value of the surface reflectivity, and a "cloud top" threshold just below the point where the cloud areas begin to vanish.

Thresholds chosen interactively as described above are somewhat arbitrary. However, if the cloud fraction is known independently, then the cloud base threshold can be adjusted to produce the correct cloud fraction. The cloud fraction of single-layer clouds such as stratocumulus and fair weather cumulus can be determined from the thermal band by the spatial coherence method (Coakley and Bretherton, 1982).

Figure 2 shows an example of a spatial coherence plot for two regions of cloudiness in a TM scene off Cape Cod – one region over the Gulf Stream and one over the Labrador Current. This is a scatter plot of the mean and standard deviation of the thermal radiance for a grid of four by four pixel arrays. Completely cloud-covered arrays give a minimum mean radiance, I_c, and a low variance; cloud-free arrays give a maximum mean radiance, I_s, and a low variance; and partially-covered arrays give an intermediate mean radiance and a high variance, forming an arch. In this example we have two concentric arches, with colder cloud tops occuring over the Gulf Stream, where the boundary layer is deeper. If f_k is the cloud fraction of pixel array k, then the radiance is $I_k = f_k I_c + (1-f_k) I_s$. Averaging over k allows us to determine the mean cloud fraction in terms of the mean radiance, along with I_c and I_s. This technique clearly fails for multi-layered clouds, where Ic is not constant.

Figures 3a to 3c show histograms of the pixel radiances in band 7 for scenes of stratocumulus, fair weather cumulus, and ITCZ clouds, respectively. In the stratocumulus case (3a) we see a peak at low reflectance from ocean surface pixels, and a second peak at high reflectance from solidly cloud-covered pixels. The cloud base threshold occurs in between, at a point such that the fraction of pixels above the threshold equals the cloud fraction determined by the spatial coherence method. The cloud top threshold is chosen at the position of the second peak. The fair weather cumulus case (3b) also shows the low reflectance "surface" peak, but rather than a second "cloud" peak we see instead a gradually decreasing number of pixels with increasing reflectivity above the cloud base threshold. The lack of any characteristic

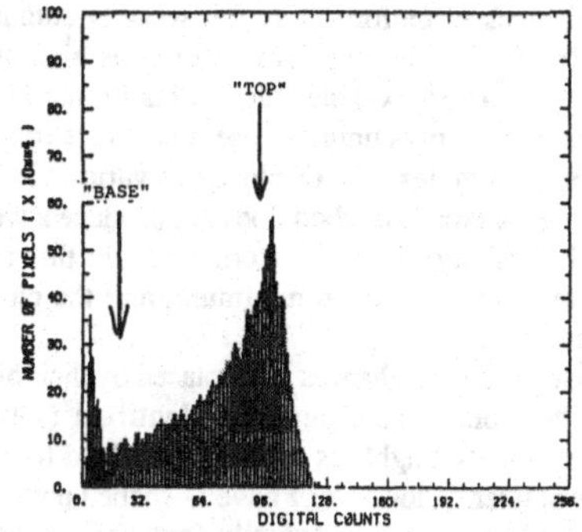

Fig. 3a: Histogram of TM Band 7 digital counts (linearly related to pixel reflectance) for a Pacific stratocumulus cloud scene. There is a peak at low reflectance from the ocean surface, and a second peak at high reflectance from the solid, relatively uniform cloud tops. An appropriate "cloud base" reflectance threshold may be chosen by dividing the distribution such that the fraction of pixels above the threshold equals the cloud fraction determined by the spatial coherence method. A "cloud top" reflectance threshold is chosen at the high reflectance peak.

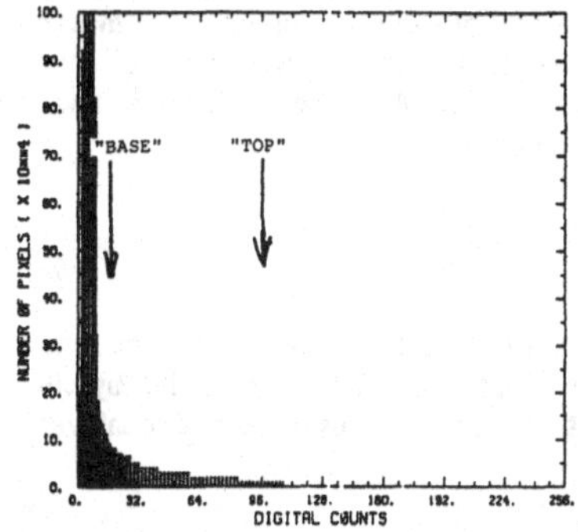

Fig. 3b: As in 3(a), but for a Pacific fair weather cumulus cloud scene. Compared to the stratocumulus case, here there are many more cloud-free pixels contributing to the low reflectivity peak, and then a gradually decreasing number of cloudy pixels with increasing reflectivity. The "cloud base" threshold is chosen as in 3(a). The lack of any peak at a characteristic "cloud top" reflectivity is due to the lack of uniformity in fair weather cumulus cloud tops. The "cloud top" threshold is then chosen by increasing the threshold until the number of contiguous areas begins to rapidly decrease.

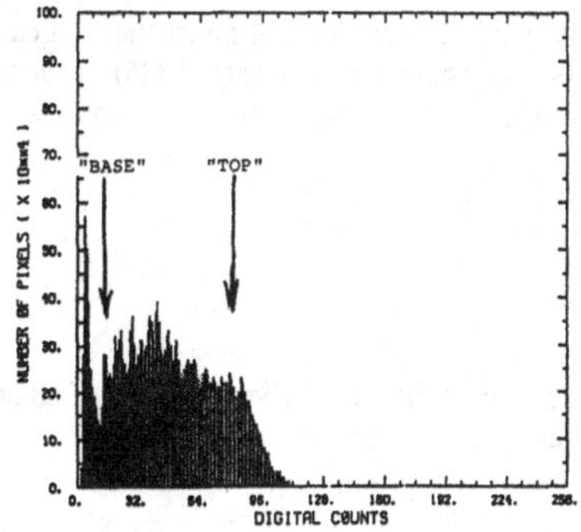

Fig. 3c: As in 3(a), but for a Pacific ITCZ cloud scene. Here there is a minimum above the surface peak, as in the stratocumulus case, but this is followed by a multiplicity of peaks at higher reflectivities, indicating a complex mix of clouds with various heights and optical thicknesses.

cloud top reflectivity is due to the lack of uniformity in fair weather cumulus cloud tops. The choice of a cloud top threshold is rather arbitrary in this case. It is chosen as high as possible without reducing the number of areas to a point insufficient for good statistics. Finally, the ITCZ case (3c) also has a surface peak followed by a minimum as in the stratocumulus case, but this is followed by a multiplicity of peaks at higher reflectivities, indicating a complex mix of clouds at various levels in the atmosphere and with various optical depths. The spatial coherence method does not produce a well-defined arch in this case, and since Ic is ill-defined the cloud fraction cannot be determined. In this case the cloud base threshold is arbitrarily chosen at the position of the histogram minimum, and the cloud top threshold is chosen just below the rapid falloff in the histogram.

Once the thresholds are defined and the radiances are replaced by their binary cloud/no-cloud values, then areas of contiguous cloud or no-cloud pixels must be identified (step 2 above). This is the most time-consuming step. Each pixel and its neighbors must be examined to decide to which cloud it belongs, whether it is part of the cloud perimeter. Clouds may have U-shaped parts, which appear distinct at first as one scans down the scene line-by-line, but are eventually found to be contiguous, so that one must scan back up and correct previous assignments. Clouds which touch an edge of the scene must be flagged, since their true area is unknown.

Eventually a file of contiguous areas is produced from each chosen TM subscene. Each record of this area file contains an area identification number, area type (cloud/no-cloud), area size in pixels and square kilometers, number of perimeter pixels of various types, center-of-mass coordinates, and an edge flag. Each such file contains thousands of records, and typically there are two or three files for each threshold and each band of a full TM scene. The statistical results presented in the following two sections are computed from this database of over 100,000 cloud areas.

3. AREAS AND PERIMITER

We first examine the distribution of cloud base areas. Since the larger areas are far less common, it is convenient to use area bins of increasing width. We do this by changing variables from area to the logarithm of the square root of the area, and then using bins of equal width. If the area is denoted as a, let

$$x = \log(\sqrt{a}) \tag{1}$$

Then the distribution of a is related to the distribution of x as

$$n(a) = n(x)\,\frac{dx}{da} = \frac{n(x)}{2a}. \tag{2}$$

How do we expect n(x) to behave in the case of scaling fractals? It can be shown that the distributions in that case must be power law (see, for example, Falconer, 1985). Mandelbrot (1983, p.117f) has shown that for a simple class of iteratively generated scaling fractals one obtains a more specific relation, termed the "Korcak law", given by

$$\int_a^\infty n(a')da' = F\,a^{-d/2}, \tag{3}$$

where d represents the fractal dimension of the cumulative perimeter of all the individual areas, and is related to the fractal dimension of individual perimeters by

$$1 \le d_p < d. \tag{4}$$

Taking the derivative of the Korcak law with respect to a, multiplying by 2a, and taking the log of the resulting expression leads to

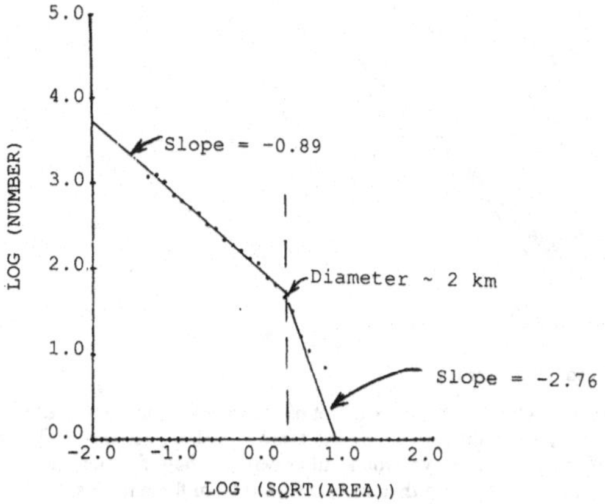

Fig. 4a: Histogram of contiguous cloud base areas, defined by a reflectance threshold, for a fair weather cumulus cloud field off the coast of South Carolina. The logarithm of the number versus the logarithm of the square root of the area is plotted, and fit to two straight lines with slopes and the location of the break determined by least squares.

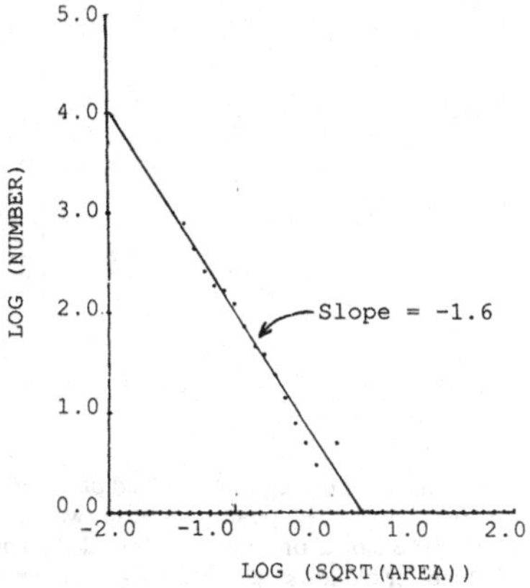

Fig. 4b: As in 4(a), but for contiguous cloud-free areas, or cloud "holes". The distribution is more peaked at small values, and no break is observed.

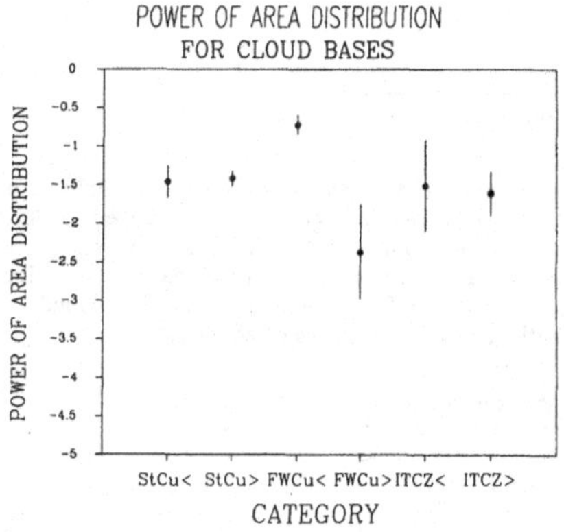

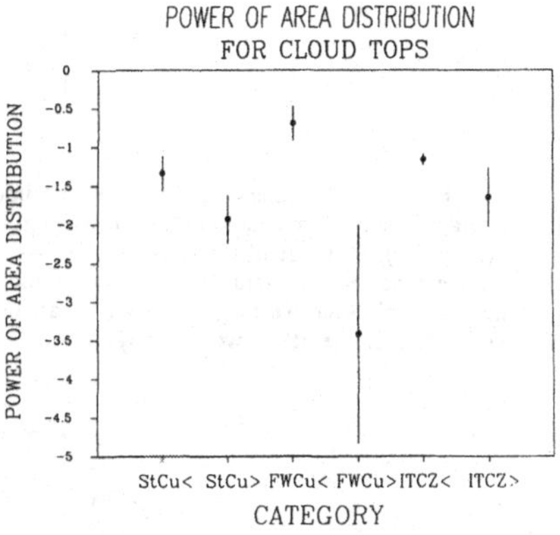

Fig. 5a: Summary of cloud base area distributions. The slopes, determined as in figure 4(a) and averaged over all available scenes of a given cloud type, are shown for the three cloud types described in figures 3(a) to 3(c). A value of the slope is obtained for smaller (<) and larger (>) areas for each cloud type. The vertical lines extend one standard deviation above and below the mean. Note that small and large clouds have significantly different slopes for both fair weather cumulus and ITCZ. Only stratocumulus are consistent with scale-invariance.

Fig. 5b: As is figure 5(a), but for cloud top areas. Comparing with figure 5(a), we see no significant dependence on threshold, except for the larger stratocumulus areas which drop off more rapidly for the higher threshold.

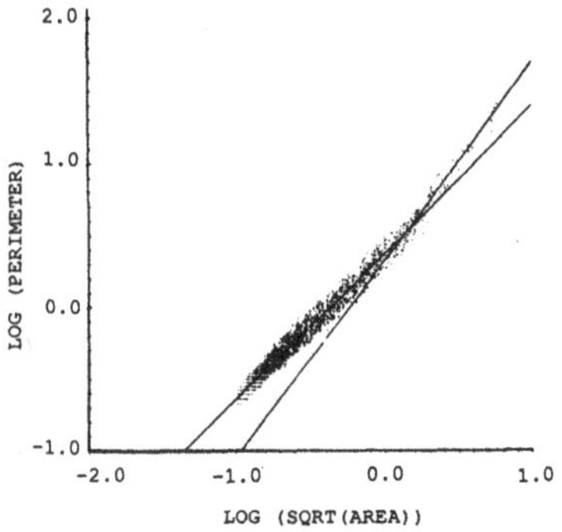

Fig. 6: Log-log plot of cloud base perimeter as a function of the square root of the area for the same scene as in figure 4(a). The points are fit to two straight lines by least squares, using the same cutoff area found in figure 4(a). The slope, or so-called "fractal dimension" of the perimeter, was found to depend on how the perimeter is defined for clouds smaller than about 12 pixels in area, and these were therefore excluded from the analysis.

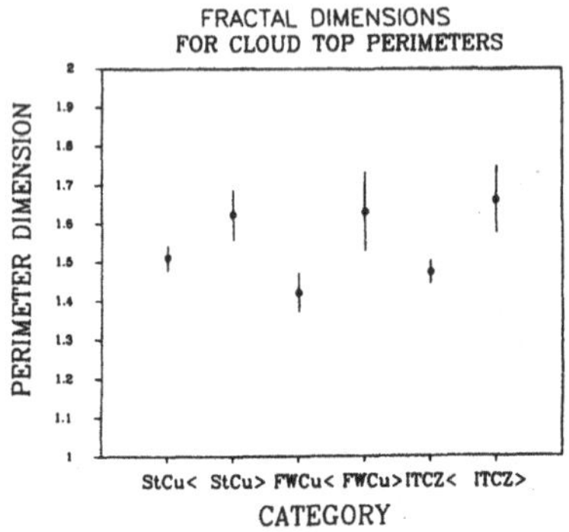

Fig. 7a: Summary of cloud base perimeter fractal dimensions. The slopes, determined as in figure 6 and averaged over all available scenes of a given cloud type, are shown for the three cloud types described in figures 3(a) to 3(c). A value of the slope is obtained for smaller (<) and larger (>) areas for each cloud type. The vertical lines extend one standard deviation above and below the mean. Note that the smaller fair weather cumulus clouds have a perimeter fractal dimension of about 4/3, while it exceeds 1.5 for the larger ones. The cloud base perimeters of the other two cloud types are consistent with scaling, with a dimension of about 1.4.

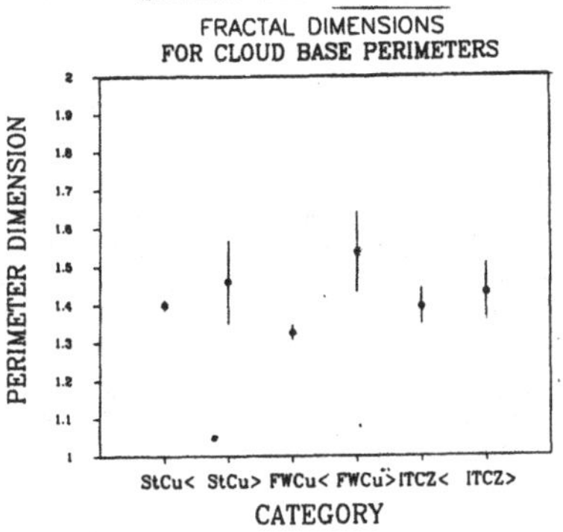

Fig. 7b: As in figure 7(a), except for cloud top perimeters. Here we see violations of scaling for all three cloud types. Comparing with figure 7(a), we see that the perimeter fractal dimension increases with threshold. This is in contrast to a Hausdorff, or covering dimension, which must decrease with threshold.

$$\log(n(x)) = -d \cdot x + \text{constant}. \quad (5)$$

How does this result compare to the TM cloud data? Figure 4a shows a plot of log(n) versus x for cloud base areas in a scene off the coast of South Carolina containing more than 10,000 fair weather cumulus clouds. (Results in this and all following figures are from TM band 7.) Also shown is a double straight line fit, with the slopes and location of the break determined by minimizing the mean square error. The great majority of clouds, those smaller than 2 km in diameter, closely follow a straight line with slope -0.89, with very small error. The population of clouds larger than 2 km falls off more rapidly, and can be fit with a straight line of slope -2.76, though with more error in the fit. The change is slope represents a clear violation of scale invariance. Similar behavior has been observed in a wide variety of fair weather cumulus cloud scenes (see also Joseph and Cahalan, 1987). Scale invariance appears to be an excellent assumption for the smaller clouds. However, since the slope exceeds -1, the simple class of scaling fractals mentioned above cannot apply.

Figure 4b shows the distribution of cloud-free areas, or cloud "holes" in the same scene. In this case a single power law fits the full range of sizes, with no significant break in slope. The slope of -1.6 is consistent with the Korcak law, and indicates that the distribution of cloud holes, unlike the clouds themselves, may be consistent with the simple iteratively generated fractals mentioned above.

Figure 5a summarizes our results for cloud base area distributions. Here we show the slopes, determined as in figure 4a, and averaged over all available scenes of a given cloud type, for the same three cloud types described in figures 3a to 3c, namely stratocumulus, fair weather cumulus, and ITCZ. For each cloud type, an average slope is shown for clouds below the breakpoint (StCu<, for example) as well as those above (StCu>). Position of the break varies from scene to scene, but is typically at about 0.5 km for fair weather cumulus, and 1 to 2 km for stratocumulus and ITCZ. In addition to the average slope in each category, the standard deviation of the slopes was also computed, and vertical lines plotted one standard deviation above and below the mean. (Each slope in the average was weighted by the number of clouds in the fit, but this weighting makes no qualitative difference in the results.)

If slopes within one standard deviation of each other are considered to be indistinguishable, we can see from figure 5a that both stratocumulus and ITCZ clouds are consistent with

$$n \sim (\sqrt{a})^{-1.5}. \quad (6)$$

However, this conclusion is somewhat premature for the ITCZ since the errors are quite large, and represent only two TM scenes. The fair weather cumulus behave as

$$n \sim \begin{cases} (\sqrt{a})^{-0.6} & (\sqrt{a}) < 0.5 \text{ km} \\ (\sqrt{a})^{-2.3} & (\sqrt{a}) > 0.5 \text{ km}. \end{cases} \quad (7)$$

Figure 5b shows a similar plot as in 5a, but for cloud tops rather than cloud bases. The only significant change in raising the threshold from cloud base to cloud top is that stratocumulus and ITCZ cloud tops are inconsistent with scaling, but behave like fair weather cumulus, with a more rapid decrease for larger sizes. Unlike fair weather cumulus, however, the slopes still lie between 1 and 2, and so may allow some kind of dimensional interpretation.

The fractal dimension of the individual perimeters, d_p, may be obtained by considering the perimeter length, p, as a function of the area a, or equivalently of x. For smooth shapes (circles, squares, etc.) $d_p=1$ and the perimeter is proportional to $(\sqrt{a})$. For a scaling fractal the perimeter is proportional to $(\sqrt{a})^{d_p}$. Thus

$$\log(p) = d_p x + \text{constant}. \quad (8)$$

Figure 6 shows a scatter plot of log(p) versus x for the same fair weather cumulus cloud scene used in figure 4a. Using the same cutoff area found in figure 4a (2 km) the points are fit to two straight lines by least squares. Clouds smaller than 12 pixels in area were excluded from the analysis, since they produced a slope sensitive to the algorithm used to estimate perimeter, while the slopes shown here are insensitive to the perimeter algorithm. The slope for those clouds between 12 pixels and 2 km is close to 4/3, while that for the larger clouds is about 1.6. Thus for the smaller clouds $d_p > d$, which again is inconsistent with the result given above for iteratively generated scaling fractals.

Figure 7a summarizes the results for the fractal dimension of cloud perimeters, d_p, and has a similar format to figure 5a. As in 5a, we see that both stratocumulus and ITCZ are consistent with a single dimension,

$$d_p \sim 1.4 \text{ to } 1.5, \tag{9}$$

while for fair weather cumulus we have

$$d_p \sim \begin{cases} 1.33 & x < 0.5 \text{ km} \\ 1.55 & x > 0.5 \text{ km}. \end{cases} \tag{10}$$

The perimeter fractal dimensions for cloud tops are summarized in figure 7b. In every category the dimension increases by about 0.1 as the threshold is increased from base to top. Note that this threshold dependence is opposite to that of the fractal dimension of area above a threshold discussed for so-called "multifractals" (Henschel and Procaccia, 1983; Grassberger, 1983; Schertzer and Lovejoy, 1983; Parisi and Frisch, 1985). This area dimension must decrease simply because the areas must cover less space as the threshold is raised. The increase in the perimeter dimension may be related to an increased intensity of turbulence higher in the cloud. If so, one expects the vertical dependence to be different for different cloud types. For stratocumulus the increase should be concentrated at cloud top, while in the ITCZ it is likely to occur through the full cloud thickness. Clearly more study is needed on this point.

4. NEAREST-NEIGHBOR SPACINGS

Properties of individual clouds such as those discussed in the preceeding section do not completely determine the radiative properties of a cloud field. It is also necessary to know something about the relative positions of the clouds, which will determine how they interact. One of the simplest measures is the distance between the center-of-mass of a given cloud and that of the closest neighboring cloud, the so-called nearest neighbor spacing, which we shall denote as s.

How would the spacing s be distributed if cloud centers were random and independently distributed in space? The answer depends on the dimension of space. (The following results are from Cahalan (1986) and references therein.) For points on a line, the spacings are distributed exponentially, with zero spacing being most likely. For points in a plane, the spacing distribution has the form $2s\exp(-s^2)$, which vanishes at zero separation, reaches a maximum at $s^2 = 1/2$, and then drops rapidly to zero. For a space of arbitrary dimension d, one finds a Weibull distribution with shape parameter d. This can be further generalized by relaxing the requirement that the coordinates be independent. If C(s), the conditional probability that there is a point at a distance s from some chosen point, given that there are none closer, is taken to vary as s^δ, then one obtains for the spacing distribution a Weibull distribution with shape parameter $d_{eff} = d + \delta$.

Thus to test whether clouds are spaced independently, or with simple power-law dependence, we need to compare the spacing distribution with a Weibull. This is most easily done for the cumulative distribution, F(s), which is the fraction of spacings greater than or equal to a given spacing. To obtain F empirically, we rank the spacings from largest to smallest, assigning rank r = 1 to the largest, rank r = 2 to the next largest, etc., then setting F equal to the rank over the total number of spacings. For a Weibull distribution with shape parameter d_{eff},

$$\ln(-\ln(F)) = d_{eff}\ln(s) + \text{constant}. \tag{11}$$

Thus, for clouds distributed independently or with power-law dependence, the double logarithm of the cumulative spacing distribution must be linear in the logarithm of the spacing.

Figure 8a shows a test of this behavior for a fair weather cumulus cloud field. Here the natural logarithm of the spacing (units = number of 30 m pixels) is plotted on the ordinate, and the double natural logarithm of F on the abscissa, so that independent points in a plane would follow the straight line with slope $1/d_{eff} = 1/2$. This appears to work reasonably well for spacings smaller than about 10 pixels (ln(s)=2.3, or about 300 m). For larger spacings the data deviates significantly from the straight line.

This non-independence of cloud locations is well known, and often referred to as cumulus cloud clumping or clustering. (For one attempt to model it, see Randall and Huffman (1980)). A simple statistical model of such clumping can be obtained from an isotropic random walk in which the step sizes are taken from a power-law probability distribution. This process is termed a "Levy flight" by Mandelbrot (1983). The power-law step size distribution yields a large number of small steps, thus generating a cluster, but large steps occur with sufficient probability to carry one from one cluster to another. Figure 8b shows the cumulative spacing distribution estimated from one realization of such a process, and plotted in the same format as figure 8a. One sees the same qualitative behavior, with the smaller intracluster spacings being nearly indistiguishable from the independent Weibull-type distribution, while the larger intercluster spacings deviate considerably.

Figure 9a summarizes the spacing distributions for cloud bases in all cloud categories. This is in the same format as in figures 5 and 7, except that the two size categories are now for spacings less than 10 pixels (StCu<, for example) and those greater than 50 pixels (StCu>). For all three cloud types the smaller spacings give a slope near 1/2, consistent with independent random spacings. The larger spacings give slopes of 2-3. Figure 9b shows that the results are essentially the same for the cloud top threshold.

5. SUMMARY AND CONCLUSIONS

We have estimated the probability distributions of cloud areas, perimeters and nearest-neighbor spacings for fair weather cumulus, stratocumulus and ITCZ clouds using thresholds for both cloud base and cloud top. The data base on which these estimates are based is summarized in Table 1, which gives the number of cloud areas contributing to the fits of the number distribution summarized in figure 5 (labelled n< and n> in Table 1), the perimeter-area fits summarized in figure 7 (p< and p> in Table 1) and the nearest neighbor distribution summarized in figure 9 (nn< and nn> in Table 1). The total number of fair weather cumulus, stratocumulus and ITCZ clouds is 121,519.

Our results may be summarized as follows: The probability distributions of cloud areas and cloud perimeters are found to be approximately power-law, with powers denoted by -d and d_p, respectively. Stratocumulus clouds conform to a single power ($d=1.5$, $d_p=1.4$) more closely than do fair weather cumulus, which exhibit a clear change in the fractal dimension, cloud bases less than about 0.5 km in diameter having a lower dimension ($d=0.6, d_p=1.33$) than the larger clouds ($d=2.3$, $d_p=1.55$). The fractal dimension also changes with the reflectivity threshold. As the threshold is raised from cloud base to cloud top, the perimeter fractal dimension increases, perhaps indicative of the increased turbulence at cloud top. In addition, scaling violations become evident at cloud top not only for fair weather cumulus but also for stratocumulus and ITCZ clouds. The distribution of nearest-neighbor spacings between cloud centers is shown to deviate from that expected for independently distributed centers due to cloud clustering. The cloud spacing distribution is qualitatively similar to that of a random walk with step sizes drawn from a power-law distribution (a "Levy flight").

Why do fair weather cumulus larger than about 0.5 km show such strikingly different scaling behavior? Perhaps this represents the largest individual convective cells, and larger cumulus are aggregates of several such cells. An observation which seems to support this view is that in searching several cumulus scenes for cloud areas having a single peak in reflectivity, we did not find any larger than about 0.5 km, while there are many larger clouds with multiple brightness peaks. If this holds up in a more objective analysis of the full data base, the question then becomes: what physically determines the maximum convective cell size?

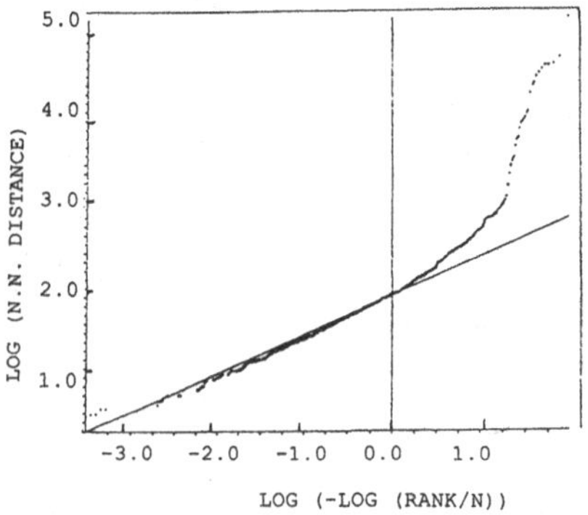

Fig. 8a: Logarithm of the distance between each fair weather cumulus cloud center and the closest neighboring cloud center plotted versus the double logarithm of the rank of the distance divided by the total number. Distances are ranked from largest to smallest, with the largest having rank 1. If the points were generated randomly, with coordinates chosen independently from a uniform distibution, then the log-double log plot would produce a straight line with slope 1/2. This random model seems to work only for clouds closer than about 0.5 km. For larger spacings, strong cloud clustering causes the plot to deviate from the straight line.

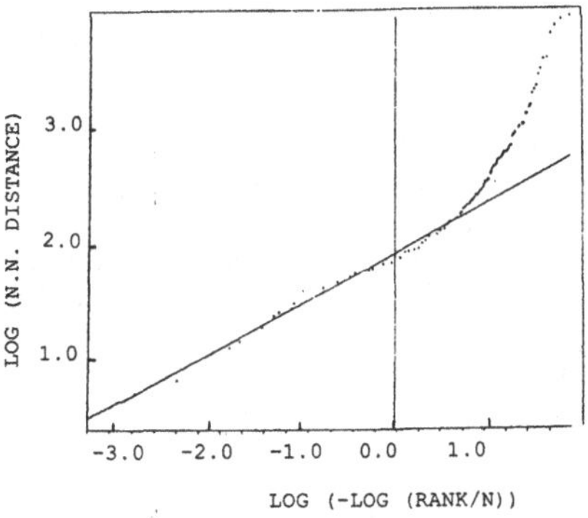

Fig. 8b: As in figure 8(a), except for data generated by a "Levy flight" in which each point is generated from the preceeding one by choosing a direction from a uniform distribution and choosing a distance from a power law distribution. The same qualitative clustering behavior is seen as in 8(a).

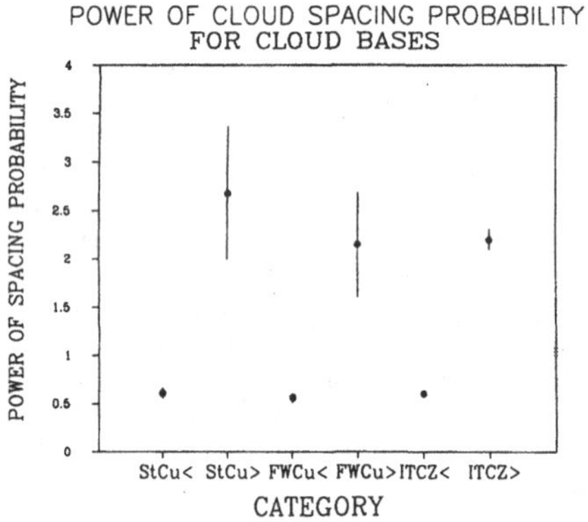

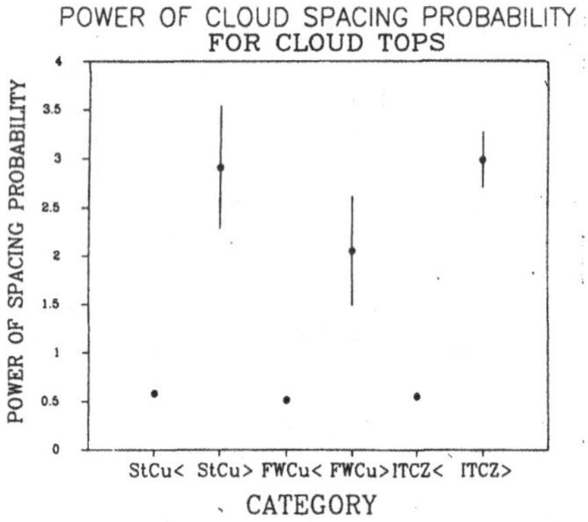

Fig. 9a: Summary of nearest neighbor spacings of cloud base areas. The slopes, determined as in figure 8(a) and averaged over all available scenes of a given cloud type, are shown for the three cloud types described in figures 3(a) to 3(c). A value of the slope is obtained for distances smaller than 0.35 km (<) and for those larger than 1.5 km (>) for each cloud type. The vertical lines extend one standard deviation above and below the mean. Note that the smaller spacings are consistent with a random distribution of points, which would have slope 0.5, while the larger spacings have slope 2-3, indicative of strong clustering.

Fig. 9b: As for figure 9(a), except for cloud top spacings. Comparing with figure 9(a), the only significant dependence on threshold occurs for the larger ITCZ clouds, which show stronger clustering for the higher threshold.

TABLE 1

		Figure 5		Figure 7		Figure 9	
		n<	n>	p<	p>	nn<	nn>
StCu	BASE	22,997	593	5,210	593	34,989	907
	TOP	25,767	1,233	6,481	1,233	41,273	1,210
FWCu	BASE	66,364	3,117	29,528	3,147	57,970	1,824
	TOP	32,417	2,272	13,721	2,272	32,204	941
ITCZ	BASE	27,063	1,385	6,206	546	12,219	412
	TOP	26,479	692	8,056	208	14,949	299

Table 1. Total number of clouds contributing to the parameters shown in figures 5, 7 and 9 for the three cloud types, for both thresholds ("base" and "top"), and for both size categories("<" and ">"). The total number of clouds analyzed may be obtained by adding n< and n> for StCu, FWCu, and ITCZ cloud bases, which gives 121,519. Note that StCu and ITCZ clouds have a greater number of contiguous cloud top areas than cloud base areas, unlike FWCu which have fewer. This is because most of the FWCu are single cells of varying thicknesses, while many of the StCu and ITCZ are multicell aggregates.

This work is part of a growing body of observational evidence on inhomogeneous cloud structure. Derr and Gunter (1982) reported searching in vain for the sort of clouds assumed by cloud radiation modelers (whether plane-parallel or a lattice of simple shapes). Plots of liquid water along the aircraft track through Arctic stratus (the quintessential plane-parallel clouds) bounce between zero and high values in an erratic fashion (Tsay and Jayaweera, 1983).

One explanation of the ragged structure of clouds has emerged from the combined observational/theoretical study of Baker and Latham (1979); see also Blyth et al. (1980). They find that the conventional assumption of homogeneous entrainment is incorrect – entrained dry air does not mix uniformly with cloudy air. Instead, it remains intact, as blobs of all sizes, which decay only slowly by invasion of cloudy air. Thus, on their model, a cloud is somewhat like a Swiss cheese, although this analogy fails to convey the essence of cloud inhomogeneity since the size spectrum of holes in Swiss cheese is rather narrow.

The implications of ragged cloud structure for climate modeling clearly need further investigation. The fractal stucture of clouds may explain the "optical depth paradox" (Wiscombe et al., 1984). On a zonal average, clouds behave as if they had optical depths less than 10 or so, otherwise the roughly 50% cloud cover would have an albedo higher than 50% (the value needed to balance the radiation budget). On the other hand, extinction coefficients on the order of 30 km^{-1} are commonly calculated by cloud radiation modelers. If such values were correct, then the average cloud would be only about 300 m thick. This apparent contradiction may be due to our misconception of clouds as homogeneous objects.

6. ACKNOWLEDGEMENTS

I am grateful to Mark Nestler of Science Applications Research, Landover, MD, for much of the computational work. This study is part of ongoing research in collaboration with Professor Joachim Joseph of Tel Aviv University and Dr. Warren Wiscombe of the Goddard Laboratory for Atmospheres. I have also benefitted from conversations with T. Bell, Harshvardhan, and W. Ridgeway.

7. REFERENCES

M. Aida , 1977: Scattering of solar radiation as a function of cloud dimensions and orientation. J. Quant. Spectrosc.Radiat. Transfer, 17, 303-310.

Baker, M., and J. Latham, 1979: The evolution of droplet spectra and the rate of production of embryonic raindrops in small cumulus clouds. J. Atmos. Sci., 36, 1612-1615.

Blackmer, R. H., and S. M. Serebreny, 1962: Dimensions and distributions of cumulus clouds as shown by U-2 photographs, Stanford Research Institute, AFCRL Rept. No. 62-609.

Blyth, A., T. Choularton, G. Fullarton, J. Latham, C. Mill, M. Smith, and I. Stromberg, 1980: The influence of entrainment on the evolution of cloud droplet spectra. II. Field experiments at Great Dun Fell, Quart. J. Roy. Met.Soc., 106, 821-840.

Cahalan, R. F., D. A. Short, and G. R. North, 1981: Cloud fluctuation statistics. Mon. Wea. Rev., 110, 26-43.

Cahalan, R. F., 1984: Climatological statistics of cloudiness. Preprints Fifth Conf. Atmos. Radiation, Baltimore, Amer. Meteor. Soc., 206-213.

Cahalan, R. F., 1986: Nearest neighbor spacing distributions of cumulus clouds, Proceeding of the 2nd International Conference on Statistical Climatology, Vienna, Austria, June, 1986.

Caughey, S. J., B. A. Crease, and W. T. Roach, 1982: A field study of nocturnal stratocumulus II, turbulence structure and entrainment. Quart. J. R. Met. Soc., 108, 125-144.

Coakley, J. A., Jr., and F. P. Bretherton, 1982: Cloud cover from high-resolution scanner data: detecting and allowing for partially filled fields of view. J. Geophys. Res., 87, 4917-4932.

Davies, R., 1978: The effect of finite geometry on the three-dimensional transfer of solar irradiance in clouds. J. Atmos. Sci., 35, 1712-1725.

Derr, V., and R. Gunter, 1982: EPOCS 1980: Summary Data Report -- Aircraft Measurements of Radiation, Turbulent Transport and Profiles in the Atmospheric and Oceanic Boundary Layers of the Tropical Eastern Pacific, NOAA Tech.Memo. ERL WPL-101, NOAA Wave Propagation Lab, Boulder, Colorado.

Falconer, K. J., 1985: The Geometry of Fractal Sets, Cambridge University Press.

Grassberger, P., 1983: Generalised dimensions of strange attractors. Phys. Lett., 97A, 227-230

Gifford, M. D., 1978: Characteristic size spectra of cumulus fields observed from satellites. Thesis available from Defense Technical Information Center, Report No. CI 78-28, ADA-049936, 79 pp..

Hahn, C. J., S. G. Warren, J. London, R. M. Chervin, and R. Jenne, 1982: Atlas of simultaneous occurence of different cloud types over the ocean. NCAR Technical Note, 212 pp., NCAR/TN-201+STR.

Harshvardhan, and J. A. Weinman, 1982: Infrared radiative transfer through a regular array of cuboidal clouds. J. Atmos. Sci., 39, 431-439.

Hentschel, H.G.E., and I. Procaccia, 1983: The infinite number of generalized dimensions of fractals and strange attractors. Physica, 8D, 435-444.

Hentschel, H.G.E., and I. Procaccia, 1984: Relative diffusion in turbulent media, the fractal dimension of clouds. Phys. Rev., A29, 1461-1476.

Hozumi, K., T. Harimaya, and C. Magono, 1982: The size distribution of cumulus clouds as a function of cloud amount. J. Met. Soc. Japan, 60, 691-699.

Joseph, J. H. , and R. F. Cahalan, 1987: Distribution of cumulus cloud sizes and nearest neighbor spacings inferred from LANDSAT, submitted to J. of Clim. App. Met.

Kunkle, K. E., E. W. Eloranto, and S. T. Shipley, 1977: Lidar observations of the convective boundary layer. J. Appl. Met., 16, 1306-1311.

Lopez, E. L., 1976: Radar characteristics of the cloud populations of tropical disturbances in the northwest Pacific. Mon. Wea. Rev., 104, 268-283.

Lopez, E. L., 1977a: The log-normal distribution and cumulus cloud populations. Mon. Wea. Rev., 105, 867-872.

Lopez, E. L., 1977b: Some properties of convective plume and fair weather cumulus fields as measured by acoustic and lidar sounders. J. Appl. Met., 16, 861-865.

Lovejoy, S., 1982: Area-perimeter relation for rain and cloud areas. Science, 216, 185-187.

Mandelbrot, B. B., 1983: The Fractal Geometry of Nature. W. H. Freeman and Co., San Francisco, 460pp..

McKee, T. B., and S. K. Cox, 1974: Scattering of visible radiaiton by finite clouds. J. Atmos. Sci., 31, 1885-1892.

McKee, T. B., and S. K. Cox, 1976: Simulated radiance patterns for finite cubic clouds. J. Atmos. Sci., 33, 2014-2020.

McKee, T. B., and J. T. Klehr, 1978: Effects of cloud shape on scattered solar radiation. Mon. Wea. Rev., 106, 399-404.

Parisi, O., and U. Frisch, 1985: A multifractal model of intermittency, p. 85-89. Proc. of the Varenna School, Italian Phys. Soc., Eds. M. Ghil, R. Benzy, G. Parisi.

Plank, V. G., 1969: The size distribution of cumulus clouds in representative Florida populations. J. Appl. Meteor., 8, 46-67.

Randall, D. A., and G. J. Huffman, 1980: A stochastic model of cumulus clumping. J. Atmos. Sci., 37, 2068-2078.

Reynolds, P. D., T. B. McKee, and K. L. Danielson, 1978: Effects of cloud size and cloud particles on satellite-observed reflected brightness. J. Atmos. Sci., 35, 160-164.

Schertzer, D. and S. Lovejoy, 1983: Onthe dimension of atmospheric motions. Preprints, IUTAM, Symp. on Turbulence and Chaotic phenomena in Fluids, Kyoto, Japan, IUTAM, 141-144.

Shenk, W. E., and V. V. Salomonson, 1971: A simulation study exploring the effects of sensor spatial resolution on estimates of cloud cover from satellites. J. Appl. Meteor., 11, 214-226.

Wiscombe, W. J., R. M. Welch, and W. D. Hall, 1984: The effects of very large drops on cloud absorption. I: Parcel models, J. Atmos. Sci., 41, 1336-1355.

FRACTAL DIMENSION ANALYSIS OF HORIZONTAL CLOUD PATTERN IN THE INTERTROPICAL CONVERGENCE ZONE

Jun-Ichi Yano
Geophysical Institute/ Laboratory for Climatic Change Research,
Kyoto University
Kyoto 606, Japan

Yoshiaki Takeuchi
Meteorological Satellite Center
Japan Meteorological Agency
Kiyose, Tokyo 204, Japan

ABSTRACT. An organized horizontal pattern of clouds of order 10^4 km over the tropical oceans,the so-called intertropical convergence zone (ITCZ), which plays an important role in the tropical atmosphere, is a fractal for the equivalent black body temperature $-10 \leq T_{BB} \leq 10°$ C, but it is not for $T_{BB} \leq -20°$ C. The fractal dimension representing a whole cloud structure of the ITCZ is determined to be $D \approx 1.5$, which is a different quantity from that determined by Lovejoy (1982) for individual clouds. The result suggests the scaling property of the dynamics of the tropical atmosphere for scales from ~1 km to ~10^3 km analysed here.

1. INTRODUCTION

Atmospheric physics has been, as in the other fields of physics, developed in a realm of dynamics, by reducing the meteorological phenomena to the dynamics of the wind-velocity field (Lorenz, 1967): in a general sense, a traditional spirit of physics is to reproduce the phenomena by solving the basic governing differential equations. This is possible at least in the higher latitudes, where the atmospheric phenomena may be understood in terms of the synoptic scale (~10^3 km) dynamics (Holton, 1979). On the other hand, the cloud systems, which are organized into a system of scale up to 10^4 km, namely the so-called intertropical convergence zone (ITCZ, fig. 1), play an important role in the tropical atmosphere (Webster, 1982). Because the clouds are complicated systems not only dynamically but also thermodynamically accompanied by processes such as the condensation, growth of raindrops, etc., a traditional approach to solving the basic governing differential equations of the system directly is quite formidable even numerically (Frank, 1983). Hence, the authors believe that another approach from the geometrical point of view is required to understand such a complicated physical system. This paper is intended to be a step to develop such an approach.

In fig. 2, we represent the two pictures that parts of the ITCZ in fig. 1 are enlarged in different scales. More specifically, the one is the enlargement of a quarter of the lower right part of the other*. A difficulty for distinction of the scales of two pictures strongly implies the self-similarity (in a statistical sense) of the ITCZ cloud pattern; i.e., the ITCZ cloud system consists of an infinitely continuing series of a hierachy of the cloud patterns, whose upper (larger) structure has a remarkable resemblance to the lower (smaller) one.

* Here, (b) is a part of (a), while (a) is an enlargement of a part just left to the center of the imagery fig. 1.

D. Schertzer and S. Lovejoy (eds.), Non-Linear Variability in Geophysics, 297–302.

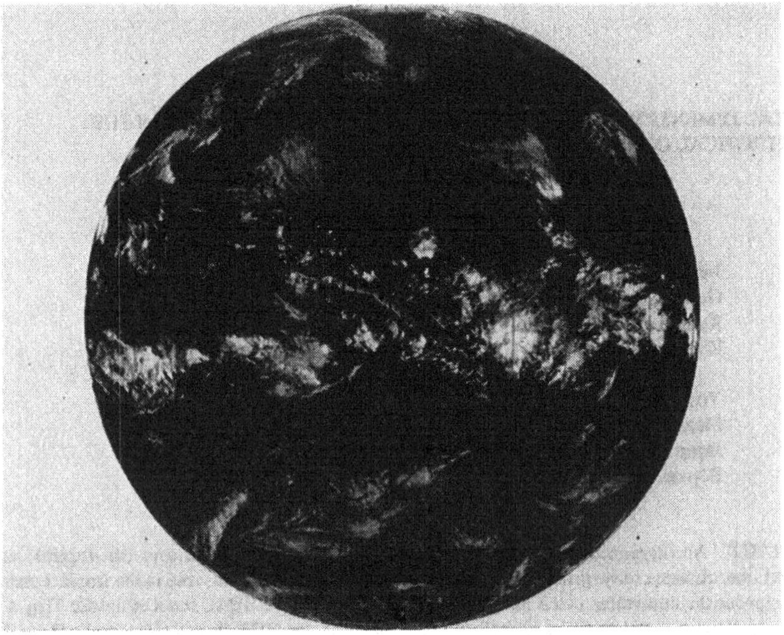

Fig. 1: The GMS-3 infrared imagery at 06Z, April 16, 1985.

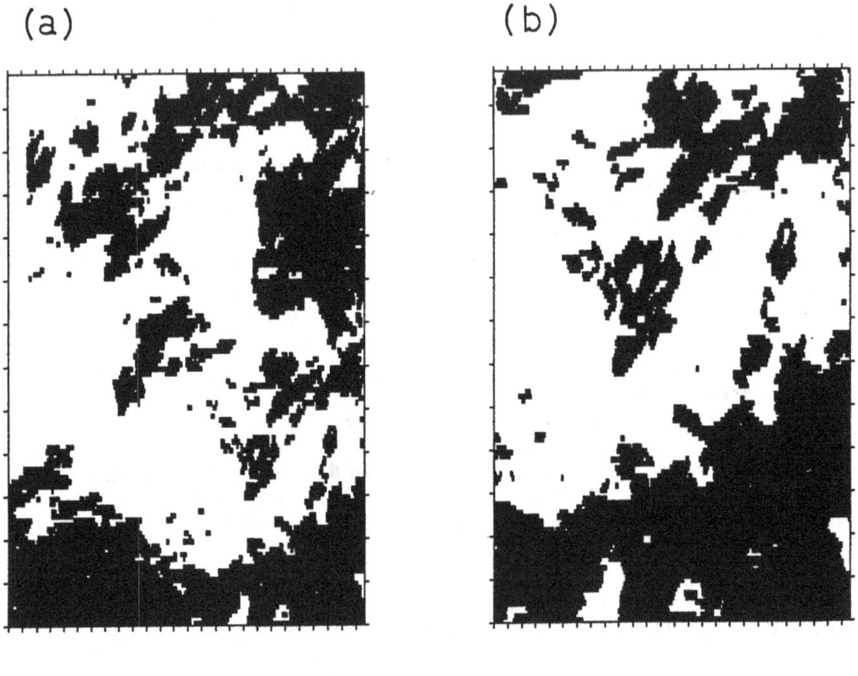

Fig. 2: The two partial enlargements of the imagery in figure 1 in same relative resolution with the regions above T_{BB} = -10°C represented by black.

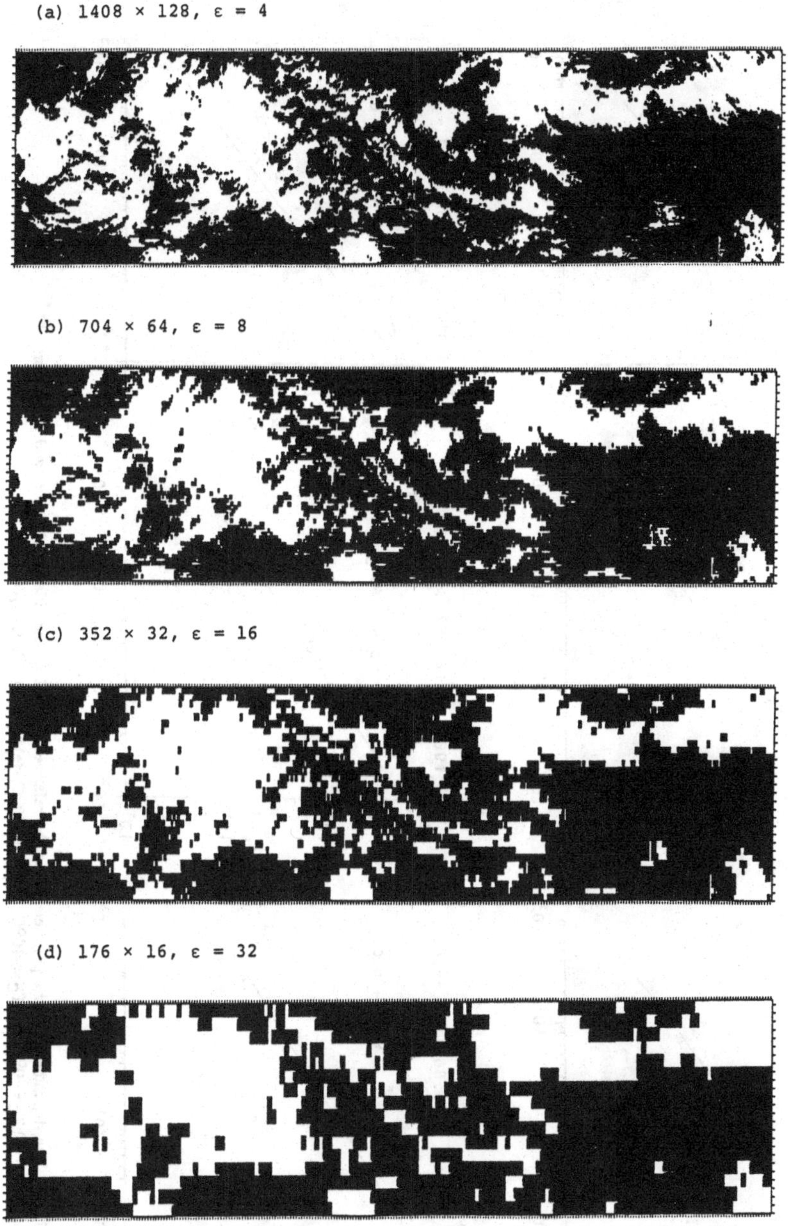

Fig. 3: The averaged imagery under the analysis in figure 1 with averaged pixel numbers: (a) 1408 x 128 (ε = 4), (b) 704 x 64 (ε = 8), (c) 352 x 32 (ε = 16), (d) 176 x 16 (ε =32). The regions above T_{BB} = -10°C are represented by black.

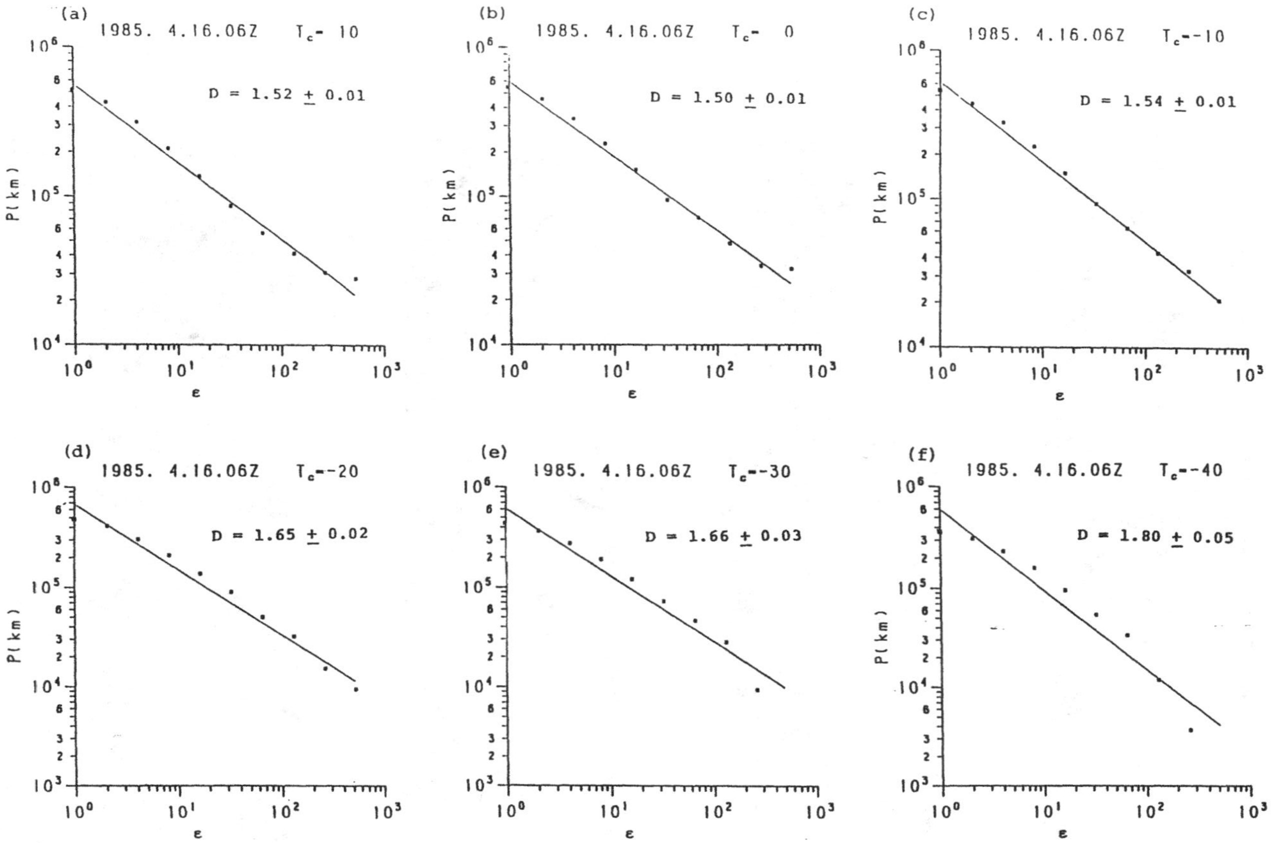

Fig. 4: The dependence of the perimeter P on the relative length ε of the sides of the averaged pixels for: (a) $T_C = 10^oC$, (b) $T_C = 0^oC$, (c) $T_C = -10^oC$, (d) $T_C = -20^oC$, (e) $T_C = -30^oC$, (f) $T_C = -40^oC$.

Studies of fractal geometry (Mandelbrot, 1983) show that self-similar geometrical figures satisfy the following relation between the length P of the perimeter and the length ε of the yardstick used for measurement:

$$P \propto \varepsilon^{1-D} \qquad (1)$$

where a constant D is called the fractal dimension. If the cloud patterns are really self-similar, they must satisfy the relation (eq. 1).

Various fractal dimension analysis methods have been proposed so far. For cloud patterns, Lovejoy (1982) demonstrated the fractal property by measuring both the perimeter P and the area A of the horizontal pattern of various individual clouds, instead of proving the relation (eq. 1) directly. He obtained the result $D \cong 1.35$ from the relation $P \propto \sqrt{A}^{D}$. Recently, Lovejoy, Schertzer and Tsonis (1987) analysed the fractal property of clouds in three-dimensional space with the use of radar rain reflectivity data by the box-counting method. We perform here another measurement of the fractal dimension of clouds by a percolation analysis method. In our analysis, instead of increasing the size of the square (or, in general, the cubic box of appropriate dimension) for counting the number of those containing the data that exceed the threshold value, we take the average of the grid-point data in sequence. The procedure is in conformity with the traditional data analysis method in meteorology, where time-averaging or space-averaging of data is taken in order to analyze the longer-period or larger-scale phenomena. It is anticipated that implications for such averaging may be also found. Note that if the field is characterised by intense regions with smaller fractal dimensions than the weaker regions (i.e the field is "multifractal"), then the temperature corresponding to a given fractal dimension will therefore depend on the scale of averaging. It would be important to use different thresholds at different scales in order to study the same fractal. The method used here does not take this effect into account and will be reliable only for warm clouds that have only weak threshold dependence (i.e that are approximately mono-dimensional).

2. ANALYSIS

We analyse here the GMS-3 infrared data at 06Z(GMT), April 16, 1985 over the tropical region of the Pacific Ocean approximately enclosed by the lines 80°E, 13°N, 15°W, and 13°S (see fig. 1 and fig. 3(a)); the domain consists of 5632 x 512 rectangular pixels of size 2 km x 5 km (measured in the scale at the center of the imagery) for longitudinal and latitudinal directions, respectively.

The infrared data are given by the equivalent black body temperature T_{BB}, which is a temperature of a corresponding black body emitting a measured infrared radiance, and can be considered as a representation of a cloud top temperature. In the following analysis, the clouds are defined as the regions where T_{BB} is below a certain critical value T_C. Instead of changing the length of the yardstick, we take the average of the units of pixels 2 x 2 in sequence, which is equivalent to increasing the length of yardstick twice. The data averaged for k times contain $11 \times 2^{(10-k)}$ by $2^{(10-k)}$ pixels with relative length of the side $\varepsilon = 2^k$, which we consider as the length of the yardstick (see fig. 3). The length of the perimeter of the regions with T_{BB} below a given critical value T_C is measured for each averaged datum along the sides of rectangular pixels.

In fig. 4, we present the results P(ε) for several values of T_C. For $-10 \le T_C \le 10$°C, the function P(ε) satisfies (eq. 1) well, with the fractal dimension D = 1.54 ±0.01, 1.50 ±0.01, and 1.52 ±0.01 for T_C = -10, 0, and 10°C, respectively, which are determined by the least square method with the possible errors indicated. Our results take a larger fractal dimension than that of Lovejoy (1982): Lovejoy has examined the fractal property of the individual clouds as isolated closed curves; we have investigated that of the cloud structure consisting of the various curves of different sizes, where a fractal property of the distribution of the individual clouds is also taken into account.

On the other hand, for T_C = -20°C, P(ε) decreases a bit faster than the value expected from (eq. 1) as ε increases. For T_C = -30 and -40°C, a deviation from the straight line is so large that the cloud patterns can be no longer considered as self-similar. Because T_{BB} is physically related to both the cloud top temperature and the thickness of the clouds, the physical interpretation of the strong dependence of P(ε) on T_{BB} is not straightforward. Nevertheless, over the tropics, the clouds of $T_{BB} \le -30$°C in most parts consist of the high-altitude cirrus clouds. Hence, we can conclude that, though the cloud patterns in the ITCZ are self-similar in general for the range of the scales between ~1 km and ~10^3 km, the high-altitude cirrus clouds are the exceptions that become more ramified on small length scales and less ramified on large length scales.

3. CONCLUSIONS

The self-similarity of the horizontal cloud pattern over the ITCZ for $-10 \leq T_{BB} \leq 10^\circ C$ also suggests the scaling property of the dynamics of tropical atmosphere. The scaling property of the dynamics also casts some doubt on the traditional data analysis procedure (e.g., Yasunari, 1981; Murakami, 1984) taking average (filter) of the original data to remove the noise.

The dependence of $P(\varepsilon)$ on T_{BB} firstly revealed here is yet little understood, and a detailed data analysis is now under progress, along with theoretical examinations. The computational algorithm employed for fractal dimension analysis (a percolation analysis method) adopted here is so simple that it would be applicable in various two-dimensional image analysis in widely different fields.

4. ACKNOWLEDGEMENTS

We express our sincere thanks to Prof. R. Yamamoto and Prof. I. Hirota of Kyoto University, and Dr. H. Takayasu of Kobe University for their invaluable comments to our study. Thanks are also due to Prof. S. Lovejoy and Prof. D. Schertzer, who offered us an opportunity to contribute this paper to the present volume.

5. REFERENCES

Frank, W. M., 1983: Mon. Wea. Rev., 111, 1859.
Holton, J.R., 1979: An Introduction to Dynamic Meteorology. Academic Press, New York, 2nd Ed., Ch.9.
Lorenz, E. N., 1967: The Nature and Theory of the General Circulation of the Atmosphere, World Meteor. Organization.
Lovejoy, S., 1982: Science, 215, 185.
Lovejoy, S., D. Schertzer, and A. A. Tsonis, 1987: Science, 235, 1036.
Mandelbrot, B.B., 1983: The Fractal Geometry of Nature, Freeman, New York.
Murakami, M., 1984: J. Met. Soc. Japan, 62, 88.
Webster, P. J., 1982: Large-Scale Dynamical Processes in the Atmosphere, eds. B. Hoskins and R. Pearce Academic Press, New York, p. 235.
Yasunari, T., 1981: J. Met. Soc. Japan, 59, 336.

RADIATIVE TRANSFER IN MULTIFRACTAL CLOUDS

A. Davis, S. Lovejoy[1], D. Schertzer
EERM/CRMD, Météorologie-Nationale,
2 Av. Rapp, Paris 75007
FRANCE

ABSTRACT. An understanding of radiative transfer in highly inhomogeneous fractal clouds is important in meteorological and climatological modelling, including possible greenhouse warming and nuclear winter cooling. In this paper, we extend previous work on monofractal clouds to the more physically relevant multifractal cases. We obtain expressions for the scaling exponents of the direct transmission in lognormal universal multifractal clouds, for multiple scattering in the corresponding plane parallel models as well as for direct transmittance in a family of microcanonical α model multifractal clouds. As for monofractal clouds, we obtain phase function independent exponents, and predict diverging thick cloud transmission ratios in comparison with (standard) non-fractal models. These results amplify our previous explanations of the so-called "albedo paradox", and highlight the importance of using the correct scaling in radiative transfer calculations.

1. INTRODUCTION

1.1 The importance of inhomogeneity in modelling radiative transfer in clouds

One of the most immediately striking visual features of clouds is their organisation into complex multifractal structures spanning many orders of magnitude in scale. Indeed, in situ and remote measurements establish that this variability extends over the virtually the entire dynamically relevant range of scales; see, e.g., Lovejoy and Schertzer (this volume) for relevant measurements from 10^3km down to 1mm. While it is true that with respect to the underlying cloud liquid water field, the associated radiation field is (non-linearly) smoothed, the striking visual appearance of clouds is in itself sufficient to demand that cloud models used in radiative transfer calculation directly incorporate the observed extreme inhomogeneity. In spite of this elementary observation, in part due to the difficulty in adequately accounting for the variability, the effects of inhomogeneity are only beginning to be taken seriously. Indeed, practically all geophysical radiative transfer calculations to date have been carried out in plane-parallel (i.e., horizontally uniform) models with vertical variability extending over very limited ranges in scale, i.e. they are also smooth in the vertical. In contrast, real clouds are highly chaotic structures shaped by turbulent flows with large variation of liquid water (hence optical density) down to the smallest observable scales; however smoothed, the levels of their multiply scattered radiation fields reflect this extreme variability.

There are at least two reasons why study of radiative transfer in highly variable media has in the past been shunned: the lack of appropriate models for spatial variability and an exagerated emphasis on the detailed angular part of the problem which aggravates the analytical and computational difficulties. The appearance of explicit multifractal cloud models (Schertzer and Lovejoy, 1987b, Wilson et al., this volume), has now made it possible to directly study radiative transfer in extremely variable media. Furthermore in a series of papers, we have studied a class of radiative transfer systems in which scattering occurs only in a finite number of discrete directions ("discrete angle" or "DA" radiative transfer), greatly simplifying the angular part of the calculations. This enabled us to obtain results in some simple monofractal clouds. In this paper we attack the problem from a different vantage: we consider the full continuous angle problem in multifractal clouds, obtaining results in a number of interesting systems.

---

[1] Physics Department, McGill University, 3600 University Street,Montréal, Québec, H3A 2T8, CANADA.

D. Schertzer and S. Lovejoy (eds.), Non-Linear Variability in Geophysics, 303–318.

1.2 Models of turbulent cloud density and radiative transfer

So far most efforts at dynamic cloud/radiation modelling have been at thermal infra-red wavelengths using statistical closure techniques (Simonin et al., 1981; Schertzer and Simonin, 1983; Coantic, 1978; and in the astrophysical literature, starting with Spiegel, 1957). The closure approach is limited primarily by its inability to properly handle intermittency (i.e., extreme variability). Another approach (e.g., Welch, 1983) is to use standard numerical modelling techniques for directly simulating all the dynamical fields and their interactions (including radiative). The difficulty here is that even the largest scale dynamical models (which have of the order a factor of only 100 in their range of scales) do not contain a sufficient range of scales to allow intermittency to build up significantly. On the one hand, the empirical cloud densities are highly intermittent (Tsay and Jayaweera, 1984) and have power law energy spectra over many orders of magnitude (King et al., 1981). On the other hand, it is precisely this scaling intermittency which leads to large radiative effects (anomalous exponents). It is therefore desirable to study stochastic cloud models which are scaling by (cascade) construction and are specifically designed to produce strong intermittency.

The simplest multiplicative cascade model is the "β-model" (Novikov and Stewart, 1964; Mandelbrot, 1974; Frisch et al., 1978); it was originally developed for modelling the energy flux density ($\varepsilon=\partial v^2/\partial t$ where v is the modulus of the velocity vector) in highly turbulent flows and, as pointed out by Schertzer and Lovejoy (1983), is the only one involving a single fractal dimension. The β-model creates a highly variable (intermittent) field by randomly concentrating the large scale flux density into sub-regions, simulating the non-linear break-up of an eddy into sub-eddies. A binomial process randomly determines if a sub-eddy is "alive" with probability λ_o^{-C} or "dead" with probability $1-\lambda_o^{-C}$. $\lambda_o>1$ is the scale ratio between successive steps of the cascade process, after N iterations the grid size is $\lambda=\lambda_o^N$ in all d spatial directions; $C\geq 0$ is the co-dimension with C=d-D where D is the fractal dimension of the live regions. In the limit of a large number of cascade steps ($\lambda\rightarrow\infty$), the flux is almost surely everywhere zero except on a fractal set with dimension D. If the eddy is alive, the flux density is increased by a constant factor λ_o^C, such that the ensemble averaged total (area integrated) flux is conserved. Radiative transfer on a simple deterministic β model was studied in Gabriel et al. (1990) and Davis et al. (1990a,b); early and/or related results on β models can be found in Gabriel et al. (1986), Lovejoy et al. (1988), Gabriel (1988), Lovejoy et al. (1989), Davis et al. (1989).

A seemingly small change yields a far more interesting cascade. In the "α-model[1]" proposed by Schertzer and Lovejoy (1984) the same binomial distribution is used but instead of being multiplied by λ_o^C and 0 the multiplicative increments of ε are taken to be $\lambda_o^{C/\alpha}>1$ and some other strictly positive constant $\lambda_o^{-C/\alpha'}$ chosen so that $\langle\varepsilon\rangle=1$ (the β-model is retrieved formally with $\alpha\rightarrow 1$, $\alpha'\rightarrow 0$). In analogy with statistical mechanics, such models are said to be "canonical", i.e. they conserve the energy flux on ensemble-average. A restrictive variant is also possible: the "micro-canonical" cascade in which energy (flux) is exactly conserved at each step. This is unphysical since, in the small scale limit, it implies flux conservation at each point, i.e. it is really "pico-canonical"; see Schertzer and Lovejoy (this volume) for more detail about these limitations. However, the microcanonical models can be easier to analyze and we consider the transfer in microcanonical α models in section 5. Cahalan (1989) also numerically simulated radiative transfer in one dimensional (horizontally varying) microcanonical α model (see section 5).

In cloud modeling, we are more interested in the (liquid water) density field (ρ) than ε (which partially describes the velocity field v). A multifractal model for ρ is given in Schertzer and Lovejoy (1987a,b) and Wilson et al. (this volume). This model is much more physically based than its additive/monofractal counterparts since it has the same symmetries (e.g. scaling) as the dynamical (partial differential) equations governing passive scalar advection in incompressible turbulence, as well as the same cascade phenomenology. This phenomenology assigns to passive scalar variance flux density ($\chi=\partial\rho^2/\partial t$ where ρ is the passive scalar density) a role analogous to that of ε (with respect to v). The quantities χ and ε are fundamental since they are exactly conserved by the non-linear terms in the dynamical equations, and are thus expected to be scale invariant down to viscous scales (typically of the order of millimeters). Once the statistically stationary conserved quantities χ and ε are given, the fluctuations in density are described by the Corrsin-Obukhov scaling law:

$$\Delta\rho \approx \phi^{1/3}\Delta x^{1/3} \tag{1.1}$$

---

[1] This "α" is quite different from and should not be confused with the Levy "α" which characterizes the generator of the multifractal and which will be used below.

where $\Delta\rho$ is the (mean) difference in density between two points separated by a distance Δx and $\phi = \varepsilon^{-1/2}\chi^{3/2}$. Various methods of numerically implementing this phenomenology are possible (Wilson et al., this volume).

2. RECAPITULATION OF SOME RESULTS OF RADIATIVE TRANSFER THEORY

2.1 The radiative transfer equation for multiple scattering

Not very much radiative transfer theory will be required in the following; however a brief summary of the necessary elements is given below. The detailed angular distribution of radiation is given by the (specific) intensity field, also called radiance, and that we shall denote $I_s(x)$. It is defined as the flux of radiant energy in direction s (at position x) per unit of solid angle and unit of area perpendicular[1] to s. The radiative transfer equation expresses energy (flux) conservation within an elementary volume around x, it reads (Chandrasekhar, 1950):

$$(s\cdot\nabla)\, I_s(x) = -\,\kappa\rho(x)\,\left\{I_s(x) - \int_{|s|=1} p(s'\rightarrow s)\, I_{s'}(x)\, d^d s'\right\} \tag{2.1}$$

where $\kappa\rho(x)$ denotes the optical density (total cross-section per unit volume) at position x; the physical interpretation of $\kappa\rho$ is the probability of (photon-cloud droplet) interaction per unit of distance. $p(s'\rightarrow s)$ is the phase function for scattering from direction s' into s (the probability of a photon incident at angle s' being scattered to angle s)[2]. d=2,3 is the dimension of space in which the scattering occurs. The (elastic) scattering probability is:

$$\int_{|s|=1} p(s'\rightarrow s)\, d^d s' = \varpi_o, \text{ for all } s \tag{2.2}$$

$\varpi_o=1$ is the important conservative case which we will consider below and $(1-\varpi_o)$ is the probability of absorption per event[3].

2.2 Direct transmission

Up until now, the only approaches to dealing with radiative transfer when the optical density field $\kappa\rho$ is mono or multifractal are either approximate or numeric. However, useful insight can be obtained by considering the direct transmittance (corresponding to those photons which undergo no scattering) which is a lower bound on the total transmittance and which can be considered as a source for the diffuse (scattered) radiation field. Consider a cloud with incident radiation $I_s^{(0)}$ in direction s, and determine the transmittance through the system in direction s by solving eq. 2.1 for I_s over the bottom of the cloud, then the direct transmittance is that which would occur through the same system without scattering, i.e. with the integral term in eq. 2.1 set equal to zero (or, equivalently, with $\varpi_o = 0$). The solution of 2.1 is an exponential which defines a direct transmittance coefficient T_d as follows:

$$T_d = \frac{I_s}{I_s^{(0)}} = e^{-\tau} \tag{2.3}$$

where τ is the optical thickness:

$$\tau = \int_{C_s} \kappa\rho\, dr \tag{2.4}$$

---

[1] In two dimensional systems, ordinary angles and unit lengths rather than areas are required.
[2] More general formulations of the multiple scattering source function, i.e. second term on r.h.s of (2.1), would incorporate the mixing of various polarizations and/or frequencies.
[3] $\varpi_o > 1$ models (neutron) fission processes.

$I_s^{(0)}$ is the radiation incident in direction s and C_s is the path in this direction from the top to the bottom of the cloud. If this distance is L, and the cloud is homogeneous, then $\tau = \kappa\rho L$. In scattering/absorbing media (when $\varpi_o > 0$), eq. 2.3 gives the distribution of (photon) free paths.

2.3 Asymptotic solution for plane parallel clouds

Another result of radiative transfer theory that will be useful is for the total (multiply scattered) transmission through plane parallel clouds (i.e., those where $\kappa\rho$ varies only in the vertical direction denoted z). The simplest way to obtain this result is to use the "two-flux" approximation (Schuster, 1905). Divide the radiation into two "beams" or "fluxes" Let $I_\pm(z)$ represent the fluxes of radiant energy (at position z) in the $\pm$z directions respectively. Energy (flux) conservation considerations (between z and z$\pm$|dz|) allow us to write:

$$\pm \frac{dI_\pm}{dz} = -\kappa\rho(z)\left(I_\pm - (tI_\pm + rI_\mp)\right) \qquad (2.5)$$

t and r are the probabilities of scattering forwards and backwards respectively ($0 \le t \le 1$, $0 \le r \le 1$ and $t+r=\varpi_o$).

In the following, we will be exclusively interested in the so-called "albedo problem" which is, for plane-parallel media, a two-point boundary value problem described by $I_+(0)=I_o$, $I_-(\tau)=0$, where τ is the total optical thickness of the layer. We are particularly interested in the radiation exiting the medium at both bottom and top boundaries; from the definition of transmission coefficient ($T_p=I_+(\tau)/I_+(0)$, "p" for plane parallel"), when we have conservative scattering ($\varpi_o=1$) we easily obtain:

$$T_p = \frac{1}{1+r\tau} \qquad (2.6)$$

in the important case of conservative ($\varpi_o=1$) scattering. For albedo ($R=I_-(0)/I_+(0)$) we find R=1-T; notice that R naturally vanishes if r=0. The functional form 2.6 occurs in the asymptotic theory for radiances in continuous angle radiative transfer (c.f. Lenoble 1977); the above formula is only asymptotically true for large[1] τ, and r is a function of angle of incidence at the cloud top as well as of the phase function. The same transmission function can also be used in an approximation[2] to the total transmission in inhomogeneous clouds (the "independent pixel approximation" - see section 5 below).

2.4 Summary of DA radiative transfer results

In a three part series of papers (Lovejoy et al., 1990a; Gabriel et al., 1990; Davis et al., 1990a) we simplified the radiative transfer problem by considering a sub-set of systems in which scattering occurs only through a finite number of discrete angles. In this case $p(s'\rightarrow s)$ is simply a finite scattering matrix, I_s a finite vector and the integro-differential equations (2.1) become a set of coupled PDE's. These systems are generalizations the two stream model above. Discrete Angle (DA) radiative transfer in fractal systems can be studied either numerically (Davis et al., 1990a), or analytically, in particular using real space renormalisation techniques (Gabriel et al., 1990). The basic result was that in scaling systems, the total transmission (T), and reflection (R) through a cloud with (average) optical thickness τ is of the following form (for $\tau \gg 1$):

$$T \approx T^* + h_T\,\tau^{-\nu_T} \qquad (2.7)$$
$$R \approx R^* - h_R\,\tau^{-\nu_R}$$

where $T\rightarrow T^*$, and $R\rightarrow R^*$ as $\tau\rightarrow\infty$ and h_T, h_R are phase function dependent prefactors, ν_T, ν_R are phase function independent exponents. If light is allowed to leak out of the sides, $\nu_T<\nu_R$, otherwise (e.g. with cyclic or reflecting side boundary conditions), $\nu_T = \nu_R$. A similar form, but with different parameters holds in the limit $\tau\rightarrow 0$. Furthermore, in homogeneous clouds (i.e., with trivial scaling), we obtain the plane

---

1 There are unimportant exponential corrections.

2 The (heuristic) justification for this approximation is that if horizontal radiative interactions are neglected, then each column of the atmosphere separately obeys eq. 2.6. Within the context of DA transfer, these horizontal fluxes vanish identically with side scattering.

parallel scaling discussed above with $\nu_T=1$ (c.f. eq. 2.6). The exponents ν_T, ν_R are said to be "universal" in the sense that they are phase function independent. This can be rigourously proved in DA systems with six orthogonal beams, however various theoretical arguments and numerical results indicate that it extends to systems with continuous angle phase functions, hence that the result eq. 2.7 is generally true in scaling systems; see Lovejoy et al. (1990a), Davis et al. (1990a) for more details. Other DA results which are likely to be quite general, include the demonstration that the problem of diffusion on fractals (studied in statistical physics) is in a different universality class from DA scattering and hence (presumably) from continuous angle scattering. We therefore expect diffusion approximations in which the diffusion equation is used in place of the radiative transfer equation to yield poor results when applied to fractal systems (Davis et al., 1990b).

3. REVIEW OF PROPERTIES OF MULTIFRACTAL PASSIVE SCALAR CLOUDS

3.1 Theoretical properties

The solution of the multiple scattering problem in fractal clouds is quite difficult. However, insight can be gained by studying the the direct transmission through fractal clouds, which as indicated in section 2.2 in a given direction is simply $e^{-\tau}$ where τ is the integral of the optical density in the corresponding direction[1]. The direct transmission is obviously a (very low) lower bound on the total transmission. Anticipating that limiting values of the direct transmission will be either 0 (the cloud is totally opaque), or 1 (the cloud is totally transparent), it is clear that the total transmission will have the same limits. Indeed, the direct transmittance exponents calculated below will provide bounds for the full (multiply scattered) transmittance exponents. In the following, we elaborate on some preliminary results presented in Lovejoy et al. (1990b). We also use a much improved notation which renders the derivations very straightforward.

Now consider a cascade process constructed from an outer unit scale down to an inner scale λ times smaller. We will be interested in the limiting scaling properties of the direct transmission. If along a certain direction, γ_ϕ is the order of the dominant singularity in the underlying conserved field ϕ (e.g. in passive scalar clouds, $\phi = \varepsilon^{-1/2}\chi^{3/2}$), and the optical density is related to this field to through the power a and via a (fractional) integration order -(b+1) (i.e. through the relation $(\kappa\rho)_\lambda = \phi_\lambda{}^a\lambda^{b+1}$, in passive scalar clouds, a = 1/3, b = -4/3) then $\tau_\lambda \approx (\kappa\rho)_\lambda\lambda^{-1}$ (recall that λ is the ratio of the outer scale (L_0) of the cascade to the scale of interest, hence $\Delta x = L_0\lambda^{-1}$ where Δx is the separation used in eq. 1.1, and the subscript λ indicates that we consider the corresponding quantity at resolution λ) and we obtain (taking $L_0 = 1$):

$$\tau_\lambda \approx \lambda^{a\gamma_\phi+b} = \lambda^{\gamma}; \qquad \gamma = a\gamma_\phi+b \tag{3.1}$$

Hence, taking powers and integrating yields a linear shift in the orders of singularities[2]. Denoting the codimension function of τ by $c(\gamma)$, then a basic multifractal relation (Schertzer and Lovejoy, 1987b) is:

$$\Pr(\tau_\lambda>\lambda^{\gamma}) \approx \lambda^{-c(\gamma)} \tag{3.2}$$

and the corresponding codimension function for ϕ is readily obtained:

$$c_\phi(\gamma_\phi) = c(\gamma) = c(a\gamma_\phi+b) \tag{3.3}$$

Similarly, we can define the corresponding multiple scaling exponent function K(h), from:

$$\langle\tau_\lambda{}^h\rangle = \lambda^{K(h)} = \int_{-\infty}^{\infty} \lambda^{h\gamma}\,\lambda^{-c(\gamma)}d\gamma \tag{3.4}$$

---

[1] In DA radiative transfer, an improved lower bound for the transfer is obtained by using the "effective" optical thickness in the above formula rather than τ, see Lovejoy et al. (1990a).

[2] For simplicity of notation, singularities, codimension and multiple scaling functions without subscripts refer to optical thickness τ.

since (to within logarithmic corrections), $\lambda^{-c(\gamma)}$ is the probability density ($=p(\gamma)=dPr/d\gamma$) and "< >" indicates statistical (ensemble) averaging. There is a similar expression for $<\phi_\lambda{}^h>$. Using the method of steepest descents to evaluate this integral, we obtain the Legendre transformation result:

$$K(h) = \max_\gamma (h\gamma - c(\gamma)); \qquad c(\gamma) = \max_h (h\gamma - K(h)) \tag{3.5}$$

An immediate consequence of this relation is that for each moment h, there is a corresponding singularity associated with the maximum: $h=c'(\gamma)$. Furthermore, $c(\gamma)$ must be strictly positive and convex, hence $c'(\gamma)$ is a monotonically increasing function. These facts will be extensively exploited below. The corresponding function for ϕ is obtained via the linear transformation of singularites (eq. 3.1), yielding:

$$K_\phi(h) = K\left(\frac{h}{a}\right) - \frac{bh}{a} \tag{3.6}$$

If ϕ is a continuous cascade process, $c(\gamma)$ will fall into the universality classes (Scherzter and Lovejoy, 1987a,b):

$$c(\gamma) = c_0\left(1 - \frac{\gamma}{\gamma_0}\right)^{\alpha'}; \qquad \frac{1}{\alpha} + \frac{1}{\alpha'} = 1 \tag{3.7}$$

where $0 \le \alpha \le 2$, is the fundamental universality parameter (note that h>0 when $\alpha < 2$, and see Schertzer and Lovejoy (this volume) for discussion of the special cases $\alpha = 0, 1$). Using 3.3, 3.6, we may obtain the corresponding expression for $c_\phi(\gamma)$. c_0, γ_0, are simply related to the two other fundamental parameters - H, C_1 - see Schertzer and Lovejoy (this volume).

3.2. Some semi-empirical estimates of $c(\gamma)$

We are now in a position to calculate the various $c(\gamma)$, K(h) functions for multifractal clouds specified by $c_\phi(\gamma)$, a, b. Below, we will treat the universal multifractal clouds with $\alpha = 2$; the other cases are slightly more involved since their extreme negative γ behaviour is not algebraic (they involve extremal Levy distributions, see Schertzer and Lovejoy (this volume)). They will be discussed elsewhere. In the case $\alpha = 2$, we have:

$$c_\phi(\gamma_\phi) = \frac{C_1}{4}\left(\frac{\gamma_\phi}{C_1} + 1\right)^2; \qquad K_\phi(h) = C_1(h^2-h) \tag{3.8}$$

hence $c_{\phi 0} = C_1/4$ and $\gamma_{\phi 0} = -C_1$ (recall that $C_1>0$, hence here $\gamma_0<0$) $\gamma_{\phi 0}$ is the most probable (space filling) singularity (note that in this case, we also have $c_{\phi 0} = -\gamma_{\phi 0}/4$). We will be particularly interested in $K(h)=C_1ah(ah-1)+bh$, and $K(1)=C_1a(a-1)+b$ which is the exponent characterizing the mean $<\tau>$.

Before proceeding to calculate the scaling of the transmittance in these cases, we first indicate briefly how we can estimate the relevant parameters empirically. Direct measurements of the fundamental multifractal parameter α for liquid water have not been made[1]. However, if we assume $\alpha = 2$ we can use the liquid water power spectrum obtained by King et al. (1981) to estimate $C_1 = -\gamma_0$. To do this, recall that the spectrum of the liquid water density is the Fourier transform of the covariance of ρ. The multiple scaling yields a correction to the 5/3 power law spectrum of ρ: $E(k) \approx k^{-5/3-B}$ where, according to King et al.'s (1981) data, $B = 0.14\pm0.05$. Since the covariance of ρ is a second order moment (and hence is proportional to $\phi^{2/3}$), we obtain $B = -K_\phi(2/3) = 2C_1/9$ hence $C_1 = 0.63\pm0.22$, and we obtain $\gamma_{\phi 0} \approx -0.63$, $c_{\phi 0} \approx 0.16$. Furthermore, using the theoretical values for a passive scalar, a = 1/3, b = -4/3 (eq. 1.1), we obtain c(0) = 4.50 and c'(0) = 3.17 (>0) and K(1) = -1.19. Therefore (as expected) as $\lambda \to \infty$, the scale is reduced and the mean optical thickness decreases; conversely for larger distances, λ decreases and $<\tau>$ increases. The large λ limit therefore tells us about the scaling of optically thin clouds. It is therefore interesting to consider a cascade constrained such that $<\tau> \approx \lambda$, i.e. such that K(1) = 1. Keeping C_1 ($= -\gamma_0$), a at their above values, we require an integration of order -(b+1) with $b = 1+C_1a(1-a) \approx 1.14$, and hence c(0) = 3.09, c'(0) = -2.21 (i.e. c'(0)<0).

[1] However Lovejoy and Schertzer (this volume) find α = 1.7, 0.63 for satellite radiances at infra-red and visible wavelengths respectively.

4. CODIMENSION FUNCTIONS AND SINGULARITIES FOR DIRECT AND PLANE PARALLEL TRANSMITTANCE AND REFLECTANCE

4.1 Discussion

The atmosphere involves many nonlinearly coupled fields. If the corresponding dynamics are scale invariant over significant ranges in scale, then we generally expect that the appropriate way of relating the various fields is via their scale/resolution independent singularities and $c(\gamma)$, $K(h)$ functions (or more fundamentally, the generators of the latter). Equations 3.1, 3.3, 3.6 give a simple illustration of the types of relations we may expect[1]. In radiative transfer through multifractal clouds, we may expect relations between the orders of singularities of the cloud density and radiation fields to be statistical. However, in the direct transmission and plane parallel cases, we have seen that deterministic, simple functional relations exist, and these will be exploited below to yield results analogous to 3.1, 3.3, 3.6.

The formulae discussed here will be for bare cascades (a cascade constructed over only a finite range of scales λ) whereas τ is clearly a dressed quantity i.e. the integral over a completed cascade over a scale λ. This fundamental distinction has been discussed at length in Schertzer and Lovejoy (1987a,b 1990a, this volume), and Lavallée et al. (this volume). The $c(\gamma)$ function for the dressed quantities is the same as for the bare quantities except for the large values of γ (where $c(\gamma)$ becomes linear). In the following, we ignore this complication whereas in section 5 this distinction is exactly accounted for exactly but in a rather special class of models. For convenience, all the definitions and analytical results of this section are summarized in tables 1-4.

For both the direct transmittance and transmittance through plane parallel layers, the relations between τ and T_d, T_p are given by eqs. 2.4 and 2.8 respectively. In order to use these to obtain relations between the corresponding orders of singularities and codimensions we introduce the following notation:

$$\tau = \lambda^{\gamma} = e^{\zeta\gamma} \qquad \zeta = \log\lambda \tag{4.1}$$

Before continuing, note that we will be interested in the limit $\lambda \to \infty$, $\zeta \to \infty$. We now consider the various relevant cases.

4.2 Direct transmittance

We will use the formula for transforming probability densities to obtain a relation between $c(\gamma)$ and $c_{Td}(\gamma_{Td})$, this will illustrate the general method employed below:

$$p_{Td}(\gamma_{Td}) = p(\gamma)\left|\frac{d\gamma}{d\gamma_{Td}}\right| \tag{4.2}$$

This implies the following general relation between codimension functions:

$$c_{Td}(\gamma_{Td}) = c(\gamma) - \frac{\log\left(\left|\frac{d\gamma}{d\gamma_{Td}}\right|\right)}{\zeta} \tag{4.3}$$

Introducing:

$$T_d = e^{-\tau} = e^{\zeta\gamma_{Td}}; \qquad \gamma_{Td} \le 0 \tag{4.4}$$

we obtain the following relation between the singularities in τ, T_d:

---

[1] A related example is the relation between various radar rain singularities discussed in Lovejoy and Schertzer (1990b).

$$\gamma = \frac{\log(-\zeta\gamma_{T_d})}{\zeta}; \qquad \gamma_{T_d} = -\frac{\exp(\zeta\gamma)}{\zeta}; \qquad \left|\frac{d\gamma_{T_d}}{d\gamma}\right| = e^{-\zeta\gamma}; \qquad \gamma_{T_d} \leq 0 \qquad (4.5)$$

Hence in the limit as $\zeta \to \infty$, all the positive orders of singularities in τ ($\gamma>0$) are mapped onto (the infinitely) strong regularity (negative singularity) in T_d ($\gamma_{T_d} \to -\infty$). Conversely, all the regularities in τ ($\gamma<0$) are mapped onto the (single) neutral singularity in T_d ($\gamma_{T_d} = 0$). We may now use eq, (4.3) to obtain the following relation between codimension functions:

$$c_{T_d}(\gamma_{T_d}) = c(\gamma) + \gamma \qquad (4.6)$$

Substituting γ_{T_d} in terms of γ in the above, and expanding $c(\gamma)$ in a Taylor series about the origin, we obtain:

$$c_{T_d}(\gamma_{T_d}) = c(0) + \frac{\log(-\zeta\gamma_{T_d})}{\zeta}(1+c'(0)) + \ldots \qquad (4.7)$$

The first term on the right hand side indicates that a single codimension ($c(0)$) dominates the behaviour of T_d. The second term on the right hand side corresponds to two prefactors in the probability density for γ_{T_d}. The first is $(-\gamma_{T_d})^{-(1+c'(0))}$, i.e. a singularity at the origin of the density of γ_{T_d} of order $(1+c'(0))$; in the probability distribution of γ_{T_d} this is of order $c'(0)$, which is regular as long as $c'(0)<0$. Conversely, when $c'(0)>0$, the probability of γ_{T_d} will be singular at the origin. The second term yields a (sub-exponential) factor $(\log \lambda)^{-(1+c'(0))}$ in the probability distribution of γ_{T_d}; the exponent $-(1+c'(0))$ is called a "subcodimension". Of all the cases we discuss below, the direct transmittance problem is the only one where the detailed consideration of these higher order terms is important.

We may now use eq. (3.5) to calculate $K_{T_d}(h)$ from $c_{T_d}(\gamma_{T_d})$, and hence to obtain the multiple scaling characteristics of T_d. However, relation (3.5) is true only when exponential factors are dominant, i.e. for $c'(0)<0$ (which corresponds to $\langle T_d^h\rangle \to 0$ as $\zeta\to\infty$; clouds become opaque). For $c'(0)>0$, either more refined analyses are required (Lovejoy et al., 1990b), or more simply, we must consider the quantity $1-T_d$ (see next subsection); we will find that $\langle T_d^h\rangle \to 1$ (the clouds become transparent in the limit). Restricting our attention to the case $c'(0)<0$, we therefore[1] obtain:

$$K_{T_d}(h) = -c(0); \qquad c'(0)<0 \qquad (4.8a)$$

Using the values discussed in the second case of section 3.2 with $c'(0) = -2.21$ (<0), $c(0) = 3.09$, we obtain $K_{T_d}(1) = -3.09$.

Eq. 4.8a shows that the direct transmittance decreases algebraically as the cloud increases in thickness. Following Lovejoy et al. (1990a), it is often convenient to introduce another exponent which is essentially a "mean field" exponent relating $\langle\tau\rangle$ to $\langle T_d\rangle$, (in analogy with eq. 2.9 for the total transmission), we obtain:

$$\langle T_d\rangle \approx \langle\tau\rangle^{-\nu_{T_d}} \qquad (4.8b)$$

where:

$$\nu_{T_d} = -\frac{K_{T_d}(1)}{K(1)} \qquad (4.8c)$$

The above result shows that the effect of the multifractal optical density field has been to greatly enhance the mean direct transmittance which is now algebraic, not exponential[2]. In the above example with $K(1) = 1$ we have $K_{T_d}(1) = -3.09$, hence $\nu_{T_d} = 3.09$. More generally from the Legendre relation 3.5, we find $-c(0) = \min(K(h))$, hence $\nu_{T_d} = -\min(K(h))/K(1)$ when $c'(0)<0$.

---

[1] Although when it holds, eq. 4.8a doesn't involve h, there is an h dependence that enters via the correction term discussed above. The exact form of this correction is given in Lovejoy et al. (1990b).

[2] The photon free path distribution will have a much longer tail.

4.3 Diffuse radiation coefficients

We saw that the exponential relation between τ and T_d mapped the continuous distribution of singularities in τ onto the two values $\gamma_{Td} = -\infty$, 0 for $\gamma>0$, $\gamma<0$ respectively. More interesting relations (useful below) may be obtained by considering the quantity $R_d = 1-T_d$, where R_d is the "diffuse radiation" coefficient (i.e. the radiation that has been scattered at least once). Introducing the corresponding orders of singularity γ_{Rd}, we obtain:

$$R_d = e^{\zeta\gamma_{Rd}} = 1 - \exp(-e^{\zeta\gamma}) \qquad \gamma_{Rd} \leq 0 \qquad (4.9)$$

This yields:

$$\gamma = \frac{\log(-\log(1-e^{\zeta\gamma_{Rd}}))}{\zeta};$$
$$\gamma_{Rd} = \frac{\log(1-\exp(-e^{\zeta\gamma}))}{\zeta} \qquad \gamma_{Rd} \leq 0 \qquad (4.10)$$
$$\left|\frac{d\gamma_{Rd}}{d\gamma}\right| = \frac{e^{\zeta\gamma_{Rd}}}{(1-e^{\zeta\gamma_{Rd}})\log(1-e^{\zeta\gamma_{Rd}})}$$

Or, approximately:

$$\gamma_{Rd} \approx \gamma - \frac{e^{\zeta\gamma}}{2\zeta}; \quad \gamma<0; \qquad \gamma_{Rd} \approx -\frac{\exp(-e^{\zeta\gamma})}{\zeta}; \quad \gamma>0$$
$$\gamma \approx \gamma_{Rd} - \frac{e^{\zeta\gamma_{Rd}}}{2\zeta}; \quad \gamma_{Rd}<0; \qquad \gamma \approx \frac{\log(-\log(-\gamma_{Rd}))}{\zeta}; \quad \gamma_{Rd} \approx 0 \qquad (4.11)$$
$$\left|\frac{d\gamma_{Rd}}{d\gamma}\right| \approx 1 - \frac{e^{\zeta\gamma_{Rd}}}{2}; \quad \gamma_{Rd}<0; \qquad \left|\frac{d\gamma_{Rd}}{d\gamma}\right| \approx \zeta\gamma_{Rd}\log(-\gamma_{Rd}); \quad \gamma_{Rd} \approx 0$$

hence as $\zeta\rightarrow\infty$; singularities in τ are mapped onto the neutral singularity in R_d, and regularities in τ are mapped onto regularities of the same order in R_d. Using these relations between γ, γ_{Rd}, and eq. (4.3), we obtain the following relation between codimension functions (valid only for $\gamma_{Rd}<0$ and ζ large enough, i.e. this formula will give information only on the regularities in τ):

$$c_{Rd}(\gamma_{Rd}) \approx c(\gamma_{Rd}); \qquad \gamma_{Rd}<0 \qquad (4.12)$$

where we have dropped the $1/\zeta$ corrections. We can now use eqs. (3.5, 4.12) to calculate the scaling exponents of $\langle R_d^h \rangle$:

$$K_{Rd}(h) = \max_{\gamma_{Rd}<0} (h\gamma_{Rd} - c(\gamma_{Rd})) = K(h) \ (> -c(0)); \qquad h<c'(0) \qquad (4.13)$$

The condition $h<c'(0)$ immediately follows from the relation $h=c'(\gamma_{Rd})$, the monotonicity of $c'(\gamma)$, and the restriction $\gamma_{Rd}<0$. Furthermore, the convexity and positivity requirements on $c(\gamma_{Rd})$ ensures that $K(h)>c(0)$, and implies that for $c'(0)>h>0$, $K(h)<0$. If $h\geq c'(0)$, the maximum in the above will depend on the limit $\gamma_{Rd}\rightarrow 0$ (i.e. the singularities in τ will dominate), and the above formula breaks down. To understand what happens, consider the two cases $c'(0)>0$, $c'(0)<0$. In the former case, for some small positive h, the above formula will hold, and $\langle R_d^h \rangle\rightarrow 0$ as $\zeta\rightarrow\infty$. This is sufficient to imply $\langle T_d^h \rangle\rightarrow 1$ for all $h>0$, in conformity with the results of subsection 4.2. Conversely, when $c'(0)<0$, we expect $\langle T_d^h \rangle\rightarrow 0$, and hence $\langle R_d^h \rangle\rightarrow 1$.

Using the above formulae, we obtain $K_{Rd}(1)=a\gamma_0+b$. Using the above parameters $a=1/3$, $b=-4/3$, $\gamma_{\phi 0}=-0.63$, recalling that $c'(0)=3.17>1$, we obtain $K_{Rd}(1) = -1.54$ which indicates that as the cloud gets thinner ($\lambda\rightarrow\infty$), that $\langle T_d \rangle$ approaches 1 as expected. Since $K_{Rd}(1) = K(1) = -1.19$, we have $v_{Rd}=K_{Rd}(1)/K(1)=1$ (i.e. R_d is a linear function of τ as expected in the case of optically thin media).

4.4 Plane parallel transmission coefficient

The plane parallel formula 2.4 leads to particularly simple mappings between singularities, codimensions and multiple scaling exponents. Following the procedure discussed above, we obtain:

$$T_p = e^{\zeta\gamma_{Tp}} = (1+ re^{\zeta\gamma})^{-1} \qquad \gamma_{Tp}<0$$
$$\gamma_{Tp} = -\frac{\log(1+re^{\zeta\gamma})}{\zeta}; \qquad \gamma = \frac{\log(r^{-1}e^{-\zeta\gamma_{Tp}}-1)}{\zeta} \tag{4.14}$$
$$\left|\frac{d\gamma_{Tp}}{d\gamma}\right| = 1 - e^{\zeta\gamma_{Tp}};$$

In the limit $\zeta\to\infty$, we obtain the following approximate formulae:

$$\gamma_{Tp} \approx -\gamma; \qquad \gamma>0 \tag{4.15}$$
$$\gamma_{Tp} \approx \frac{re^{\zeta\gamma}}{\zeta}; \qquad \gamma<0$$

Hence, the singularities in τ are mapped onto the corresponding regularities in T_p, and the regularities in τ are mapped onto the neutral singularity $\gamma_{Tp}=0$. Considering only the case $\gamma_{Tp}<0$, we therefore obtain the following relation between codimension functions:

$$c_{Tp}(\gamma_{Tp}) = c(-\gamma_{Tp}); \qquad \gamma_{Tp}<0 \tag{4.16}$$

Finally, using 3.5, we find the following relation between multiple scaling exponents:

$$K_{Tp}(h) = \max_{\gamma_{Tp}<0}(h\gamma_{Tp} - c(-\gamma_{Tp})) = K(-h)\ [>-c(0)]; \qquad h<-c'(0) \tag{4.17}$$

In particular, we see that as long as $c'(0)<0$, then the above is true for some positive h, and the corresponding moments will tend to 0 as $\zeta\to\infty$. As in the above, if $h>-c'(0)$, then it is the regularities in τ that dominate ($\gamma_{Tp}=0$), and the above formula breaks down. Again, rather than attempt a complicated analysis of this case, it is much simpler to analyse $R_p=1-T_p$; i.e., the plane parallel reflection coefficient. If we now use eq. 4.17 to determine v_{Tp}, we obtain[1]: $v_{Tp} \leq 1$ which is a simple consequence of the concavity of the function τ^{-1}. Another relation worth mentioning is that by comparing eq. 4.8a and 4.17, we have $K_{Tp}(h)\geq K_{Td}(h)$ for all $h>0$. This follows from the fact that T_p takes into account multiple scattering, and hence $T_p\geq T_d$. Finally, the relations are all phase function (r) independent: the latter only affect the prefactors but not the exponents.

4.5 Plane parallel reflection coefficient

This case is very easy to deal with since:

$$R_p = e^{\zeta\gamma_{Rp}} = (1+ r^{-1}e^{-\zeta\gamma})^{-1} \tag{4.18}$$

which is identical to eq. 4.14 for T_p except that r is replced by r^{-1}, and γ by $-\gamma$. We therefore immediately obtain:

$$\gamma_{Rp} \approx \gamma; \qquad \gamma<0 \tag{4.19}$$
$$\gamma_{Rp} \approx \frac{r^{-1}e^{-\zeta\gamma}}{\zeta}; \qquad \gamma>0$$

---

[1] This follows from: $v_{Tp} = -K(-1)/K(1) = (-K_\phi(-a)+b)/(K_\phi(a)+b) \leq 1$ since $K_\phi(-a)\geq -K_\phi(a)$ (due to the concavity of $K_\phi(h)$). In the above example with $K(1) = 1$ (i.e. $a=1/3$, $b=1.14$), we obtain $K(-1) = -0.86$, hence $v_{Tp} = 0.86<1$.

The regularities of τ are mapped onto the corresponding regularities of R_p, and the singularities of τ onto the neutral singularity of R_p (γ_{Rp}=0) as in the case for R_d, with comments analogous to those at the end of subsection 4.4. The fact that the results for R_d and R_p are the same follows from the fact that both are linear funcitons of τ for $\tau \ll 1$.

4.6 Summary of definitions and results

Definitions
$\tau = \phi^a \lambda^b$
$T_d = e^{-\tau}$
$T_p = (1 + r\tau)^{-1}$
$R_d = 1 - T_d = 1 - e^{-\tau}$
$R_p = 1 - T_p = (1 + r^{-1}\tau^{-1})^{-1}$

Table 1: Summary of the definitions discussed in the text.

	Regularities ($\gamma<0$)	Singularities ($\gamma>0$)
γ_ϕ:	$(\gamma-b)/a$	$(\gamma-b)/a$
γ_{Td}:	0	$-\infty$
γ_{Tp}:	0	$-\gamma$
γ_{Rd}, γ_{Rp}:	γ	0

Table 2: The mapping of singularities in τ onto the various singularities discussed in the text.

Codimension formulae	Range of validity
$c_\phi(\gamma_\phi) = c(a\gamma_\phi + b)$	$-\infty \le \gamma \le \infty$
$c_{Td}(\gamma_{Td}) = c(0)$	$\gamma_{Td}<0$
$c_{Tp}(\gamma_{Tp}) = c(-\gamma_{Tp})$	$\gamma_{Tp}<0$
$c_{Rd}(\gamma_{Rd}) = c(\gamma_{Rd}),\ c_{Rp}(\gamma_{Rp}) = c(\gamma_{Rd})$	$\gamma_{Rd}<0,\ \gamma_{Rp}<0$

Table 3: The codimension formulae discussed in the text.

Multiple scaling exponents	Conditions of validity
$K_\phi(h) = K(h/a) - bh/a$	$-\infty \le h \le \infty$
$K_{Td}(h) = -c(0)$	$c'(0)<0$
$K_{Tp}(h) = K(-h)$	$h<-c'(0)$
$K_{Rd}(h) = K_{Rp}(h) = K(h)$	$h<c'(0)$

Table 4: The multiple scaling formulae discussed in the text.

5. APPLICATIONS TO A CLASS OF MICROCANONICAL α MODEL CLOUDS

We are now in a position to exploit the above results to obtain the direct transmission through a simple class of "α model" discrete microcanonical[1] cascades for the optical density. As noted in Schertzer and Lovejoy (1987a,b), this discrete cascade is outside the scope of the universality classes which are obtained for continuous cascades. We discuss only the simplest case in two dimensions (see fig. 1), in which each eddy is broken up into four sub-eddies, each with one half the size of the parent eddy (λ_0=2 per step). For the moment, we consider the unnormalized case in which the sum of the two multiplicative factors on the left and right are fixed (=$\lambda_0^{\gamma_1}$, $\lambda_0^{\gamma_2}$ respectively, - the factors which share a column can be randomly

[1] Recall that in microcanonical cascades, there is strict conservation at each step, i.e. in fig. 3, a+b+c+d = constant. In canonical cascades we have the weaker restriction <a+b+c+d> = constant. Actually the model described here is more strongly microcanonical than is usual since each column at each step is microcanonical, i.e. a+c=W_1, b+d =W_2. where W_1, W_2 are constants.

In this model, the introduction of the microcanonical constraint is really an artifice designed to avoid the problem of "bare"/"dressed" properties by ensuring that the spatial averages of τ are equal to the ensemble averages.

chosen as long as they respect this constraint). This constraint is such that as the cascade proceeds, if we are only interested in the integral of the optical density (τ) over columns, then the latter is modulated by one of the above factors at each step.

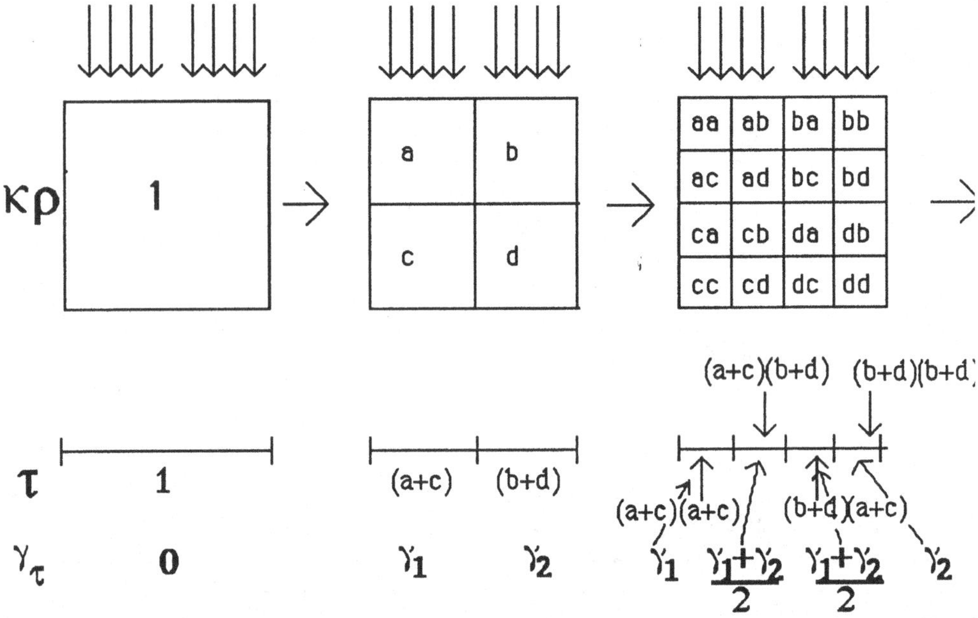

Fig. 1: The unit square with unit optical density ($=\kappa\rho$) is broken up into four sub-eddies at each step in the cascade, each sub-eddy being modulated by the factors a,b,c,d which are here deterministically arranged as shown in the middle figure (the first step of the cascade). The expressions in each box indicate the local optical density. The one dimensional cascade shown below it shows the evolution of the vertical integral, the optical thickness (τ) with the corresponding expression for the optical density. On the line below, we indicate the corresponding orders of singularity of τ ($\gamma_\tau=\log\tau/\log\lambda$) with $\lambda_0^{\gamma_1}=a+c$, $\lambda_0^{\gamma_2}=b+d$, $\lambda_0=2$. Closer inspection reveals that as far as the τ cascade is concerned, that a,b,c,d can be random, as long as they are constrained so that a+c, b+d are constants. The arrows at the top indicate the incident radiation.

Calculating the multiple scaling for the optical thickness[1] (K(h)), we obtain:

$$K(h) = \frac{\log \langle\tau^h\rangle}{\log \lambda} = \frac{\log \left(\lambda_0^{h\gamma_1-1} + \lambda_0^{h\gamma_2-1}\right)}{\log \lambda_0} \qquad (5.1)$$

(since the probabilities are both $1/2 = \lambda_0^{-1}$) hence, for a conserved cascade in τ (i.e. K(1)=0 so that $\langle\tau\rangle$ is independent of λ), we require that γ_1, γ_2 be constrained such that $\lambda_0^{\gamma_1-1}+\lambda_0^{\gamma_2-1}=1$. Similarly, since $\tau_\lambda=(\kappa\rho)_\lambda\lambda^{-1}$ the condition for a conserved cascade[2] in $\kappa\rho$ is $\lambda_0^{\gamma_1-2}+\lambda_0^{\gamma_2-2}=1$.

According to our previous analysis, the two fundamental cases of interest are $c'(0)<0$, $c'(0)>0$, and we will be also interested in the value of c(0). Although the Legendre transform of eq. 5.1 cannot be performed analytically for all values of γ, it is straightforward in the special case of interest, $\gamma=0$. Expressions for c(0), c'(0) are obtained using the fact that $c(0)=K(h_0)$ where $h_0=c'(0)$ is the value that yields the minimum K(h). We therefore have:

[1] Lovejoy et al. (1990b) uses a quadratic approximation to 5.1 and gives results corresponding to those below to within this approximation.

[2] It is straightforward to calculate the corresponding multiple scaling function for $\kappa\rho$, but to do this, we must introduce the (joint) probabilities for a,b,c,d.

$$c'(0) = h_0 = -\frac{\log(-\gamma_1) - \log(\gamma_2)}{(\gamma_2-\gamma_1)\log\lambda_0} \qquad (5.2)$$

where we have taken[1] $\gamma_1<0$, $\gamma_2>0$. The sign of $c'(0)$ is therefore the opposite of $\gamma_1+\gamma_2$. In particular whenever $\gamma_1+\gamma_2 = 0$ we have simultaneously $c'(0)=h_0=c(0)=0$. The two cases are shown in fig. 2. Using the above results for the transmission and reflection coefficients, we have:

$c'(0)\geq 0,\ \gamma_1+\gamma_2\leq 0$:

This case[2] yields:

$$K_{Rd}(h) = K_{Rp}(h) = K(h) \qquad h<c'(0) \qquad (5.3)$$

$c'(0)\leq 0,\ \gamma_1+\gamma_2\geq 0$:

This case[3] yields:

$$K_{Td}(h) = -c(0) \qquad h<-c'(0) \qquad (5.4)$$
$$K_{Tp}(h) = K(-h)$$

We can now conveniently display the combined results for K_{Td}, K (fig. 2), where we see that there are three regions corresponding to $<\tau>\rightarrow 0$, $<T_d>\rightarrow 1$ (thin transparent clouds), $<\tau>\rightarrow\infty$, $<T_d>\rightarrow 1$ (thick but transparent clouds), and $<\tau>\rightarrow\infty$, $<T_d>\rightarrow 1$ (thick opaque clouds). The intersection of the curves $K_{Td}(1)=0$ and $K(1)=0$ occurs at $\gamma_1=\gamma_2=0$ and corresponds to multiplicative factors of 1 in each column - i.e., this is the (only) non-fractal case (however $\gamma_1=\gamma_2$ yields a non-fractal τ).

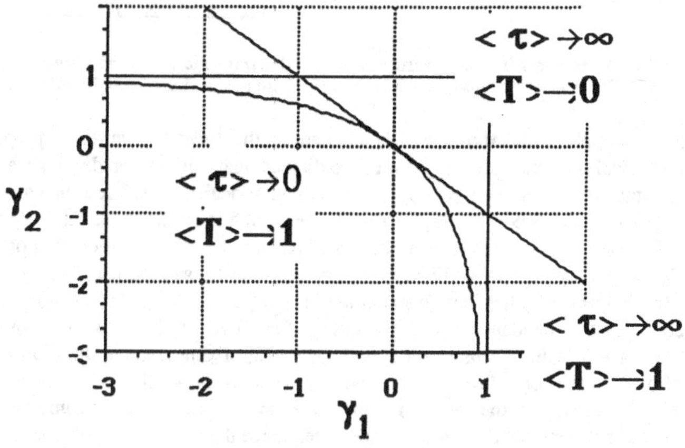

Fig. 2: A diagram of the γ_1, γ_2 plane showing the three different regions characterize the scaling of the optical thickness and direct transmission. The straight line is the line $c'(0)=c(0)=K_{Td}(1)=0$, and the curved line is $K(1)=0$.

As an example of the above[4], we may consider the $D = \log3/\log2 = 1.58\cdots$ monodimensional cloud discussed in Gabriel et al. (1990), Davis et al. (1990a) and shown in fig. 3; it corresponds to $\gamma_2=1$, $\gamma_1=0$,

---

[1] If $\gamma_1, \gamma_2 > 0$, $c'(0) = -\infty$, $\gamma_1, \gamma_2 < 0$, $c'(0) = \infty$; in both cases, $c(0) = 0$.

[2] In Lovejoy et al. (1990b) a formula for $<T_d^h>$ only valid for $c_0= -\gamma_0/4$ was used to obtain $K_{Rd}(1)$. The latter formula is erroneous since here this special condition is not satisfied; 5.3 is correct.

[3] Here T_p is obviously not a plane parallel transmittance, but the "independent pixel" approximation transmittance. It is also the exact solution for DA phase functions with no side scattering, only forward/backward.

[4] Many (but not all) random microcanonical β models are special cases of the above: the total number of alive eddies per column must be fixed.

and $c'(0) = -\infty$, so using eq. 5.7, we find $K_{Td}(1) = -c(0) = K(-\infty) = -1$, hence $\nu_{T_d} = 1.71$. We therefore have $\langle T_d \rangle \to 0$, $\langle \tau \rangle \to \infty$ the cloud is thick and opaque. For comparison, numerical multiple scattering results on this cloud are $\nu_T = 0.41$ for the case with cyclic boundary conditions, and 0.51 for open sides (see Davis et al. (1990a) for details). Both exponents are smaller than 1.71 as required since the direct transmittance is a lower bound on the total transmittance. Finally, we can compare these results to analytic renormalization estimates (Gabriel et al., 1990) which yields .60 for the case with cyclic boundary conditions[1]. In order to make the model transparent in the limit, we must divide τ by at least a factor $\sqrt{2}$ (i.e. we decrease γ_1, γ_2 by 1/2), yielding $K_{Td}(1)=0$, and $K(1)=\log_2(3/2\sqrt{2}) = 0.08\cdots$ which implies that τ increases without bound. We also find $K_{Tp}(1) = 0.41$, $\nu_{T_p} = 0.71$ (which is <1, as expected, and is an improved bound on ν_T).

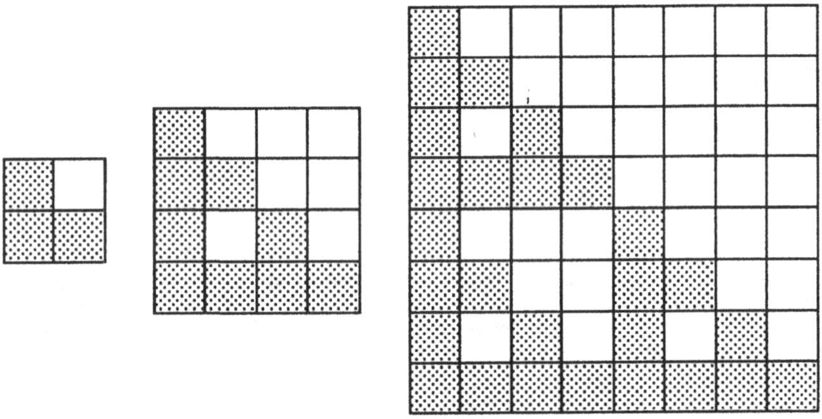

Fig. 3: The first three steps in the construction of the deterministic monofractal used discussed at length in Gabriel et al. (1990), Davis et al .(1990a), which is a special case of the above model with $\gamma_1=1$, $\gamma_2=0$.

As a final example of the above model, we examine the direct transmittance properties of Cahalan's (1989) microcanonical α model. His model was one-dimensional - the optical density was assumed constant in vertical columns, i.e. he assumed a=c, b=d. He also numerically considered the case a=c=1.3, b=d=0.7 which yields $\gamma_2=\log_2 1.3=0.379\cdots$, $\gamma_1=\log_2 0.7=-0.575\cdots$, which is on the curve $K(1)=0$ in fig. 2, and yields $c'(0)=0.497$, $c(0)=0.017$, hence we use formula 5.3 which will be valid for moments of R_d, R_p for h up to 0.497. We see that for the case of vertically incident radiation considered here, that we expect $\langle T \rangle \to 1$ as $\langle \tau \rangle \to \infty$, i.e. for "overhead" illumination his cloud is transparent in the limit $\lambda \to \infty$. However, Cahalan only examined the case of incidence at 60^0, and obtained an increase of $\langle T \rangle$ with λ, but apparently tending towards a finite nonzero value. This result is perhaps not surprising since in the limit $\lambda \to \infty$, photons will traverse an infinite number of columns before exiting, whereas the vertically incident directly transmitted photons examined here stay in the same column. As $\lambda \to \infty$ most columns become very thin, while rare, sparsely distributed columns get thicker to compensate, hence the results $\langle T \rangle \to 1$ for vertical incidence but another result for non-vertical incidence. This artificial dependence on incident angle, should disappear if vertically varying $\kappa\rho$ fields (such as the slightly more general model used here) were employed.

6. METEOROLOGICAL IMPLICATIONS AND CONCLUSIONS

Elsewhere we have already shown that the scaling exponents characterizing the mean transmission and reflection properties of clouds depends directly on the scaling of the cloud optical density field. These results (summarized in eq. 2.7) were essentially monofractal because we only considerd the scaling of the means, not the higher order moments. Even so, they indicated that in comparing thick homogeneous clouds and equally thick fractal clouds that the corresponding exponents will be different leading to diverging

---

[1] With open boundary conditions, we no longer have $\nu_T = \nu_R$ and the renormalization method estimates ν_R not ν_T.

transmission ratios as $\tau \rightarrow \infty$. The monofractal results were therefore sufficient to solve the so-called "albedo paradox" (Wiscombe et al., 1984). This "paradox" is simply the observation that if horizontally homogeneous (plane parallel) cloud models are used to infer cloud liquid water from measurements of cloud albedo, that discrepancies of factors of 10 with respect to in situ measurements are common. Other implications concerned the interpretation of satellite pictures, the estimation of cloud radiation budgets and global climate models.

In this paper we obtained results on multifractal clouds obtaining in many cases explicit analytical formulae for relating the multiple scaling of the cloud optical density field with the transmitted intensities. The cases examined were for the direct transmittance, the total multiple scattering through multifractal plane parallel clouds and the corresponding quantities through a microcanonical α model cloud. The direct transmittance is considerably easier to calculate than the total transmittance and yields exponents which provide bounds on the multiple scattering exponents. As in Discrete Angle radiative transfer, the exponents are all phase function independent and the corresponding transmitted fields will be multifractal. There is no reason to suspect these general results to change when multiple scattering is included.

7. ACKNOWLEDGMENTS

We acknowledge stimulating and helpful discussions with G.L. Austin, R. Davies, R. Devaux, P. Gabriel, I. Graham, M. Grant, A. Saucier, G. Sèze and L. Smith.

8. REFERENCES

Cahalan, R.F., 1989: Overview of fractal clouds. Advances in Remote Sensing (RSRM'87), pp 371–389, Edited by A. Deepak et al., A. Deepak, Hampton, Va..

Coantic, M.F., 1978: Interaction Between Turbulence and Radiation. AGARDograph, 232, 175-236.

Chandrasekhar, S., 1950: Radiative Transfer. Oxford University Press (reprinted by Dover Publ., New York, 1960), pp xiv+393.

Davis, A., P. Gabriel, S. Lovejoy, D. Schertzer, 1989: Asymptotic Laws for Thick Clouds, Dimensional Dependence - Phase Function Independence. IRS'88: Current Problems in Atmospheric Radiation, Eds. J. Lenoble and J. F. Geleyn, Deepak Publ., 103-106.

Davis, A., P. Gabriel, S. Lovejoy, D. Schertzer, G.L. Austin, 1990a: Discrete Angle Radiative Transfer - Part III: Numerical Results and Meteorological Applications. J. Geophys. Res., 95, 11729-11742.

Davis, A., S. Lovejoy, D. Schertzer, P. Gabriel, G.L. Austin, 1990b: Discrete Angle Radiative Transfer Through Fractal Clouds. Proceedings of the 7th Conf. on Atmos. Rad. (San Fransisco, 23-27 July), A.M.S. Boston.

Frisch, U., P.L. Sulem, M. Nelkin, 1978: A Simple Dynamical Model of Intermittent Fully Developed Turbulence. J. Fluid Mech., 87, 719-724.

Gabriel, P., S. Lovejoy, G.L. Austin, D. Schertzer, 1986: Radiative Transfer in Extremely Variable Fractal Clouds. 6th Conference on Atmospheric Radiation, AMS, 12-16/5, Williamsburg, Va.

Gabriel, P., 1988: Radiative transfer in extremely variable clouds, Ph.D. Thesis, McGill U., Montréal.

Gabriel, P., S. Lovejoy, A. Davis, D. Schertzer, G.L. Austin, 1990: Discrete Angle Radiative Transfer - Part II: Renomalization Approach for Homogeneous and Fractal Clouds. J. Geophys. Res. 95, 11717-11728.

King, W.D., C.T. Maher, G.A. Hepburn 1981: Further Performance Tests on the CSIRO Liquid Water Probe. J. Atmos. Sci., 38, 195-200.

Lovejoy, S. and B. Mandelbrot, 1985: Fractal Properties of Rain and a Fractal Model. Tellus, 37A, 209-232.

Lovejoy, S., D. Schertzer, 1985: Generalized Scale Invariance in the Atmosphere and Fractal Models of Rain. Wat. Resour. Res., 21, 1233-1250.

Lovejoy, S., D. Schertzer, 1986: Scale Invariance, Symmetries, Fractals and Stochastic Simulation of Atmospheric Phenomena. Bull AMS, 67, 21-32.

Lovejoy, S., P. Gabriel, G.L. Austin, and D. Schertzer 1988: Modeling the scale dependence of visible satellite images by radiative transfer in fractal clouds, Proceedings, 3rd Conference on Satellite Meteorology and Oceanography, Am. Meteorol. Soc., Annaheim, Calif., Jan. 31 to Feb. 5.

Lovejoy, S., P. Gabriel, D. Schertzer, and G.L. Austin, 1989: Fractal clouds with discrete angle radiative transfer, in IRS'88: Current Problems in Atmospheric Radiation, edited by J. Lenoble and J. F. Geleyn, pp. 99–102, A. Deepak, Hampton, Va.

Lovejoy, S., D. Schertzer, 1990a: Multifractal Analysis Techniques and Rain and Cloud Fields from 10^{-3} to 10^6m. Scaling, Fractals and Non-Linear Variability in Geophysics, edited by D. Schertzer and S. Lovejoy, Kluwer, Dotrecht/Boston, this volume.

Lovejoy, S., D. Schertzer, 1990b: Multifractals, universality classes, satellite and radar measurements of clouds and rain. J. Geophys. Res., 95, 2021-2034.

Lovejoy, S., A. Davis, P. Gabriel, D. Schertzer, G.L. Austin, 1990a: Discrete Angle Radiative Transfer. Part I: scaling and similarity, universality and diffusion. J. Geophys. Res. 95, 11699-11715.

Lovejoy, S., A. Davis, D. Schertzer, 1990b: Radiative transfer in multifractal clouds: observations and theory, Proceedings of the 7th Conf. on Atmos. Rad., (San Fransisco, 23-27 July), A.M.S., Boston.

Mandelbrot, B., 1974: Intermittant Turbulence and Self-similar Cascades: Divergence of High Moments and Dimension of the Carrier. J. Fluid Mech., 62, 331-350.

Schertzer, D., O. Simonin 1983: A Theoretical Study of Radiative Cooling in Homogeneous and Isotropic Turbulence. Turbulent Shear Flow, vol. 3, Eds. L.J.S. Bradbury et al., 262-274.

Schertzer, D., S. Lovejoy, 1983: On the Dimension of Atmospheric Motions. Preprint Vol., IUTAM Symposium on Turbulence and Chaotic Phenomena in Fluids, 141-144.

Schertzer, D., S. Lovejoy, 1984: On the Dimension of Atmospheric Motions. Turbulence and Chaotic Phenomena in Fluids, Ed. Tatsumi, Elsevier North-Holland, New York, 505-508.

Schertzer, D., S. Lovejoy, 1987a: Singularités Anisotropes et Divergence de Moments en Cascades Multiplicatives. Annales Math. Qué., 11, 139-181.

Schertzer, D., S. Lovejoy, 1987b: Physically Based Rain and Cloud Modeling by Anisotropic, Multiplicative Turbulent Cascades. J. Geophys. Res., 92, 9693-9714.

Schertzer, D., S. Lovejoy, 1990a: Nonlinear geodynamical variability: Multiple singularities, universality and observables. Scaling, Fractals and Non-Linear Variability in Geophysics, edited by D. Schertzer and S. Lovejoy, Kluwer, Hingham, Mass., this volume.

Schertzer, D., S. Lovejoy, 1990b: Nonlinear variability in geophysics: multifractal anlaysis and simulations. Fractals :the physical origin and consequences, L. Pietronero ed., pp. 49–79.

Schuster, A., 1905: Radiation through a Foggy Atmosphere. Astrophys. J., 21, p1.

Simonin, O., M.F. Coantic, D. Schertzer, 1981: Effet du Rayonnement Infrarouge sur la Turbulence de Temperature dans l'Atmosphère: Structure Spectrale et Taux de Dissipation. C.R. Acad Sci., 293, 245-248.

Spiegel, E.A., 1957: The Smoothing of Temperature Fluctuations by Radiative Transfer. Astrophys. J., 126, 202-207.

Tsay, S. and K. Jayaweera, 1984: Characteristics of Arctic Stratus Clouds. J. Climate Appl. Meteor., 23, 584-596.

Wilson, J., D. Schertzer, S. Lovejoy, 1990: Physically Based Rain and Cloud Modelling. Scaling, Fractals and Non-Linear Variability in Geophysics, edited by D. Schertzer and S. Lovejoy, Kluwer, Hingham, Mass., this volume.

Wiscombe, W.J., R.M. Welch, W.D. Hall, 1984: The Effects of Very Large Drops on Cloud Absorption,.Part 1: Parcel Models. J. Atmos. Sci., 41, 1336-1355.